DICTIONARY
OF
METALS

A—Hearth. B—Heap. C—Slag-vent. D—Iron mass. E—Wooden mallets.
F—Hammer. G—Anvil.

Iron making in the sixteenth century. Source: Georgius Agricola, *De Re Metallica*, translated by Herbert Clark Hoover and Lou Henry Hoover, Dover Publications, Inc., New York, 1950, p 422

DICTIONARY
OF
METALS

Editor
Harold M. Cobb

**The Materials
Information Society**

ASM International®
Materials Park, Ohio 44073-0002
www.asminternational.org

Prepared under the direction of the ASM International Technical Book Committee (2011–2012), Bradley J. Diak, *Chair.*

ASM International staff who worked on this project include Scott Henry, Senior Manager, Content Development and Publishing; Karen Marken, Senior Managing Editor; Amy Nolan, Content Developer; Victoria Burt, Content Developer; Sue Sellers, Editorial Assistant; Bonnie Sanders, Manager of Production; Madrid Tramble, Senior Production Coordinator; and Kelly Sukol, Production Coordinator.

Library of Congress Control Number: 2012940641

ISBN-13: 978-1-61503-978-4
ISBN-10: 1-61503-978-3

SAN: 204-7586

ASM International®
Materials Park, OH 44073-0002
www.asminternational.org

Printed in the United States of America

Dictionary of Metals
H.M. Cobb, editor

Contents

Preface

The story of metals is undeniably entwined with the history of humanity, as evidenced by the division of the ages by the well-known Stone Age, Bronze Age, Iron Age, and what some have called the Steel Age and the Stainless Steel Age. Metals, more than any other material, have had the greatest influence on the development of civilization from prehistoric times.

Metals were used for horseshoes, tools, knives, cook pots, cups and plates, nails, chains, cannon balls, and coins. That was when metals were made in small enough crucibles so that one or two men could lift them for casting. In the middle of the 19th century, Henry Bessemer showed how to make steel by the ton so that a long piece of metal could be made into a pipe, sheet, wire, or beam, leading eventually to the manufacture of machinery, ships, refineries, power plants, skyscrapers, and airplanes.

Every metal has its own exciting story that tells of its unique properties, the obstacles encountered in producing it, and its advantages and special uses. Metals are all around us and, for the most part, are taken for granted. But what would we do without that tiny amount of silicon in every computer chip produced, or the stainless steel in the utensils on our kitchen tables?

The *Dictionary of Metals* includes descriptions of metals and terms relating to metals. It also includes a considerable amount of the history, starting with the seven metals of antiquity. Each of the 73 metallic elements has a discussion that includes the discoverer and date, the naming of the metal and its meaning, major applications, and the significance of discovery. Charts show their physical properties. An appendix includes a timeline of important events in the history of metals and metallurgy.

In addition to the basic metals, hundreds of alloys are described, as well as common names such as mild steel, cartridge brass, wrought iron, sterling silver, Muntz metal, Alclad, rare earth metal, metalloids, and killed steel.

Most of the elements in the periodic table of chemical elements are classified as metals. In fact, 73 of the 92 naturally-occurring elements are metals. Metals have been divided into two classes, with *ferrous* implying all metals and alloys that are principally iron, while the *nonferrous* consists of all others. This seemingly one-sided division is not so strange when it is realized that about half of all of the alloys are ferrous, of which most are some type of steel. In all, it is estimated that there are at least 25,000 alloys.

Because each metallic element consists of atoms containing different numbers of subatomic particles, it follows that each metal must be different from all the others, having properties that make each metal unique. The properties of metals are classified as physical and mechanical. The physical properties are the basic characteristics, while the mechanical properties are those that can be determined by deforming or breaking a specimen.

The *Dictionary of Metals* was compiled in a comprehensive manner, and as such it brings together terms from dozens of authoritative publications, introducing new terms and preserving the old.

I wish to acknowledge the following staff of ASM International for their work on this book: Vicki Burt, Scott Henry, Steve Lampman, and Amy Nolan in the Content Department, and Kelly Sukol and Madrid Tramble in the Production Department. I especially thank my son, Bruce Warren Cobb, and Evelyn Dorothy Roberts for assistance with typing; to my cousin, Anne Cobb Moore for assistance with translation; to Susan Frederick for historical research at the public library in Exton, Pennsylvania; and my wife Joan Inman Cobb for proofreading, construction and many suggestions.

This book includes both original work and carefully selected terms from the *ASM Materials Engineering Dictionary,* the *ASM Handbook* series, and other references as listed in the Bibliography. Terms are supplemented by illustrations and tables, and Technical Notes provide concise overviews of the properties, compositions, and applications of selected metals, and direct readers to more detailed information.

The book will be of primary interest to engineers, metallurgists, chemists, professors of Materials Engineering, technicians, librarians, and historians. For these and other interested audiences, in addition to meeting the obvious need of having metallurgical definitions at one's fingertips, it was written with the intent of being an engaging volume that actually can be read as a book.

Harold M. Cobb

About the Author

Harold M. Cobb graduated from Yale University in 1943 with a degree in metallurgical engineering. He has had a long and varied metallurgical career that began with 22 years at companies that produced aircraft gas turbines, propeller blades, helicopters, and fuel elements for nuclear submarines.

He then was on the staff of the American Society for Testing and Materials (ASTM) for 18 years as Group Manager of the Metals Division, where he was an ex-officio member of 12 metals committees. He was a member of ISO Committee TC 17 on Steel and Secretary of Subcommittee 12 on Steel Sheet for 15 years.

In 1970, Cobb was a principal developer of the Unified Numbering System (UNS) for metals and alloys. He developed the individual numbering systems for three categories: Miscellaneous Steels, the K series; Cast Steels, the J series; and Steels Specified by Mechanical Properties, the D series. He also became the number assigner for the D and K series of steels and is still responsible for that activity.

For many years he has been an editorial consultant for the ASTM/SAE publication *Metals and Alloys in the Unified Numbering System (UNS)*. Cobb created and now maintains the Index of Common Names and Trade Names of that book, an index that has grown to approximately 20,000 entries.

He has edited 24 books on metals, and in 2010 authored *History of Stainless Steel*. He has authored several dozen articles, including "Development of the Unified Numbering System for Metals," "The Naming and Numbering of Stainless Steels," and "What's in a Name?" He has worked at various times as an editorial consultant for ASM International, ASTM, the Association for Iron & Steel Technology, the Society of Automotive Engineers, and the Specialty Steel Industry of North America.

Cobb also is a guest editor for *Stainless Steel World*, a Dutch magazine, creating a monthly feature story under the byline "Cobb's Corner."

He holds a patent on a process for manufacturing fuel elements for nuclear submarines. He is a Life Member of ASM International, and a member of ASTM and Committee A-1 on Steel, Stainless Steel, and Related Alloys. Cobb and his wife reside in Kennett Square, Pennsylvania.

Dictionary of Metals
H.M. Cobb, editor

Introduction

Without doubt, none of the arts is older than agriculture, but that of the metals is not less ancient; in fact they are at least equal and coeval, for no mortal man ever tilled a field without implements. In truth, in all the works of agriculture, as in other arts, implements are used which are made of metals which could not be made without the usage of metals; for this reason the metals are of the greatest necessity to man . . . for nothing is made without tools.
—*Vannoccio Biringuccio, Italy, 1540*

Contents

The contents of the book fall into the following categories:

- Detailed descriptions of each of the 73 metallic elements, including the date of discovery, the discoverer, the meaning and source of the name, and principal applications.
- Tables in the Appendix showing the physical properties of each element and its abundance in the earth's crust and in seawater.
- Descriptions of alloys and groups of alloys, often with sources for further information.
- Definitions of metallurgical terms, with references.
- Descriptions of test methods, with references to ASTM tests.
- Historical notes on the prominent men and women in the field of metallurgy.
- Descriptions and illustrations of notable metal structures and applications.
- A separate Metals History Timeline of metals, metallurgy, and notable events and people.

The Earliest Discoveries

The field of metals and metallurgy begins with the seven metals of antiquity, dating from the Bronze and Iron Ages: gold, silver, copper, iron, lead, tin, and mercury. They were found in ores, present primarily as oxides, sulfides, or silicates, and occasionally in pure form as nuggets. The metals were discovered by ancient man inadvertently roasting around a fire these mineral-laden rocks. This led to the building of furnaces to recover the metals, which were made into items such as ornaments, utensils, farming implements, knives, axes, spears, and swords.

In those early days, men associated metals with the gods and the planets. Brilliant, yellow gold became associated with the sun, silver with the moon, and

Mars, the god of war, was for the strong metal, iron. Dark lead was associated with the planet Saturn; the word *saturnine* means heavy, grave, gloomy, dull— the opposite of *mercurial*. Mercury, also known as quicksilver, fell, naturally, to Mercury, the planet that moved fastest across the sky.

The Ages of Man

In 1818, the Danish archeologist Christian Jürgensen Thomsen (1788–1865) divided the ages of man into the Stone, Bronze, and Iron Ages, illustrating the significance of these materials during certain periods of history. His work was not published until 1836 when it appeared in Copenhagen in *Ledetraad til Nordisk Oldkyndighed* (*Guideline to Scandinavian Antiquity*). He did not assign dates, but the dating of artifacts suggests the beginning of the Bronze Age from 3300 to 1200 B.C., depending upon the location. Iron implements found in the Indian subcontinent dating back to circa 1800 B.C. indicate the earliest beginning of the Iron Age in any region. (If Thomsen were alive today, he might well create a Steel Age that started in 1855 with Henry Bessemer's great invention for mass production of steel from molten pig iron.)

A copper-arsenic bronze alloy may have been the first alloy discovered, because copper and arsenic often occurred in the same mineral. When copper was smelted, it also would be mixed with tin to form a preferred type of bronze. The bronze alloys were harder than copper and better suited for tools and weapons.

When iron ore was discovered, it was heated to a red heat with charcoal and the spongy mass, called a *bloom*, was squeezed and hammered to remove some of the silica slag. The 1–2% of slag remaining in the mass consisted of long thin strings that contributed to the properties of the metal that became known first as *iron*, and in the 18th century as *wrought iron*, apparently to ensure it was not confused with *cast iron*. The product also had approximately 0.02% C. This, then, was the *iron* of the Iron Age. There never was a pure form of iron, except what was produced in modern times for experimental purposes. But the iron worked just fine. It eventually would be the stuff of the Eiffel Tower. In addition to its strength and malleability, iron possesses good corrosion resistance.

Although the term *wrought iron* is still often used for some products made today, it is not the wrought iron of old; it is ordinary steel and will have no corrosion resistance. In 1969, the last wrought iron producer in the United States, the A.M. Byers Company of Pittsburgh, closed its doors. Their process was labor intensive and they could not compete with other metals.

With the discovery of all seven of the ancient metals likely having occurred no later than 1000 B.C., it is interesting to note that over 2000 years passed until the next metal was discovered. About 1250 A.D., a German monk by the name of Albertus Magnus discovered arsenic. About 1450, bismuth was recognized as a new metal by Basilius Valentinus of Erfut, Germany. He called it "wismut," which the early mineralogists Latinized to "bismutum." In *De Re Metallica,* published in 1556, Georgius Agricola mentioned that zinc was identified as a metal in India in the 13th century.

During the last half of the 18th century, a dozen metals were discovered that included, in chronological order, cobalt, nickel, manganese, molybdenum, tellurium, tungsten, uranium, zirconium, titanium, yttrium, beryllium, and chromium. It was many years before these metals were actually put to use.

The 19th century, with the Industrial Revolution in full swing, saw the discovery, in chronological order, of niobium, tantalum, iridium, palladium, rhodium, potassium, sodium, boron, barium, calcium, magnesium, strontium, cerium, lithium, cadmium, selenium, silicon, aluminum, thorium, vanadium, lanthanum, erbium, terbium, ruthenium, cesium, rubidium, thorium, indium, and gallium.

In the early part of the 20th century, the final eight metals were discovered (not counting the trans-uranium elements): europium, lutetium, hafnium, rhenium, technetium, promethium, francium, and astatine.

The Naming and Numbering of Metals

Each of the elements was called a different name in every language. Iron, for example, was *ferrum* in Latin, *hierro* in Spanish, *fer* in French, and *eisen* in German. But scientists in virtually every country have agreed on the same chemical symbol for each metal, with the exception of a disagreement over whether one element should be called columbium (Cb) or niobium (Nb). The naturally occurring metallic elements also have atomic numbers in accordance with the periodic table of elements, ranging from atomic number 3 for lithium to number 92 for uranium.

However, the identification system for alloys, which are mixtures of two or more chemical elements, one of which is a metal, has been far more complicated, so much so that it requires large reference books such as *Stahlschlüssel* (*Key to Steel*), published in Germany, just to list all of the steels with their names, numbers, and chemical compositions, for each of approximately 20 countries.

What Is a Metal?

There is no generally agreed-upon definition of metal, and there probably does not need to be. Several approaches to the subject are listed subsequently.

In *De Re Metallica*, published in 1556, German mineralogist Georgius Agricola wrote the following description, which was translated from Latin in 1912 by future U.S. president Herbert Clark Hoover and his wife, Lou Henry Hoover:

> Metal body, by nature either liquid or somewhat hard. The latter may be melted by the heat of fire, but when it is cooled down again and lost all heat it becomes hard again and resumes its proper form. In this respect it differs from the stone which melts in the fire, for although the latter regains its hardness, yet loses its pristine form and properties. Traditionally there are six different kinds of metals, namely gold, silver, copper, iron, tin, and lead. There are really others, for quicksilver is a metal, although the Alchemists disagree with us on this subject, and bismuth is also. The ancient Greek writers seem to have been ignorant of bismith [sic], wherefore Ammonius rightly states that there are many species of metal, animals, and plants that are unknown to us. Stibium when melted in the crucible and refined has as much right to be regarded as a proper metal as is accorded to lead by writers. If, when smelted, a certain amount be added to tin, a bookseller's alloy is produced from which the type is made that is used by those who print books on paper. Each metal has its own form which it preserves when separated from those metals which were mixed with it. Therefore neither electrum nor stannum is a real metal, but rather an alloy of two metals. Electrum is an alloy of gold and silver, stannum of lead and silver. And yet if silver is to be taken from stannum, then lead remains and not stannum. Whether brass, however, is found as native metal or not, cannot be ascertained by any surety. We only know of the artificial brass, which consists of copper tinted with the color of the mineral calamine. And yet if any should be dug up, it would be a proper metal. Black and white copper seem to be different from the red kind. Metal, therefore, is by nature either solid, as I have stated, or fluid, as in the unique case of quicksilver. But enough now, concerning the simple kinds.

In 1965, British metallurgist Donald Birchon inserted the following statement in his *Dictionary of Metallurgy*:

> Metal. There is no rigourous definition of a metal. Earlier attempts to define it in terms of ductility, lustre, conductivity, etc., all fail due to anomalies or inability to be exclusive, and the chemical definition of a substance whose hydroxide is alkaline fails since it does not define materials having commonly accepted "metallic" properties.
>
> A more acceptable approach is a crystalline material, in which the ions are connected indirectly through the field of free electrons surrounding

them. Each ion attracts as many neighboring ions as it can, giving a close-packed structure of short bonds, therefore good strength and relatively high density, associated with good electrical and thermal conductivity, ductility, and reflectivity.

The Nature of Metals

Everything on earth is composed of one or more of the 92 natural elements that fall into two groups: metals and nonmetals. The nonmetals consist of the gases hydrogen, oxygen, and nitrogen plus the inert gases argon, krypton, xenon, and neon. The other nonmetallic elements are carbon, phosphorus, sulfur, chlorine, bromine, fluorine, radon, and iodine.

The 73 naturally occurring metals are derived primarily from minerals in the earth's crust in the form of oxides, silicates, or sulfides. Iron, for example, is found as iron oxide in several different minerals. Iron is separated from the oxide by heating at a high temperature with coke, which provides carbon that combines with the oxygen of the ore to allow molten iron to be produced. Many of the smelting procedures are very complex.

Each metal is unique because its atoms are composed of a specific number of protons, neutrons, and electrons. This means that the properties of each metal are different. The density of each metal, for example, is different. Lead is approximately four times as heavy as aluminum. The melting points are different and vary from -38.89 °C (-38 °F) for mercury to 3410 °C (6170 °F) for tungsten.

The thermal conductivities are mostly different, with copper and silver being almost twice as heat conductive as aluminum. The strengths of the metals vary considerably, and some metals are much stronger than others at elevated temperatures. Metals also have enormous differences with respect to their corrosion resistance in various environments.

The unique properties of each metal give rise to special uses. Copper wire, for example, is ideal for conducting electricity. Aluminum is the ideal metal for beverage cans, and lead is the best possible metal for storage batteries for cars. Tungsten and molybdenum are ideal for the fine filaments in incandescent bulbs. Pure chromium metal is as brittle as glass but is applicable for the very thin chrome plate finish on steel. Gold is the most highly prized metal for its appearance and the most precious to own.

The Development of Alloys

Alloys are mixtures of two or more chemical elements, one of which is a metal. For example, iron, a metal, combined with 1% or less of carbon, creates alloys of steel. Each of the metals described previously can be combined to make dozens or hundreds of alloys. One of the most interesting things in the field of metallurgy is that iron and carbon can be mixed to form the greatest number of alloys of any metal. These are primarily steels and cast irons. Some of the steels are just iron and carbon alloys, but there are hundreds of other alloys that have been created by adding small or large amounts of chromium, molybdenum, vanadium, tungsten, and other elements. The total number of steels is approximately 5000. The so-called *tool steels* are used to forge, roll, press, and cut steels themselves and all the other metals and alloys, in addition to providing the tools and equipment necessary to produce almost everything.

A286. An Fe-Ni-Cr superalloy with additions of molybdenum and titanium; an age-hardenable alloy with good strength and oxidation resistance up to approximately 700 °C (1300 °F). It is also known as type 660 and UNS S66286. The alloy is described in ASTM A453.

A5. A French coinage alloy, 90% Al, 5% Ag, 5% Cu.

abnormal steel. Fully-hardened or case-hardened steels that are of the correct carbon content and have been heat treated in the normal way but contain soft spots. In an annealed specimen, the structure in these areas consists of coarse and irregular pearlite, and at the eutectoid composition, massive cementite is found in the austenite grain boundaries with ferrite on either side. Such abnormality appears to result from the use of aluminum as a deoxidizer during the manufacture of the steel and may be detected by the *McQuaid-Ehn Test.*

Abros. A corrosion-resistant alloy, 88% Ni, 10% Cr, 2% Mn.

Abyssinian gold. A mock gold, typically 88% Cu, 11.5% Zn, 0.5% Au.

accelerated corrosion test. See *salt spray test.*

ACI alloys. A designation system developed by the Alloy Casting Institute in the United States in 1940 for stainless steel and high-nickel casting alloys. The system is divided into a C Group for corrosion-resisting alloys primarily intended for use below 650 °C (1200 °F), and an H Group for heat-resisting alloys to be used primarily at temperatures above 650 °C (1200 °F).

A second letter, from A to Z, denotes the approximate amounts of chromium and nickel. The number following the first letter, for the C grades, indicates the carbon limit in hundredths of a percent. For the H grades, the number indicates the middle of the carbon range in hundredths of a percent. One or more letters following the carbon content code number signify the principal alloying element(s) in a grade, using a mixture of code letters and chemical symbols.

For example, CA-15 is the ACI designation for the corrosion-resisting 12%

Cr alloy with 0.15% maximum carbon, and HW is the designation for a Ni-Fe-Cr alloy having a total nickel and chromium content of 68% and a carbon content of 0.50%.

The ACI system is administered by Subcommittee 18 on Steel Castings of ASTM's Committee A-1 on Steel.

acicular ferrite. A highly substructured nonequiaxed ferrite formed upon continuous cooling by mixed diffusion and shear mode of transformation that begins at a temperature slightly higher than the transformation temperature range for upper bainite. It is distinguished from bainite in that it has a limited amount of carbon available; thus, there is only a small amount of carbide present.

acid bottom and lining. The inner bottom and lining of a melting furnace, consisting of materials such as sand, siliceous rock, or silica brick that give an acid reaction at the operating temperature.

acid bronze. A leaded nickel-tin bronze of variable composition, usually 8–10% Sn, 2–17% Pb, 0–2% Zn, 0–2% Ni, bal Cu, which casts well because it is really a leaded nickel-containing gun metal.

acid embrittlement. A form of hydrogen embrittlement that may be induced in some metals by acid treatment.

acid steel. Steel melted in a furnace with an acid bottom and lining under a slag containing an excess of an acid substance such as silica.

actinide metals. A series of chemically similar radioactive elements ranging from atomic number 89 (actinium) through 103 (lawrencium).

actinium. A radioactive metal having atomic number 89, atomic weight 227, and symbol Ac. It is extracted with thorium from pitchblende. The metal was erroneously discovered and named by André Debierne for *Aktis* or *Aktinos,* the Greek words for *beam* or *ray.* The actual element was obtained in 1902 by Friedrich Otto Giesel.

activation. The changing of a passive surface of a metal to a chemically active state. Contrast with *passivation.*

activation energy. The energy required to initiate a rate process (e.g., plastic flow, diffusion, chemical reaction).

adhesive wear. The removal of material from a surface by the welding together and subsequent shearing of minute areas of the two surfaces that slide across each other under pressure. In advanced stages this may lead to *galling* or *seizing.*

admiralty brass. An α-brass, nominally 70% Cu, 29% Zn, 1% Sn. An alloy exhibiting good corrosion resistance in seawater; its resistance to *dezincification* is improved by the addition of 0.02–0.05% As.

Adnic alloy. A copper-nickel alloy, typically 70% Cu, 29% Zn, 1% Sn, used for condenser tubes due to its high resistance to corrosion and impingement attack.

Advance. A 56–60% Cu, bal Ni alloy (i.e., similar to Constantan and Eureka) of high electrical resistivity (approximately 48 microhm-cm), very small temperature coefficient of resistivity up to 450 °C, excellent corrosion resistance, and high thermal electromotive force (see *thermocouple*). It is used for standard resistances, potentiometers, and thermocouples.

Aerospace Material Specifications (AMS). Specifications by SAE International that have been used worldwide since 1939 for the design and production of aircraft components and systems. The specifications have four-digit numbers, such as AMS 4000, which covers aluminum alloy 1060.

age hardening. Hardening by aging, usually after rapid cooling or cold working. See *aging*.

age softening. Spontaneous decrease of strength and hardness that takes place at room temperature in certain strain-hardened alloys, especially those of aluminum.

aging. A change in the properties of certain metals and alloys that occurs at ambient or slightly elevated temperatures after hot working or heat treatment (quench aging in ferrous alloys, natural or artificial aging in ferrous or nonferrous alloys) or after a cold working operation (strain aging). The change in properties is often, but not always, due to a phase change (precipitation), but never involves a change in the chemical composition of the metal or alloy. See also *age hardening, artificial aging, interrupted aging, natural aging, overaging, precipitation hardening, precipitation heat treatment, progressive aging, step aging, strain aging*.

AgION antimicrobial coating. A silver-containing coating on flat rolled stainless and carbon steel, introduced in 2002 by AK Coatings, a wholly-owned subsidiary of AK Steel, in Middletown, Ohio. The coating suppresses a broad array of destructive microbes including bacteria, molds, and fungi. The steel sheet can be used in construction, for food equipment and appliances, and for heating, ventilating, and air conditioning systems.

Agricola, Georgius. 1494–1555. Author of *De Re Metallica,* a book cataloging the state of the art of mining, refining, and smelting metals. Agricola was a German born in Glauchau in Saxony. His real name was Georg Pawer. Agricola was the Latinized version of his name, Pawer (Bauer), meaning farmer. He was a brilliant scholar who, at the age of 20, was appointed Rector extraordinarius of Greek at Great School of Zwickau. He studied medicine, physics, and chemistry in Leipzig in 1516 and pursued a doctorate degree in Italy from 1524 to 1526.

After returning to Zwickau in 1527, Agricola was appointed town physician of Joachimsthal, which was in the very center of the greatest mining area of Eastern Europe. He proceeded to spend all of his spare time roaming the countryside interviewing miners and smelters, writing everything down and drawing some 300 sketches. Over a 20-year period he produced *De Re Metallica*. He accomplished this at a time when only ten metallic elements were known—the seven metals of antiquity plus the more recent discoveries of arsenic, antimony, and bismuth. His 600-page textbook remained the classic guide to mining and smelting for 200 years.

In 1912, Herbert Clark Hoover, a mining engineer who became president of the United States in 1928, and his wife Lou Henry, a Latin scholar, undertook the first English translation of *De Re Metallica*. Hoover added copious footnotes to explain much of the text. In 1986, Dover Publications, Inc., New York, republished an unabridged version of the translation that included the original 289 woodcuts.

agricultural steels. Spades, forks, hoes, and plows usually are made of 0.25–0.5% C steel. A cast iron weld deposit often is applied to plows as an inexpensive hard facing, which, being chilled by the mass of the plow, forms a white or mottled iron.

AIME. The American Institute of Mining Engineers (now the American Institute of Mining, Metallurgical, and Petroleum Engineers), founded in 1871 by a group of 22 mining engineers in Wilkes-Barre, Pennsylvania, in the heart of the anthracite coal mining district. In 1906, the organization was incorporated in New York and the headquarters were established in New York City. In 1946, AIME was divided into three professional societies: mining, metallurgical, and petroleum engineering. They began publishing three journals, including the *Journal of Metals*. In 1956, the organization became the American Institute of Mining, Metallurgical, and Petroleum Engineers, but the name remained AIME.

air classification. The separation of metal powder into particle-size fractions by means of an air stream of controlled velocity; an application of the principle of *elutriation*.

aircraft-quality alloy steel. Alloy steel sheet of sufficiently high quality for use in highly stressed parts of aircraft, missiles, and similar applications involving stringent requirements. This quality requires exacting steelmaking, conditioning, and processing controls. Electric furnace melting and vacuum degassing sometimes are required.

Internal soundness, uniformity of chemical composition, austenitic grain size (5 or finer), and good surface are primary requirements. The quality is normally furnished in the annealed, spheroidized annealed, and normalized conditions. Aircraft-quality steel sheet

can be specified to *Aerospace Material Specifications (AMS)*.

air-hardening steel. A steel of alloy content sufficient to ensure hardening throughout the section during slow cooling in still air from the austenitizing temperature. Clearly dependent on section size, the usual limit is a ruling section of 5.72 cm (2.25 in.).

AISI. The American Iron & Steel Institute, organized in New York City on March 31, 1908.

AISI alloy steels. AISI alloys are designated as shown in Table 1.

AISI carbon steels. Carbon steels codified and numbered by the AISI system for numbering carbon steels, which was established about 1917. Carbon steels were codified according to these rules:

Carbon steels with 1.00% maximum manganese are identified with a four-digit

Table 1 AISI alloy designations

Designation	Approximate percentages of identifying elements
13xx	Mg 1.75
40xx	Mo 0.25
41xx	Cr 0.50, 0.80 or 0.95; Mo 0.12, 0.16, 0.20 or 0.30
43xx	Ni 1.83, Cr 0.50 or 0.80; Mo 0.25
46xx	Ni 1.83, Mo 0.25
47xx	Ni 0.85 or 1.05; Cr 0.45 or 0.55; Mo 0.20 or 0.55
48xx	Ni 3.50, Mo 0.25
51xx	Cr 0.80, 0.88, 0.93, or 1.00
51xxx	Cr 1.03
52xxx	Cr 1.45
61xx	Cr 0.60 or 0.95; V 0.13 or 0.15 min.
86xx	Ni 0.55, Cr 0.50, Mo 0.20
87xx	Ni 0.55, Cr 0.50, Mo 0.25
88xx	Ni 0.55, Cr 0.50, Mo 0.35
92xx	Si 2.00 or Si 1.40 and Cr 0.70; or Si 1.00 and Cr 0.55
50Bxx	Cr 0.28 or 0.50
51Bxx	Cr 0.80
81Bxx	Ni 0.30, Cr 0.45, Mo 0.12
94Bxx	Ni 0.45, Cr 0.40, Mo 0.12

B denotes boron steel

system from 1005 to 1095, wherein the final two digits signify the nominal carbon content from 0.05–0.95%. Resulfurized carbon steels are identified with a four-digit system from 1110 to 1152, wherein the last two digits signify the carbon content. Carbon steels with over 1.00% Mn are identified with a four-digit number from 1117 to 1590, wherein the last two digits signify the carbon content. Rephosphorized and resulfurized steels are identified with a four-digit system from 1212 to 1215 plus 12L14, a leaded steel, wherein the last two digits signify the carbon content.

AISI steel numbers. The American Iron & Steel Institute's numbering system for carbon and alloy steels (e.g., 1010 and 4130) that specifies steels by hardenability (e.g., 4130H), stainless steels (e.g., type 410), and tool steels (e.g., M-1). AISI became the principal developer in the United States of such numbering systems beginning in approximately 1918. AISI also assumed the responsibility for assigning numbers, but not to proprietary grades of steel. The AISI numbers became widely used in North America and abroad, where some standards organizations, including those in England, Australia, and Japan, adopted them.

AISI discontinued assigning numbers in the 1960s, which led to the development of the Unified Numbering System (UNS) for Metals and Alloys in the United States. See *UNS*.

AIST. The Association for Iron & Steel Technology, a nonprofit organization dedicated to advancing the technical development, production, processes, and applications of iron and steel. AIST was founded in 2004 by a merger of the Association of Iron & Steel Engineers and the Iron & Steel Society. The Iron & Steel Society was a society organized by AIME (American Institute of Mining, Metallurgical, and Petroleum Engineers) in 1974. AIST is the current publisher of the classic *Making, Shaping and Treating of Steel*.

aitch metal. A brass of *Muntz metal* type (60% Cu, 40% Zn) with the addition of 1–2% Fe, combining good casting properties with strength.

AK 410Cb. A proprietary stainless steel alloy manufactured by the AK Steel Corporation of Middletown, Ohio. Approximately 0.15% Nb is added to type 410 stainless steel for greater ease of heat treatment, higher strength, and improved grain size control.

Alclad. A composite wrought product comprised of an aluminum alloy core having on one or both surfaces a metallurgically bonded aluminum or aluminum alloy coating that is anodic to the core and thus protects the core against corrosion.

alkali metal. A metal in Group IA of the periodic system—namely lithium, sodium, potassium, rubidium, cesium, and francium. They form strongly alkaline hydroxides, hence the name.

alkaline cleaner. A material blended from alkali hydroxides and alkaline salts such as borates, carbonates, phosphates, or silicates. The cleaning action may be enhanced by the addition of surface active agents and special solvents.

alkaline earth metal. A metal in Group IIA of the periodic system—namely, beryllium, magnesium, calcium, strontium, barium, and radon—so called because the oxides or "earths" of calcium, barium, and strontium were found by the early chemists to be alkaline in reaction.

alligatoring. The longitudinal splitting of flat slabs in a plane parallel to the rolled surface (Fig. 1). Also called fish-mouthing.

Fig. 1 Alligatoring in a rolled slab

allotriomorphic crystal. A crystal whose lattice structure is normal but whose external surfaces are not bounded by regular crystal faces; rather, the external surfaces are impressed by contact with other crystals or another surface such as a mold wall, or are irregularly shaped because of nonuniform growth. Compare with *idiomorphic crystal*.

alloy. A substance having metallic properties and being composed of two or more chemical elements of which at least one is a *metal*.

alloying element. An element that is added to a metal (and remains with the metal) to effect changes in properties.

alloy plating. The co-deposition of two or more metallic elements.

alloy powder. A powdered metal in which each particle is an alloy.

alloy steel. A steel, other than a stainless steel, conforming to a requirement that one or more of the following elements, by mass percent, have a minimum content equal to or greater than 0.30 for aluminum; 0.0008 for boron; 0.30 for chromium; 0.30 for cobalt; 0.40 for copper; 0.40 for lead; 1.65 for manganese; 0.08 for molybdenum; 0.30 for nickel; 0.06 for niobium (columbium); 0.60 for silicon; 0.05 for titanium; 0.30 for tungsten (wolfram); 0.10 for vanadium; 0.05 for zirconium; or 0.10 for any other alloying element, except sulfur, phosphorus, carbon, and nitrogen (ASTM A941). Table 2 lists compositions for typical alloy steels.

all-weld-metal test specimen. A test specimen wherein the portion being tested is composed wholly of weld metal.

alpha-beta (α-β) brass. A copper alloy with 35–45% Z. Also called duplex brass. The α phase is a face-centered cubic structure, and the β phase is a body-centered cubic structure.

alpha (α) brass. A copper alloy having less than 38% Z, which is malleable and used to make inexpensive jewelry. It is a one-phase face-centered cubic structure.

alpha (α) iron. The body-centered cubic form of pure iron, stable at 910 °C (1670 °F).

alternate immersion test. A corrosion test in which the specimens are intermittently immersed in and removed from a liquid medium at definite time intervals.

Alumel. A nickel-base alloy containing approximately 2.5% Mn, 1% Al, and 1% Si, used chiefly as a component of a pyrometric thermocouple (e.g., Chromel-Alumel).

aluminizing. Forming of an aluminum or an aluminum alloy coating by dipping, hot spraying, or diffusion.

aluminum. Atomic number 13, atomic weight 27, symbol Al for *alumen* the

Table 2 Chemical compositions for typical alloy steels

Steel	Composition, wt%(a)								
	C	Si	Mn	P	S	Ni	Cr	Mo	Other
Low-carbon quenched and tempered steels									
A 514/A 517 grade A	0.15–0.21	0.40–0.80	0.80–1.10	0.035	0.04	...	0.50–0.80	0.18–0.28	0.05–0.15 Zr(b) 0.0025 B
A 514/A 517 grade F	0.10–0.20	0.15–0.35	0.60–1.00	0.035	0.04	0.70–1.00	0.40–0.65	0.40–0.60	0.03–0.08 V 0.15–0.50 Cu 0.0005–0.005 B
A 514/A 517 grade R	0.15–0.20	0.20–0.35	0.85–1.15	0.035	0.04	0.90–1.10	0.35–0.65	0.15–0.25	0.03–0.08 V
A 533 type A	0.25	0.15–0.40	1.15–1.50	0.035	0.04	...	...	0.45–0.60	...
A 533 type C	0.25	0.15–0.40	1.15–1.50	0.035	0.04	0.70–1.00	...	0.45–0.60	...
HY-80	0.12–0.18	0.15–0.35	0.10–0.40	0.025	0.025	2.00–3.25	1.00–1.80	0.20–0.60	0.25 Cu 0.03 V 0.02 Ti
HY-100	0.12–0.20	0.15–0.35	0.10–0.40	0.025	0.025	2.25–3.50	1.00–1.80	0.20–0.60	0.25 Cu 0.03 V 0.02 Ti
Medium-carbon ultrahigh-strength steels									
4130	0.28–0.33	0.20–0.35	0.40–0.60	...	...	...	0.80–1.10	0.15–0.25	...
4340	0.38–0.43	0.20–0.35	0.60–0.80	...	...	1.65–2.00	0.70–0.90	0.20–0.30	...
300M	0.40–0.46	1.45–1.80	0.65–0.90	...	...	1.65–2.00	0.70–0.95	0.30–0.45	0.05 V min
D-6a	0.42–0.48	0.15–0.30	0.60–0.90	...	...	0.40–0.70	0.90–1.20	0.90–1.10	0.05–0.10 V
Carburizing bearing steels									
4118	0.18–0.23	0.15–0.30	0.70–0.90	0.035	0.040	...	0.40–0.60	0.08–0.18	...
5120	0.17–0.22	0.15–0.30	0.70–0.90	0.035	0.040	...	0.70–0.90	...	...
3310	0.08–0.13	0.20–0.35	0.45–0.60	0.025	0.025	3.25–3.75	1.40–1.75	...	...
Through-hardened bearing steels									
52100	0.98–1.10	0.15–0.30	0.25–0.45	0.025	0.025	...	1.30–1.60	...	...
A 485 grade 1	0.90–1.05	0.45–0.75	0.95–1.25	0.025	0.025	0.25	0.90–1.20	0.10	0.35 Cu
A 485 grade 3	0.95–1.10	0.15–0.35	0.65–0.90	0.025	0.025	0.25	1.10–1.50	0.20–0.30	0.35 Cu

(a) Single values represent the maximum allowable. (b) Zirconium may be replaced by cerium. When cerium is added. The cerium/sulfur ratio should be approximately 1.5/1. based on heat analysis.

Latin word for *alum*, an aluminum-phosphorus-silicate compound used medically as an astringent or styptic.

Aluminum was isolated by Friedrich Wohler in 1827, although an impure form was prepared by Hans Christian Oersted two years earlier. The element already was known as one component of a base called alum or alumen. It is the third most abundant element, making up 8.3% of the earth's crust, and the second most abundant metal, after silicon. Its density is approximately one-third that of steel. (See *Appendix 3* for other physical properties.)

The metal actually was called aluminium in the United States until 1925, when the American Chemical Society changed the name to aluminum in their publications. Today, all countries use the *aluminium* spelling except the United States and Canada. Aluminum is too reactive to occur in nature as a free metal but it occurs in over 270 combination forms. The principal source of the metal is bauxite, a mineral discovered in the French town of Baux, near Arles, France. Initially, aluminum was considered to be a precious metal because it was so costly to refine. Napoleon III, Emperor of France, is said to have given a banquet where the honored guests were given aluminum utensils to use, while the other guests had to make do with gold.

In 1884, when the Washington Monument was completed, it was decided to top it with a pyramid-shaped cap of cast aluminum that weighed approximately six pounds and was the largest aluminum casting ever made. At that time it was as expensive as silver. The method of producing aluminum metal by the electrolysis of alumina dissolved in cryolite was discovered by Charles Martin Hall in the United States in 1886 and by Paul Heroult in France in 1888, a process that drastically cut the cost of producing aluminum. Aluminum and its alloys, with their light weight and other unique properties, have become the most widely used metals after steel. Approximately one-fourth of all aluminum produced is used for packaging and beverage cans; a fifth is used in the construction of airplanes and buses; and the remainder is used in the building and construction industry, and for electric transmission lines, kitchenware, and hardware. See *Technical Note 1*.

aluminum brass. A copper-zinc alloy with added aluminum to improve corrosion resistance.

amalgam. An alloy of mercury with one or more other metals.

americium. A chemical element having atomic number 95, atomic weight 243, and the symbol Am, named for America, which, in turn, was named for Amerigo Vespucci, the Italian explorer. This trans-uranium radioactive element was identified by Glenn T. Seaborg, Albert Ghiorso, and co-workers in 1944.

amorphous. Not having a crystal structure; noncrystalline.

amphoteric. Possessing both acidic and basic properties.

AMS. See *Aerospace Material Specifications*.

anchorite. A zinc-iron phosphate coating for iron and steel.

anelasticity. The property of solids by virtue of which strain is not a single-value function of stress in the low-stress range where no permanent set occurs.

anion. A negative ion.

TECHNICAL NOTE 1

Aluminum and Aluminum Alloys

ALUMINUM, also called aluminium in England, is a white metal with a bluish tinge obtained chiefly from bauxite. It is the second most abundant metallic element on earth. Aluminum metal is produced by first extracting alumina (aluminum oxide) from bauxite by a chemical process. The alumina is then dissolved in a molten electrolyte, and an electric current is passed through the bath, causing the metallic aluminum to be deposited on the cathode.

The unique combinations of properties provided by aluminum and its alloys make it suitable for a broad range of uses—from soft, highly ductile wrapping foil to the most demanding engineering applications. Its low density and strength-to-weight ratio are its most useful characteristics. It weighs only about 2.7 g/cm^3, approximately one-third as much as the same volume of steel, permitting design and construction of strong, lightweight structures—particularly advantageous for aerospace, aircraft, and land- and waterborne vehicles. Aluminum has high resistance to corrosion in atmospheric environments, in fresh and salt water, and in many chemicals. It has no toxic reactions and is highly suitable for processing, handling, storing, and packaging of foods and beverages. The high electrical and thermal conductivities of aluminum account for its use in many applications. It is also non-ferromagnetic, a property of importance in the electrical and electronic industries.

Mill products constitute the major share (~80%) of total aluminum product shipments, followed by casting and ingot other than for castings. In decreasing order of current market size, the major application categories are containers and packaging (27.4%), transportation (21.1%), building and construction (17.8%), electrical (9.1%), consumer durables (8.0%), machinery and equipment (5.9%), and others, including exports (<12%).

In addition to being available in wrought and cast form, aluminum alloys are also produced by powder metallurgy processing. Aluminum is also used extensively in *metal-matrix composites*.

Nominal compositions of selected aluminum alloys
See also wrought and cast aluminum alloy designations

Alloy	Nominal composition, %	Alloy	Nominal composition, %
Wrought aluminum alloys(a)		2618	Al-2.3Cu-1.6Mg-1.0Ni-l.1 Fe-0.07Ti
1100	0.12Cu-99.00Al (min)	3003	Al-0.12Cu-1.2Mn
1230	99.30Al (min)	5052	Al-2.5Mg-0.25Cr
2014	Al-0.8Si-4.4Cu-0.8Mn-0.5Mg	5083	Al-0.6Mn-4.45Mg-0.15Cr
2024	Al-4.4CU-0.6Mn-1.5	5086	Al-0.45Mn-4.0Mg-0.15Cr
2025	Al-0.8Si-4.5Cu-0.8Mn	5454	Al-0.8Mn-2.7Mg-0.12Cr
2117	Al-2.6Cu-0.35Mg	5456	Al-0.8Mn-5.1Mg-0.12Cr
2218	Al-4.0Cu-1.5Mg-2.0Ni	5457	Al-0.3Mn-1.0Mg
2219	Al-6.3Cu-0.3Mn-0.06Ti-0.1 V-0.18Zr	5657	Al-0.8Mg

(continued)

(a) Wrought alloys are identified by Aluminum Association designations. (b) Casting alloys are identified first by Aluminum Association designations (without decimal suffixes) and then, parenthetically, by industry designations.

TECHNICAL NOTE 1 (*continued*)

Nominal compositions of selected aluminum alloys (*continued*)
See also wrought and cast aluminum alloy designations

Alloy	Nominal composition, %	Alloy	Nominal composition, %
Wrought aluminum alloys(a) (continued)		A356 (A356)	Al-7.0Si-0.3Mg-0.2Fe max
6061	Al-0.6Si-0.27Cu-1.0Mg-0.2Cr	A357 (A357)	Al-7.0Si-0.5Mg-0.15Ti
6063	Al-0.4Si-0.7Mg	380 (380)	Al-9.0Si-3.5Cu
6151	Al-0.9Si-0.6Mg-0.25Cr	384 (384)	Al-12.0Si-3.8Cu
6351	Al-1.0Si-0.6Mn-0.6Mg	392 (392)	Al-19.0Si-0.6Cu-0.4Mn-1.0Mg
7004	Al-0.45Mn-1.5Mg-4.2Zn-0.15Zr	413 (13)	Al-12.0Si
7039	Al-0.27Mn-2.8Mg-0.2Cr-4.0Zn	443 (43)	Al-5.0Si
7072	Al-1.0Zn	B443 (43)	Al-5.0Si-0.3Cu max
7075	Al-1.6Cu-2.5Mg-0.3Cr-5.6Zn	C443 (A43)	Al-5.0Si-2.0Fe max
7079	Al-0.6Cu-0.2Mn-3.3Mg-0.2 Cr-4.3Zn	520 (220)	Al-10.0Mg
		D712 (D612, 40E)	Al-0.6Mg-5.3Zn-0.5Cr
7178	Al-2.0Cu-2.7Mg-0.3Cr-6.8Zn	850 (750)	Al-1.0Cu-1.0Ni-6.5Sn
Aluminum casting alloys(b)		**Aluminum alloy filler metals and brazing alloys**	
201 (KO-1)	Al-4.7Cu-0.6Ag-0.3Mg-0.2Ti	ER2319	Al-6.2Cu-0.30Mn-0.15Ti
222 (122)	Al-10.0Cu-0.2Mg	ER4043	Al-5.2Si
224 (···)	Al-5.0Cu-0.4Mn	ER5356	Al-0.12Mn-5.0Mg-0.12 Cr-0.13Ti
238 (138)	Al-l0.0Cu-4.0Si-0.3Mg		
A240 (A140)	Al-8.0Cu-0.5Mn-6.0Mg-0.5Ni	5456	Al-0.8Mn-5.lMg-0.l2Cr
242 (142)	Al-4.0Cu-1.5Mg-2.0Ni	R-SG70A	Al-7Si-0.30Mg
295 (195)	Al-4.5Cu-0.8Si	4047 (BAlSi-4)	Al-12Si
308 (A108)	Al-4.5Cu-5.5Si	4245	Al-10Si-4Cu-10Zn
319 (319)	Al-3.5Cu-6.0Si	4343 (BAlSi-2)	Al-7.5Si
A332 (A132)	Al-12.0Si-0.8Cu-1.2Mg-2.5Ni	No. 12 brazing	3003 alloy, 4343
354 (354)	Al-9.0Si-1.8Cu-0.5Mg	sheet	cladding on both sides
355 (355)	Al-1.3Cu-5.0Si-0.5Mg		
356 (356)	Al-7.0Si-0.3Mg		

(a) Wrought alloys are identified by Aluminum Association designations. (b) Casting alloys are identified first by Aluminum Association designations (without decimal suffixes) and then, parenthetically, by industry designations.

---

anistropy. The characteristic of exhibiting different values of a property in different directions with respect to a fixed reference system in the material.

Anka. The first austenitic stainless steel, an alloy with 15% Cr and 11% Ni, produced by Brown of Bayley's Steel Works in Sheffield, England.

annealing. A generic term denoting a treatment consisting of heating to and holding at a suitable temperature followed by cooling at a suitable rate, used primarily to soften metallic materials, but also to simultaneously produce desired changes in other properties or in microstructure. The purpose of such changes may be, but is not confined to, improvement of machinability, facilitation of cold work, improvement of mechanical or electrical properties, and/or increase in stability of dimensions. When the term is used unqualifiedly, full annealing is implied. Where applied only for the relief of stress, the process is properly called *stress relieving* or stress-relief annealing.

In ferrous alloys, annealing usually is done above the upper critical temperature, but the time-temperature cycles vary widely both in maximum temperature attained and the cooling rate employed, depending on composition, material condition, and results desired. When applicable, the following process names should be used: *black annealing, blue annealing, box annealing, bright annealing, cycle annealing, flame annealing, full annealing, graphitizing,* in-process annealing, *isothermal annealing, malleablizing,* orientation annealing, *process annealing, quench annealing, spheroidizing,* and *subcritical annealing.*

In nonferrous alloys, annealing cycles are designed to: (a) remove all or part of the effects of cold working (recrystallization may or may not be involved); (b) cause substantially complete coalescence of precipitates from solid solution in relatively coarse form; or (c) both, depending on composition and material condition. Specific process names in commercial use are *final annealing, full annealing, intermediate annealing, partial annealing, recrystallization annealing,* stress-relief annealing, and *anneal to temper.*

annealing carbon. Fine, apparently amorphous carbon particles formed in white cast iron and certain steels during prolonged annealing. Also called *temper carbon.*

annealing twin. A *twin* formed in a crystal during recrystallization.

anneal to temper. A final partial anneal that softens a cold worked nonferrous alloy to a specified level of hardness or tensile strength.

anode. The electrode where electrons leave an operating system such as a battery, an electrolytic cell, an x-ray tube, or a vacuum tube. In the first of these, it is negative; in the other three, positive. In a battery or electrolytic cell, it is the electrode where oxidation occurs. See *cathode.*

anode copper. Special-shaped copper slabs resulting from the refinement of *blister copper* in a reverbatory furnace, used as anodes in electrolytic refinement.

anode corrosion. The dissolution of a metal acting as an *anode.*

anode effect. The effect produced by polarization of the *anode* in electrolysis. It is characterized by a sudden increase in voltage and a corresponding decrease in amperage due to the anode becoming virtually separated from the electrolyte by a gas film.

anodic cleaning. *Electrolytic cleaning* in which the work is the anode. Also called *reverse-current cleaning.*

anodic coating. A film on a metal surface resulting from an electrolytic treatment at the anode.

anodic pickling. *Electrolytic pickling* in which the work is the anode.

anodic protection. Imposing an external electrical potential to protect a metal from corrosive attack.

anodizing. Forming a *conversion coating* on a metal surface by anodic oxidation; most frequently applied to aluminum.

anolyte. The electrolyte adjacent to the *anode* in an *electrolytic cell.*

antiferromagnetic material. A material wherein interatomic forces hold the elementary atomic magnets (electron spins) of a solid in alignment, a state similar to that of *ferromagnetic material* but with the difference that equal numbers of elementary magnets (spins) face in opposite directions and are antiparallel, causing the solid to be weakly magnetic, that is, paramagnetic, instead of ferromagnetic.

antimony. A chemical element having atomic number 51 and atomic weight 122. The chemical symbol, Sb, is from the Latin *stibium,* but the word *antimony* is from the Greek *anti* + *monos,* meaning a metal not found alone. The lustrous metal with a bluish-silvery-white appearance was discovered near the year 1500. Antimony and its compounds are highly toxic.

Antimony is used as an alloying element to harden lead. During solidification the metal expands slightly, which is advantageous for making lead shot and type metal. The most important use is the Babbitt bearing alloys (see *Babbitt metal*), which consist of tin, antimony, and copper with or without lead. A major use is in solders.

antipitting agent. An addition agent for electroplating solutions to prevent the formation of large pits or pores in the electrodeposit.

AOD process. Argon-oxygen-decarburization process. A secondary refining process for steel whereby these gases are injected into the molten metal through submerged side-mounted tuyeres.

The process was invented in 1954 by William A. Krivsky (1927–2006) at the Metals Research Laboratory of the Union Carbide Corporation in Niagara Falls, New York. At that time there was no practical way of reducing the carbon content of 18-8 stainless steel below 0.03%, which could prevent the susceptibility of the metal to intergranular corrosion. The use of oxygen to decarburize the molten steel in the arc furnace was not practical because oxygen overheated the steel, causing damage to the furnace lining. Krivsky experimented by melting a small heat of stainless steel while diluting the flow of oxygen with the inert gas argon. He repeated the experiment, which was highly successful, several times. Fully mindful that he had just made one of the most important metallurgical discoveries, he applied for a patent

and proceeded to scale up his experiment. His experiments with larger heats were not successful, and the project was turned over to the Joslyn Stainless Steel Company in Ft. Wayne, Indiana. They solved the problem only after building a special refining vessel and experimenting with various designs over a period of 12 years. Joslyn built a 15-ton refining vessel in 1968, which produced the following results:

1. The productivity was doubled.
2. The lower refining temperatures reduced refractory lining costs.
3. Elimination of the refining and finishing operations from the melting furnace more than offset the operating cost of the AOD.
4. The yield of the metallic alloying elements increased.
5. Lead was kept at very low levels of no more than 0.007%.
6. Sulfur, oxygen, hydrogen, and nitrogen levels were lower than previously possible.
7. The consistency, control, and reproducibility improved ductility, fatigue strength, and machinability of many alloys.
8. It was possible to add nitrogen within very close limits for the production of new nitrogen-hardening alloys.
9. The refinements of the process opened the possibility of developing new alloys.

The AOD process ended the susceptibility of stainless steels to intergranular corrosion. It opened the way to development of the duplex class of stainless steels and improved foundry operations. Within ten years there were over 100 AOD vessels in operation worldwide. Figure 2 shows a schematic of an AOD vessel.

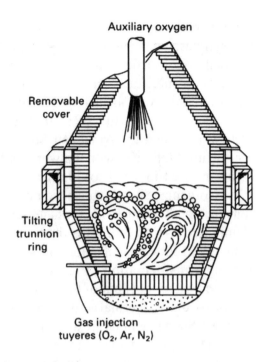

Auxiliary oxygen

Removable cover

Tilting trunnion ring

Gas injection tuyeres (O_2, Ar, N_2)

Fig. 2 Schematic of an argon-oxygen-decarburization (AOD) vessel

apparent density. (1) The weight per unit volume of a metal powder, in contrast to the weight per unit volume of the individual particles. (2) The weight per unit volume of a porous solid, where the unit volume is determined from external dimensions of the mass. Apparent density is always less than the true density of the material itself.

Ar_{cm}, Ar_1, Ar_3, Ar_4, Ar', Ar''. Defined under *transformation temperature.*

arbitration bar. A test bar, cast with a heat of material, used to determine chemical composition, hardness, tensile strength, and deflection and strength under transverse loading in order to establish the acceptability of the casting.

arc brazing. A brazing process in which the heat required is obtained from an electric arc.

arc cutting. Cutting processes that melt the metal to be cut with the heat of an arc between an electrode and the base metal. See *metal arc cutting, gas tungsten arc cutting,* and *plasma arc cutting.*

arc furnace. A furnace in which material is heated either directly by an electric arc between the electrode and the work, or indirectly by an arc between two electrodes adjacent to the material.

arc melting. Melting metal in an electric arc furnace.

arc time. The time the arc is maintained in making an arc weld. Also known as *weld time.*

arc voltage. The voltage across any electric arc (e.g., across a welding arc).

arc welding (AW). Welding processes that fuse metals together by heating them with an arc, with or without the application of pressure, and with or without the application of filler metal.

argon-oxygen-decarburization process. See *AOD process.*

arsenic. A chemical element with atomic number 33 and atomic weight 75. The chemical symbol, As, is from the Greek *arsenikon,* meaning yellow orpiment. Arsenic was discovered in the 12th century by the German monk Albertus Magnus. (It is remarkable that after the discovery of the seven metals of antiquity it took approximately two thousand years to discover the next metal, which happened to be arsenic.) By 1641, arsenious oxide ore was being reduced by charcoal. Mispickel is one of the most common ores. Arsenical copper (tough pitch) contains 0.35–0.55% arsenic, which, at elevated temperatures, increases the tensile strength to withstand flame heat and scaling better than high-conductivity copper. Arsenic is used for lead alloys to promote hardness and decrease segregation. In brasses it is used to improve resistance to dezincification. It is the toxic compound of some insecticides and antifouling paints. Arsenic is one of the metalloids, and along with boron and silicon, is the least metallic of the metals.

arsenical brass. A copper-zinc alloy with arsenic.

artifact. A feature of artificial character (such as a scratch or a piece of dust on a metallographic specimen) that can be erroneously interpreted as a real feature. During inspection, an artifact often produces a *false indication.*

artificial aging. Aging above room temperature. See *aging* and *natural aging.*

ASME. The American Society of Mechanical Engineers, a professional society focused on mechanical engineering. It was organized in 1880 in response to numerous steam boiler pressure vessel failures. The organization is known for setting codes and standards for mechanical devices.

ASM International. An organization for the materials profession that had its beginnings in 1913 when a group of heat treaters met to discuss heat treating. The American Society for Steel Treaters (ASST) was officially organized in Cleveland, Ohio in 1920. In 1933, the Society was reorganized to have a broader scope, including the entire field of metals and metallurgy, and was named the American Society for Metals. The organization was headquartered in Cleveland, Ohio. In 1986, ASM changed its scope to include virtually all engineering materials and changed the name of the organization to ASM International.

astatine. An element having atomic weight 85 and atomic weight 210. The chemical symbol, At, stems from the Greek *astatos,* meaning unstable. The element was synthesized in 1940 by Dale Corson, Kenneth MacKenzie, and Emilio Segre.

ASTM. The American Society for Testing and Materials, established in Philadelphia, Pennsylvania, in 1898. It originally was named the American Society for Testing Materials, with the word *and* being inserted in 1970 for clarification. It was organized to establish voluntary standards and test methods for materials. Committee A-1 on Steel was the first of more than 150 technical committees.

athermal transformation. A reaction that proceeds without benefit of thermal fluctuations—that is, thermal activation is not required. Such reactions are diffusionless and can take place with great speed when the driving force is sufficiently high. For example, many martensitic transformations occur athermally on cooling, even at relatively low temperatures, because of the progressively increasing driving force. In contrast, a reaction that occurs at constant temperature is an *isothermal transformation*; thermal activation is necessary in this case and the reaction proceeds as a function of time.

atmospheric corrosion test sites. Test sites set up across the country in 1932 by ASTM (American Society for Testing Materials) as part of a program for the atmospheric corrosion testing of metal panels. The test sites were considered to be representative of one of three atmospheric conditions: industrial, marine, and rural. The sites selected were at Kure Beach, North Carolina; Newark-Kearney, New Jersey; Point Reyes, California; State College, Pennsylvania; and Panama Canal Zone. Over the years, metals were tested that included various tempers, coatings, and new alloys.

Summary reports of the corrosion results were published by ASTM. Some tests ran as long as 20 or 30 years and provided much important data on the life of coatings and materials.

atmospheric riser. A riser that uses atmospheric pressure to aid feeding. Essentially a *blind riser* from which a small core or rod protrudes, the function of the core or rod being to provide an open passage so that the molten interior of the riser will not be under a partial vacuum when metal is withdrawn to feed the casting, but will always be under atmospheric pressure. Often called *Williams riser.*

atomic fission. The breakup of the nucleus of an atom, in which the combined weight of the fragments is less than that of the original nucleus, the difference being converted to a very large energy release.

atomic hydrogen welding. An arc welding process that fuses metals together by heating them with an electric arc maintained between two metal electrodes enveloped in a stream of hydrogen. Shielding is provided by the hydrogen, which also carries heat by molecular dissociation and subsequent recombination. Pressure may or may not be used. (This process is now of limited industrial significance.)

atomic number. The number of protons in an atom nucleus, which determines the individuality of the atom as a chemical element.

atomic percent. The number of atoms out of a total of 100 representative atoms of a substance.

atomization. The dispersion of a molten metal into small particles by a rapidly moving stream of gas or liquid.

attenuation. The fractional decrease of the intensity of an energy flux, including the reduction of intensity resulting from geometrical spreading, absorption, and scattering.

attritious wear. Wear of abrasive grains in grinding such that the sharp edges gradually become rounded. A grinding wheel that has undergone such wear usually has a glazed appearance.

ausforming. The deformation of metastable austenite within controlled ranges of temperature and time that avoids formation of nonmartensitic transformation products.

austempering. A heat treatment for ferrous metals in which a part is quenched from the austenitizing temperature at a rate fast enough to avoid formation of ferrite or pearlite and then held at a temperature just above M_s until transformation to bainite is complete.

austenite. A solid solution of one or more elements in face-centered cubic iron. Unless otherwise designated (such as nickel austenite), the solute is generally assumed to be carbon. Figure 3 shows equiaxed austenite grains and annealing twins in an austenitic stainless steel. Austenite was discovered by Floris Osmond, a French metallurgist, in 1895 and named for William Chandler Roberts-Austen.

austenitic grain size. The size attained by grains of steel when heated to the austenitic region. This may be revealed by appropriate etching of cross sections after cooling to room temperature.

austenitic steel. An alloy steel whose structure is normally austenitic at room temperature.

austenitizing. Forming austenite by heating a ferrous alloy into the transformation range (partial austenitizing) or above the transformation range (complete austenitizing). When used without

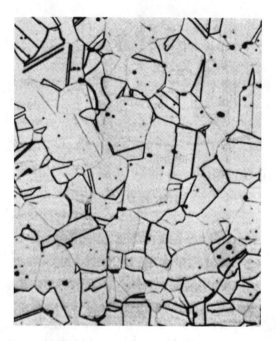

Fig. 3 Equiaxed austenite grains and annealing twins in an austenitic stainless steel

qualification, the term implies complete austenitizing.

autofrettage. Prestressing a hollow metal cylinder by the use of momentary internal pressure exceeding the yield strength.

autogenous weld. A fusion weld made without the addition of filler metal.

automatic brazing. Brazing with equipment that performs the brazing operation without constant observation and adjustment by a brazing operator. The equipment may or may not load and unload the work.

automatic press. A press in which the work is fed mechanically through the press in synchronism with the press action. It also is provided with built-in electrical and pneumatic control equipment.

automatic welding. Welding with equipment that performs the welding operation without adjustment of the controls by an operator. The equipment may or

may not load and unload the work. Compare with *machine welding*.

automation press. See *automatic press*.

autoradiography. An inspection technique in which radiation spontaneously emitted by a material is recorded photographically. The radiation is emitted by radioisotopes that are (a) produced in a metal by bombarding it with neutrons, (b) added to a metal such as by alloying, or (c) contained within a cavity in a metal part. The technique serves to locate the position of the radioactive element or compound.

auxiliary anode. In electroplating, a supplementary anode positioned so as to raise the current density on a certain area of the cathode and thus obtain better plate distribution.

Avesta 253 MA. A fully austenitic stainless steel containing small additions of rare earth metals that provide oxidation resistance and good creep strength. Despite the relatively low levels of chromium and nickel, this steel can, in many cases, replace more highly alloyed steels and nickel-base alloys. The steel has a very high scaling temperature in air. It is designated as S30815 in ASTM specification A240. Other designations for 253MA include EN 1.4835, EN 1.4893, 253RE, F45, Outokumpu 253MA, RA 253MA, S15, and Sirius S15.

Avogadro's number. The number of atoms (or molecules) in a mole of substance, which equals 6.02252×10^{23} per mole.

AWS. The American Welding Society, founded in 1919 to advance the science, technology, and application of welding and related joining disciplines.

axis of weld. A line through the length of a weld perpendicular to the cross section at its geometric center.

Babbitt, Isaac. 1799–1862. A goldsmith born in Taunton, Massachusetts. Babbitt made the first *Britannia metal* in the United States in 1824 to compete with imports of utensils made of the alloy, which was similar to pewter and very popular at the time. Babbitt became superintendent of South Boston Iron Works in Boston, Massachusetts, in 1834. He produced the first American brass cannon and then invented a tin-base bearing alloy that became known as *Babbitt metal.* It was used in steam engines and in railroad car axle boxes. For this work he received an award of $20,000 by the U.S. Congress. His alloy, one of the white metals, contained tin and small amounts of antimony and copper. Compositions other than Babbitt's have been made that also are called Babbitt metals.

Babbitt metal. A nonferrous bearing alloy originated by Isaac Babbitt in 1839. Currently the term includes several tin-base alloys consisting mainly of various amounts of copper, antimony, tin, and lead. There also are lead-base Babbitt metals. Table 3 lists compositions of tin-base Babbitt alloys.

back draft. A reverse taper on a casting pattern or a forging die that prevents the pattern or forged stock from being removed from the cavity.

backstep sequence. A longitudinal welding sequence in which the direction of general progress is opposite the direction in which the individual increments are welded (Fig. 4).

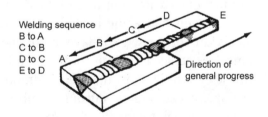

Fig. 4 Backstep sequence

back weld. A weld deposited at the back of a single-groove weld.

baghouse. A chamber containing bags for filtering solids out of gases.

Table 3   Compositions of tin-base Babbitt alloys

Designation	Sn(a)	Sb	Pb max(b)	Cu	Fe max	As max	Bi max	Zn max	Al max	Total other max
										Nominal composition, %
ASTM B 23 alloys										
Alloy 1	91.0	4.5	0.35	4.5	0.08	0.10	0.08	0.005	0.005	0.05 Cd(c)
Alloy 2	89.0	7.5	0.35	3.5	0.08	0.10	0.08	0.005	0.005	0.05 Cd(c)
Alloy 3	84.0	8.0	0.35	8.0	0.08	0.10	0.08	0.005	0.005	0.05 Cd(c)
Alloy 11	87.5	6.8	0.50	5.8	0.08	0.10	0.08	0.005	0.005	0.05 Cd(c)
SAE alloys										
SAE 11	86.0	6.0–7.5	0.50	5.0–6.5	0.08	0.10	0.08	0.005	0.005	0.20
SAE 12	88.0	7.0–8.0	0.50	3.0–4.0	0.08	0.10	0.08	0.005	0.005	0.20
Intermediate lead-tin alloys										
Lead-tin Babbit	75	12	9.3-10.7	3	0.08	0.15	...	...	...	...
ASTM B 102, Alloy PY1815A	65	15	17-19	2	0.08	0.15	...	0.01	0.01	...

(a) Desired minimum in ASTM alloys; specified minimum in SAE alloys. (b) Maximum unless a range is specified. (c) Total named elements, 99.80%

bainite. A metastable aggregate of *ferrite* and *cementite* resulting from the transformation of *austenite* at temperatures below the *pearlite* range but above M_s. Bainite formed in the upper part of the bainite transformation range has a feathery appearance; bainite formed in the lower part of the range has an acicular appearance resembling that of tempered martensite.

bake-hardenable steels. Steels in a special product class with controlled interstitial solute levels and aging behavior. These steels are processed to have moderate aging resistance to permit forming while the steel is in its most ductile condition. Aging occurs largely during a subsequent thermal treatment, for example, during paint curing, which results in desirable hardening of the final part for better durability.

baking. (1) Heating to a low temperature in order to remove gases. (2) Curing or hardening surface coatings such as paints by exposure to heat. (3) Heating to drive off moisture, as in baking of sand cores after molding.

balance. (1) (dynamic) Condition existing where the principal inertial axis of a body lies on its rotational axis. (2) (static) Condition existing where the center of gravity of a body lies on its rotational axis.

ball burnishing. (1) Same as *ball sizing*. (2) Removing burrs and polishing small stampings and small machined parts by *tumbling* in the presence of metal balls.

ball mill. A machine consisting of a rotating hollow cylinder partially filled with metal balls (usually hardened steel or white cast iron) or sometimes pebbles; used to pulverize crushed ores or other substances such as pigments.

ball sizing. Sizing and finishing a hole by forcing a ball of suitable size, finish, and hardness through the hole or by using a burnishing bar or broach consisting of a series of spherical lands of gradually increasing size coaxially arranged. Also called *ball burnishing*, and sometimes ball broaching.

banded structure. A segregated structure consisting of alternating nearly parallel bands of different composition, typically aligned in the direction of primary hot working. Figure 5 illustrates hot rolled 1022 steel showing severe banding.

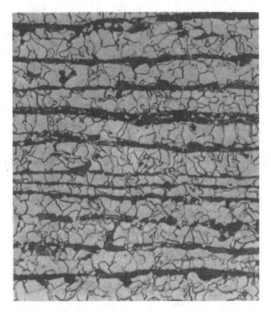

Fig. 5 Hot rolled 1022 steel showing severe banding

band mark. An indentation in carbon steel sheet or strip caused by external pressure on the packaging band around cut lengths or coils; it may occur in handling, transit, or storage.

bands. (1) Hot rolled steel strip, usually produced for rerolling into thinner sheet or strip. Also known as hot bands or band steel. (2) See *electron bands*.

bar. (1) An obsolete unit of pressure equal to 100 kPa. (2) An elongated rolled metal product that is relatively thick and narrow; most bars have simple, uniform cross sections such as rectangular, square, round, oval, or hexagonal. Also known as *barstock*.

bare electrode. A filler-metal arc welding electrode in the form of a wire or rod having no coating other than incidental to the drawing of the wire or to its preservation.

bar folder. A machine in which a folding bar or wing is used to bend a metal sheet whose edge is clamped between the upper folding leaf and the lower stationary jaw into a narrow, sharp, close, and accurate fold along the edge. It also is capable of making rounded folds such as those used in wiring. A universal folder is more versatile in that it is limited in width only by the dimensions of the sheet.

barium. A metal having atomic number 56, atomic weight 137.3, and symbol Ba. Barium is one of the three alkaline earth metals, the other two being calcium and strontium. The metal was first isolated by Sir Humphry Davy in 1808, although its existence had been suspected earlier by Carl Wilhelm Scheele in the compound barytes, also called heavy spar, which was barium sulfate. The name *barytes* was taken from the Greek *barys,* meaning heavy. The word *barium,* therefore, came from a word meaning heavy, although, with a specific gravity of 3.66, it was only a little denser than aluminum at 2.7. Barium resembles calcium in its behavior. It is spontaneously inflammable in moist air. Alloys with magnesium or aluminum are used for getters (i.e., materials that scavenge the last traces of gas) for vacuum tubes and similar sealed components. A 0.2% Ba-Ni alloy is used for spark plugs because it is said to reduce the voltage required for sparking. Small additions of barium are used to stiffen lead.

bark. The decarburized layer just beneath the scale that results from heating steel in an oxidizing atmosphere.

Barkhausen effect. The sequence of abrupt changes in magnetic induction occurring when the magnetizing force acting on a ferromagnetic specimen is varied.

barrel cleaning. Mechanical or electrolytic cleaning of metal in rotating equipment.

barrel finishing. Improving the surface finish of metal objects or parts by processing them in rotating equipment along with abrasive particles that may be suspended in a liquid.

barreling. Convexity of the surfaces of cylindrical or conical bodies often produced unintentionally during upsetting or as a natural consequence during compression testing.

barrel plating. Plating articles in a rotating container, usually a perforated cylinder that operates at least partially submerged in a solution.

barstock. An elongated rolled metal product that is relatively thick and narrow; most bars have simple, uniform cross sections such as rectangular, square, round, oval, or hexagonal. Also known as *bar*.

basal plane. A plane perpendicular to the principal axis (*c* axis) in a tetragonal or hexagonal structure.

base. (1) The surface on which a single-point tool rests when held in a tool post. Also known as *heel*. See sketch accompanying *single-point tool*. (2) A chemical substance that yields hydroxyl ions (OH⁻) when dissolved in water.

base bullion. Crude lead containing recoverable silver, with or without gold.

base metal. (1) The metal present in the largest proportion in an alloy; brass, for example, is a copper-base alloy. (2) The metal to be brazed, cut, soldered, or welded. (3) After welding, that part of the metal which was not melted. (4) A metal that readily oxidizes, or that dissolves to form ions. Contrast with *noble metal* (2).

basic bottom and lining. The inner bottom and lining of a melting furnace, consisting of materials such as crushed burned dolomite, magnesite, magnesite bricks, or basic slag that give a basic reaction at the operating temperature.

basic oxygen furnace. A large tiltable vessel lined with basic refractory material that is the principal type of furnace for modern steelmaking. After the furnace is charged with molten pig iron (which usually comprises 65–75% of the charge), scrap steel, and fluxes, a lance is brought down to the surface of the molten metal and a jet of high-velocity oxygen impinges on the metal. The oxygen reacts with carbon and other impurities in the steel to form liquid compounds that dissolve in the slag and gases that escape from the top of the vessel. Figure 6 is a schematic of a basic oxygen furnace vessel.

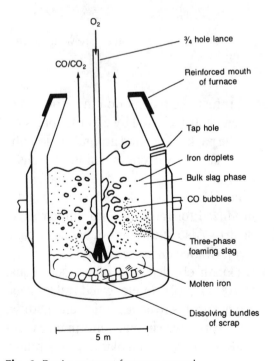

Fig. 6 Basic oxygen furnace vessel

basic steel. Steel melted in a furnace with a *basic bottom and lining* and under a slag containing an excess of a basic substance such as magnesia or lime.

basis metal. The original metal to which one or more coatings are applied.

batch. See *lot*.

Bauschinger effect. For both single-crystal and polycrystalline metals, any change in stress-strain characteristics that can be ascribed to changes in the microscopic stress distribution within the metal, as distinguished from changes caused by strain hardening. In the narrow sense, the process whereby plastic deformation in one direction causes a reduction in yield strength when stress is applied in the opposite direction.

Bayer process. A process for extracting alumina from bauxite ore before the electrolytic reduction. The bauxite is digested in a solution of sodium hydroxide, which converts the alumina to soluble aluminate. After the "red mud" residue has been filtered out, aluminum hydroxide is precipitated, filtered out, and calcined to alumina.

beach marks. Progression marks on a fatigue fracture surface that indicate successive positions of the advancing crack front. The classic appearance is of irregular elliptical or semi-elliptical rings, radiating outward from one or more origins. Beach marks (also known as clamshell marks or tide marks) typically are found on service fractures where the part is loaded randomly, intermittently, or with periodic variations in mean stress or alternating stress.

beading. Raising a ridge or projection on sheet metal.

bead weld. See preferred term *surfacing weld*.

bearing steels. *Alloy steels* used to produce rolling-element bearings. Typically, bearings have been manufactured from both high-carbon (1.00%) and low-carbon (0.20%) steels. The high-carbon steels are used in either a through-hardened or a surface induction-hardened condition. Low-carbon bearing steels are carburized to provide the necessary surface properties while maintaining desirable core properties. Tables 4 and 5 list compositions of typical bearing steels.

bearing stress. The shear load on a mechanical joint (such as a pinned or riveted joint) divided by the effective bearing area. The effective bearing area of a riveted joint, for example, is the sum of the diameters of all rivets times the thickness of the loaded member.

Table 4   Nominal compositions of high-carbon bearing steels

Grade	Composition, %					
	C	Mn	Si	Cr	Ni	Mo
AISI 52100	1.04	0.35	0.25	1.45	...	...
ASTM A 485-1	0.97	1.10	0.60	1.05	...	...
ASTM A 485-3	1.02	0.78	0.22	1.30	...	0.25
TBS-9	0.95	0.65	0.22	0.50	0.25 max	0.12
SUJ 1(a)	1.02	<0.50	0.25	1.05	<0.25	<0.08
105Cr6(b)	0.97	0.32	0.25	1.52	...	...
SHKH15-SHD(c)	1.00	0.40	0.28	1.48	<0.30	...

(a) Japanese grade. (b) German grade. (c) Russian grade

Table 5　Carburizing bearing steels

Grade	Composition, %					
	C	Mn	Si	Cr	Ni	Mo
4118	0.20	0.80	0.22	0.50	...	0.11
5120	0.20	0.80	0.22	0.80	...	...
8620	0.20	0.80	0.22	0.50	0.55	0.20
4620	0.20	0.55	0.22	...	1.82	0.25
4320	0.20	0.55	0.22	0.50	1.82	0.25
3310	0.10	0.52	0.22	1.57	3.50	...
SCM420	0.20	0.72	0.25	1.05	...	0.22
20MnCr5	0.20	1.25	0.27	1.15	...	...

Beilby layer. A layer of metal disturbed by mechanical working presumed to be without regular crystalline structure (amorphous); originally applied to grain boundaries.

belt grinding. Grinding with an abrasive belt.

bend allowance. The bend of the arc of the neutral axis between the tangent points of a bend.

bend angle. The angle through which a bending operation is performed. Figure 7 shows the terms used in bend testing.

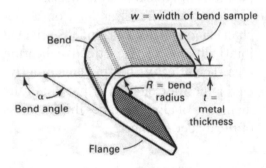

Fig. 7 Terms used in bend testing

bender. A die impression, tool, or mechanical device designed to bend forging stock to conform to the general configuration of die impressions to be subsequently used.

bending brake. A *press brake* used for bending.

bending moment. The algebraic sum of the couples or the moments of the external forces, or both, to the left or right of any section on a member subjected to bending by couples or transverse forces, or both.

bending rolls. Two or three rolls with an adjustment for imparting a desired curvature in sheet or strip metal.

bend radius. (1) The inside radius of a bend section. (2) The radius of a tool around which metal is bent during fabrication.

bend tangent. A tangent point at which a bending arc ceases or changes.

bend test. A test for determining relative ductility of metal that is to be formed (usually sheet, strip, plate, or wire) and for determining soundness and toughness of metal (after welding, for example). The specimen usually is bent over a specified diameter through a specified angle for a specified number of cycles.

beneficiation. Concentration or other preparation of ore for smelting.

bentonite. A colloidal claylike substance derived from the decomposition of volcanic ash chiefly composed of the minerals of the montmorillonite family. Western bentonite is slightly alkaline; southern bentonite usually is slightly acidic.

berkelium. A chemical element having atomic number 97, atomic weight 247, and the symbol Bk. Synthesized by neutron bombardment in 1949 by Glenn T. Seaborg, Albert Ghiorso, and co-workers, it is named for the town of Berkeley, California, where it was discovered.

beryllium. A chemical element having atomic number 4, atomic weight 9, and the symbol Be, from the Greek *beryllos,* for which beryl, the ore, is named. The element was identified by Nicolas-Louis Vauquelin in 1797 and isolated by Friedrich Wöhler in 1828. The element also was widely known by the name glucinium (Gl) because of the sweet taste of many of its salts. This name, however, was abandoned in 1949 because of the popular usage of beryllium.

beryllium-copper. Copper-base alloys containing not more than 3% Be. Available in both cast and wrought form, these alloys rank high among copper alloys in attainable strength while retaining useful levels of electrical and thermal conductivity. Applications for these alloys include electronic components (connector contacts), electrical equipment (switch and relay blades), antifriction bearings, housings for magnetic sensing devices, and resistance welding contacts. Table 6 lists compositions of commercial beryllium-copper alloys.

beryllium-nickel. Age-hardenable nickel-base alloys containing up to 2.75% Be. Beryllium-nickel alloys are used primarily as mechanical and electrical/electronic components. Cast alloys are used in molds and cores for glass and polymer molding, diamond drill bit matrices, and cast turbine parts. Table 7 lists compositions of commercial beryllium-nickel alloys.

bessemer process. A process for making steel by blowing air through molten pig iron contained in a refractory lined vessel so as to remove by oxidation most of the carbon, silicon, and manganese. The process is obsolete.

Bessemer, Sir Henry. 1813–1898. Henry Bessemer of Sheffield, England, discovered and patented the first practical

Table 6   Compositions of commercial beryllium-copper alloys

UNS number	Composition, wt%							
	Be	Co	Ni	Co + Ni	Co + Ni + Fe	Si	Pb	Cu
Wrought alloys								
C17200	1.80–2.00	...	...	0.20 min	0.6 max	...	...	bal
C17300	1.80–2.00	...	...	0.20 min	0.6 max	...	0.20–0.6	bal
C17000	1.60–1.79	...	...	0.20 min	0.6 max	...	...	bal
C17510	0.2–0.6	...	1.4–2.2	...	...	...	...	bal
C17500	0.4–0.7	2.4–2.7	...	...	...	...	...	bal
C17410	0.15–0.50	0.35–0.60	...	...	...	...	...	bal
Cast alloys								
C82000	0.45–0.80	...	...	2.40–2.70	...	...	...	bal
C82200	0.35–0.80	...	1.0–2.0	...	...	...	...	bal
C82400	1.60–1.85	...	...	0.20–0.65	...	...	...	bal
C82500	1.90–2.25	...	...	0.35–0.70	...	0.20–0.35	...	bal
C82510	1.90–2.15	...	...	1.00–1.20	...	0.20–0.35	...	bal
C82600	2.25–2.55	...	...	0.35–0.65	...	0.20–0.35	...	bal
C82800	2.50–2.85	...	...	0.35–0.70	...	0.20–0.35	...	bal

Note: Copper plus additions, 99.5% min

Table 7   Nominal compositions of commercial beryllium-nickel alloys

Product Form	Alloy	Composition, %			
		Be	Cr	Other	Ni
Wrought	N03360	1.85–2.05	...	0.4–0.6 Ti	bal(a)
Cast	M220C	2.0	...	0.5 C	bal
Cast	41C	2.75	0.5	...	bal(b)
Cast	42C	2.75	12.0	...	bal(b)
Cast	43C	2.75	6.0	...	bal(b)
Cast	44C	2.0	0.5	...	bal(b)
Cast	46C	2.0	12.0	...	bal(b)
Cast	Master	6	...	...	bal(c)

(a) 99.4 Ni + Be + Ti + Cu min, 0.25 Cu max. (b) 0.1 C max. (c) Master alloys with 10, 25, and 50 wt% Be are also available.

method for making steel in 1855. Bessemer's father, Anthony, was a highly successful inventor whose son followed in his footsteps, making inventions one after the other. One of his most successful early inventions was a gold paint that he made from brass powder using a secret process. He patented a method for making continuous glass plate, which was not a success, but he learned a great deal about making furnaces that would be useful in developing his steel-making process.

Before Bessemer's process, there were only two metallic construction materials available: cast iron and wrought iron. Steel could be made only by the crucible method in 60-pound casts—a very expensive, time-consuming process that could not be used to make large or long plates. Cast iron proved to be brittle and subject to failures in bridges and railway tracks. Bessemer steel became the material of choice for most construction work, and the process was the first step in the development of the steel industry and the beginning of the Steel Age.

beta (β) brass. A brass with 45–50% Zn that is the hardest and strongest of the brasses and can only be hot worked. Beta brass is suitable for castings.

beta ray. A ray of electrons emitted during the spontaneous disintegration of certain atomic nuclei.

beta structure. A Hume-Rothery designation for structurally analogous body-centered cubic phases (similar to β brass) or electron compounds that have ratios of three valence electrons to two atoms. Not the same as a β phase on a constitution diagram.

Betts process. A process for the electrolytic refining of lead in which the electrolyte contains lead fluosilicate and fluosilicic acid.

biaxiality. In a *biaxial stress* state, the ratio of the smaller to the larger principal stress.

biaxial stress. A state of stress in which only one of the *principal stresses* is zero, the other two usually being in tension.

billet. (1) A solid semifinished round or square product that has been hot worked by forging, rolling, or extrusion; usually smaller than a *bloom*. (2) A general term for wrought starting stock in making forgings or extrusions.

billet mill. A primary rolling mill used for making steel billets.

binary alloy. An alloy containing only two component elements.

binder. (1) In founding, a material, other than water, added to foundry sand to bind the particles together, sometimes with the use of heat. (2) In powder metallurgy, a cementing medium; either a material added to the powder to increase the green strength of the compact, which is expelled during sintering; or a material (usually of relatively low melting point) added to a powder mixture for the specific purpose of cementing together powder particles that alone would not sinter into a strong body.

bipolar electrode. An *electrode* in an *electrolytic cell* that is not mechanically connected to the power supply, but is placed in the electrolyte, between the *anode* and *cathode*, so that the part nearer the anode becomes cathodic and the part nearer the cathode becomes anodic. Also called *intermediate electrode.*

bipolar field. A longitudinal magnetic field that creates two magnetic poles within a piece of material. Compare with *circular field.*

biscuit. (1) An upset blank for drop forging. (2) A small cake of primary metal (such as uranium made from uranium tetrafluoride and magnesium by bomb reduction). Compare with *derby* and *dingot.*

bismuth. A chemical element having atomic number 83, atomic weight 209, and symbol Bi. Bismuth initially was not recognized as a metal and was confused with lead. It was mentioned for the first time in 1472 in a German document where it was called "wismuth." In 1753 Claude Geoffroy found it to be a metal. Bismuth is one of the nine metalloids. It is used in fusible metals such as Wood's metal, which is 50% Bi with 25% Pb, 12.5% Cd, and 12.5% Sn, an alloy with a melting range of 72 to 70 °C (162 to 158 °F).

There is a major use in the pharmaceutical industry for bismuth subsalicylate products such as Pepto-Bismol.

black annealing. Box annealing or pot annealing of ferrous alloy sheet, strip, or wire. See *box annealing.*

blackheart malleable. See *malleable cast iron.*

blacking. Carbonaceous materials such as plumbago, graphite, or powdered carbon used to coat pouring ladles, molds, runners, and pig beds.

black light. Electromagnetic radiation not visible to the human eye. The portion of the spectrum generally used in fluorescent inspection falls in the ultraviolet region between 330 and 400 nm, with the peak at 365 nm.

black oxide. A black finish on a metal produced by immersing it in hot oxidizing salts or salt solutions.

blade-setting angle. See preferred term *cone angle.*

blank. (1) In forming, a piece of sheet material, produced in cutting dies, that usually is subjected to further press operations. (2) A pressed, presintered, or fully sintered powder metallurgy compact, usually in the unfinished condition and requiring cutting, machining, or some other operation to produce the final shape. (3) A piece of stock from which a forging is made, often called a *slug* or *multiple.*

blank carburizing. Simulating the carburizing operation without introducing carbon. This usually is accomplished by using an inert material in place of the carburizing agent, or by applying a suitable protective coating to the ferrous alloy.

blankholder. The part of a drawing or forming die that holds the workpiece against the draw ring to control metal flow.

blanking. Producing desired shapes from metal to be used for forming or other operations, usually by punching.

blank nitriding. Simulating the nitriding operation without introducing nitrogen. This usually is accomplished by using an inert material in place of the nitriding agent, or by applying a suitable protective coating to the ferrous alloy.

blast furnace. A shaft furnace in which solid fuel is burned with an air blast to smelt ore in a continuous operation. Where the temperature must be high, as in the production of pig iron, the air is preheated. Where the temperature can be lower, as in the smelting of copper, lead, and tin ores, a smaller furnace is economical, and preheating of the blast is not required. Figure 8 is a schematic of a blast furnace.

blasting. Cleaning or finishing metals by impingement with abrasive particles moving at high speed and usually carried by gas or liquid or thrown centrifugally from a wheel.

blemish. A nonspecific quality control term designating an imperfection that mars the appearance of a part but does not detract from its ability to perform its intended function.

blending. In powder metallurgy, the thorough intermingling of powders of the same nominal composition (not the same as *mixing*).

blind riser. A *riser* that does not extend through the top of the mold.

blister. A raised area, often dome-shaped, resulting from (a) loss of adhesion between a coating or deposit and the basis metal or (b) delamination under the pressure of expanding gas trapped in a metal near a subsurface zone. Very small blisters may be called pinhead blisters or pepper blisters.

blister copper. An impure intermediate product in the refining of copper, produced by blowing copper *matte* in a converter, the name being derived from the large blisters on the cast surface that result from the liberation of SO_2 and other gases.

block brazing. An obsolete brazing process in which the joint is heated using hot blocks.

blocker. The impression in the dies (often one of a series of impressions in a single die set) that imparts to the forging an intermediate shape, preparatory to forging

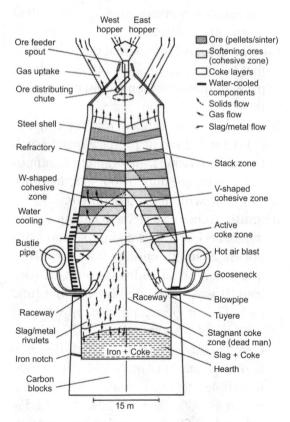

Fig. 8 Principal zones and component parts of an iron blast furnace

the final shape. Also called *blocking impression*.

blocker-type forging. A forging that approximates the general shape of the final part with relatively generous *finish allowance* and radii. Such forgings are sometimes specified to reduce die costs where only a few forgings are desired and the cost of machining each part to the final shape is not excessive.

blocking. In forging, a preliminary operation performed in closed dies, usually hot, to position metal properly so that in the finish operation the dies will be filled correctly.

blocking impression. Same as *blocker*.

block sequence. A welding sequence in which separated lengths of a continuous multiple-pass weld are partly or completely built up in cross section before intervening lengths are deposited. Compare with *cascade sequence*.

bloom. (1) A semifinished hot rolled product, rectangular in cross section, produced on a blooming mill. See also *billet*. For steel, the width of a bloom is not more than twice the thickness, and the cross-sectional area usually is not less than approximately 230 cm^2 (36 in.^{2}). Steel blooms are sometimes made by forging. (2) A visible exudation or efflorescence on the surface of an electroplating bath. (3) A bluish fluorescent cast to a painted surface caused by a deposition of a thin film of smoke, dust, or oil. (4) A loose, flowerlike corrosion product that forms when certain metals are exposed to a moist environment.

bloomer. The mill or other equipment used in reducing steel ingots to blooms.

blooming. The process for reducing cast iron into malleable iron (wrought iron).

blooming mill. A primary rolling mill used to make blooms.

blotter. In grinding, a disk of compressible material, usually blotting-paper stock, used between the grinding wheel and its flanges to avoid concentrated stresses.

blowhole. A hole in a casting or a weld caused by gas entrapped during solidification.

blowpipe. A welding or cutting torch.

blue annealing. Heating hot rolled ferrous sheet in an open furnace to a temperature within the transformation range and then cooling it in air, in order to soften the metal. The formation of a bluish oxide on the surface is incidental.

blue brittleness. Brittleness exhibited by some steels after being heated to some temperature within the range of approximately 200 to 370 °C (400 to 700 °F), particularly if the steel is worked at the elevated temperature. Killed steels are virtually free of this kind of brittleness.

blue dip. A solution containing a mercury compound, once widely used to deposit mercury on a metal by immersion, usually prior to silver plating.

bluing. Subjecting the scale-free surface of a ferrous alloy to the action of air, steam, or other agents at a suitable temperature, thus forming a thin blue film of oxide and improving the appearance and resistance to corrosion. This term ordinarily is applied to sheet, strip, or finished parts. It also is used to denote the heating of springs after fabrication in order to improve their properties.

board hammer. A type of forging hammer in which the upper die and ram are attached to "boards" that are raised to the striking position by power-driven rollers and then fall by gravity. See *drop hammer*.

bog iron. Iron ore deposits in bogs and swamps. The ores consist primarily of hydrated ferric oxide ($2Fe_2O_3 \cdot H_2O$). Iron from such sources was used during the American Revolution. Sites were developed in Maryland, New Jersey, and notably at the Saugus Iron Works in Massachusetts for making iron from such ores.

bolster. A plate to which dies may be fastened, the assembly being secured to the top surface of a press bed. In mechanical forging, such a plate also is attached to the ram.

bond. (1) In grinding wheels and other relatively rigid abrasive products, the material that holds the abrasive grains together. (2) In welding, brazing, or soldering, the junction of joined parts. Where filler metal is used, it is the junction of the fused metal and the heat-affected base metal. (3) In an adhesive bonded or diffusion bonded joint, the line along which the faying surfaces are joined together.

book mold. A split permanent mold hinged like a book.

boron. A chemical element having atomic number 5, atomic weight 11, and symbol B. The element takes its name from the Arabic boron-oxide powder, *boraq,* meaning white, from which we have the word *borax.* Boron was first isolated in 1808 by Louis-Joseph Gay Lussac and Louis Thénard. Boron forms very hard carbides with chromium and silicon, making these compounds highly useful abrasives. Boron on the order of 0.003%, when used to deoxidize steel, confers hardenability equivalent to several hundred times that of chromium or nickel. Boron is alloyed with stainless steel for nuclear components because of its very high neutron capture cross section. Boron is one of the nine metalloids and, along with arsenic and bismuth, has the least metallic characteristics of these metals.

bort. Industrial diamond.

bosh. (1) The section of a blast furnace extending upward from the tuyeres to the plane of maximum diameter. (2) A lining of quartz that builds up during the smelting of copper ores and that decreases the diameter of the furnace at the tuyeres. (3) A tank, often with sloping sides, used for washing metal parts or for holding cleaned parts.

boss. A relatively short protrusion or projection from the surface of a forging or casting, often cylindrical in shape.

bottom board. A flat base for holding the *flask* in making sand molds.

bottom drill. A flat-ended twist drill used to convert a cone into a cylinder at the bottom of a drilled hole.

bottoming tap. A tap with a *chamfer* of 1 to 1½ threads in length.

bottom pipe. An oxide-lined fold or cavity at the butt end of a slab, bloom, or billet, formed by folding the end of an ingot over on itself during primary rolling. Bottom pipe is not *pipe,* in that it is not a shrinkage cavity; in that sense, the term is a misnomer. Bottom pipe is similar to extrusion pipe. It normally is discarded when the slab, bloom, or billet is cropped following primary reduction.

bowing. Deviation from flatness.

box annealing. Annealing a metal or alloy in a sealed container under conditions that minimize oxidation. In box annealing a ferrous alloy, the charge usually is heated slowly to a temperature below the transformation range, but sometimes above or within it, and then is cooled slowly; this process also is called close annealing or pot annealing. See *black annealing*.

boxing. Continuing a fillet weld around a corner as an extension of the principal weld. Also called an end return.

brake. A device for bending sheet metal to a desired angle.

brale. A diamond penetrator of specified spheroconical shape used with a Rockwell hardness tester. This penetrator is used for the A, C, D, and N scales for testing hard metals. Figure 9 shows a brale indenter used in Rockwell hardness testing.

Fig. 9 Brale indenter used in Rockwell hardness testing

brass. A copper alloy with up to 50% Zn, which may include small amounts of other alloys, such as aluminum, lead, tin, and manganese, for special purposes. Brass is noted for its beauty and corrosion resistance. There are three types of brass, according to their crystalline structures. The *alpha (α) brasses,* with up to 30% Zn, are the most ductile. The *alpha-beta*

(α-β) brasses have from 30–43.5% Zn; and the *beta (β) brasses* include those from 43.5–50% Zn. The hardness and strength increase with increasing zinc content, while the cost of the alloy decreases. A group of alloys known as gilding metals are α brasses with 5–20% Zn. They are valued for their pleasing colors, which vary from red to golden as the zinc is increased. Gilding metal is used for inexpensive jewelry and items to be gold plated. The alloys are very ductile. *Cartridge brass* contains from 28–32% Zn and is the most ductile. As indicated by the name, it is possible to deep draw cartridge cases using intermediate annealing. Basis brass or *common brass* is the least expensive, and ranges from 35–38% Zn. *Muntz metal*, also known as 60/40 or yellow brass, contains 40% Zn. It is an α-β brass, the strongest, and must be hot worked. Brass is widely used for decoration and for applications where low friction is required, such as for locks, valves, and gears. It also is used for plumbing and electrical applications. Because it is nonsparking, brass is used for fittings and tools around flammable gases or liquids. Brass is the metal of choice for musical instruments such as trumpets and horns. See also *copper, Technical Note 4*.

Artifacts consisting of copper with 23% Zn have been found dating at least to 1000 B.C., but brass was not in common use until approximately a thousand years later, in the time of the Romans. It is one of the strangest stories in the history of metallurgy because zinc, one of the constituents of brass, was unknown and, in fact, not even identified until 1526. Roman artisans became engaged in a process that they described as the coloring of copper. Pieces of copper, along with powdered calamine and charcoal, were packed in a clay crucible with a tight-fitting lid and baked in a furnace for 24 hours at a temperature much too low to melt the copper. However, the copper turned a beautiful golden color; the Romans called it "aurichalcum," which translates to "golden copper." We call it brass. The colored copper was melted and made into ornaments, dress armor, utensils, and *sesterces* (Roman coins).

Unknown to the Romans was the fact that calamine actually was zinc carbonate, which, in the furnace, vaporized the zinc that was absorbed by the copper to form a copper-zinc alloy. The "calamine brass" process remained the preferred method of making brass for many years after the discovery and production of metallic zinc in the 18th century.

Brass became popular for church ornaments and for thin plates in floors inscribed to commemorate the dead. One of the principal commercial uses of brass was for brass pins used for carding in wool processing.

Because of its ease of manufacturing, good machinability, and good corrosion resistance, brass became widely used for precision instruments including clocks, watches, sextants, and surveying instruments.

By 1900, approximately 10,000 Americans were employed in the brass industry.

(The word *brass* is from the Old English word *braes*. When "brass" is mentioned in the King James version of the *Bible*, scholars now advise that the interpretation should be "bronze" or "copper.")

braze welding. A method of welding by using a filler metal having a liquidus above 450 °C (840 °F) and below the solidus of the base metals. Unlike *brazing*, in braze welding, the filler metal is not distributed in the joint by capillary attraction.

brazing. Welding processes that join solid materials together by heating them to a suitable temperature and using a filler metal having a liquidus above 450 °C (840 °F) and below the solidus of the base materials. The filler metal is distributed between the closely fitted surfaces of the joint by capillary attraction.

brazing alloy. See the preferred term *brazing filler metal*.

brazing filler metal. A nonferrous filler metal used in *brazing* and *braze welding*.

brazing sheet. Brazing filler metal in sheet form or flat rolled metal clad with brazing filler metal on one or both sides.

breakdown. (1) An initial rolling or drawing operation, or series of such operations, for the purpose of reducing a casting or extruded shape prior to the finish reduction to desired size. (2) A preliminary press-forging operation.

breaking stress. Same as *fracture stress* (1).

breaks. Creases or ridges usually in "untempered" or aged material where the yield point has been exceeded. Depending on the origin of the breaks, they may be termed *cross breaks*, *coil breaks*, edge breaks, or *sticker breaks*.

Brearley, Harry. 1871–1948. A self-made analytical chemist and metallurgist credited with the invention of stainless steel in 1908. Brearley was born into a poor family in Sheffield, England, and left school at the age of 12. Eventually he became manager at the Brown Firth Research

Laboratory, a joint venture between Thos. Firth & Sons and John Brown & Co. While experimenting to find an improved steel for gun barrels, he discovered a high-chromium steel that did not rust. In 1913, he produced a commercial heat of the alloy that contained 12.86% Cr and 0.24% C. The alloy was found to be ideal for making stainless steel knife blades and came to be called "cutlery steel" or chromium stainless steel. Brearley received patents for the alloy in the United States and Canada. He promoted the steel, organizing the Firth Brearley Stainless Steel Syndicate in England and the American Stainless Steel Company in Pittsburgh. He is generally recognized as the discoverer of chromium stainless steel, although others had found virtually the same alloy but had never manufactured it for commercial purposes. Brearley received the Bessemer zaward bestowed on him in 1920.

bridge die. A two-section extrusion die capable of producing tubing or intricate hollow shapes without the use of a separate mandrel. Metal separates into two streams as it is extruded past a bridge section, which is attached to the main die section and holds a stub mandrel in the die opening; the metal then is rewelded by extrusion pressure before it enters the die opening. Compare with *porthole die.*

bridging. (1) Premature solidification of metal across a mold section before the metal below or beyond it solidifies. (2) Solidification of slag within a cupola at or just above the tuyeres. (3) Welding or mechanical locking of the charge in a downfeed melting or smelting furnace. (4) In powder metallurgy, the formation of arched cavities in a powder mass. (5) In soldering, an unintended solder

connection between two or more conductors, either securely or by mere contact. Also called a *crossed joint* or *solder short.*

bright annealing. Annealing in a protective medium to prevent discoloration of the bright surface.

bright dip. A solution that produces, through chemical action, a bright surface on an immersed metal.

brightener. An agent or combination of agents added to an electroplating bath to produce a lustrous deposit.

bright finish. A high-quality finish produced on ground and polished rolls. Suitable for electroplating.

bright plate. An electrodeposit that is lustrous in the as-plated condition.

bright range. The range of current densities, other conditions being constant, within which a given electroplating bath produces a bright plate.

Brillouin zones. See *electron bands.*

Brinell hardness test. A test for determining the hardness of a material by forcing a hard steel or carbide ball of specified diameter into it under a specified load. The result is expressed as the Brinell hardness number, which is the value obtained by dividing the applied load in kilograms by the surface area of the resulting impression in square millimeters. Figure 10 is a schematic of the Brinell hardness test.

brinelling. Evenly spaced dents in a raceway of a rolling-element bearing that occur when the bearing assembly is subjected to a force or impact load great enough to cause the rolling elements to indent the raceway surface. Also called true brinelling. Compare with *false brinelling.*

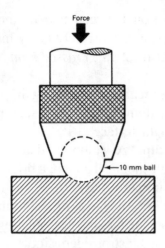

Force

←—10 mm ball

Fig. 10 Schematic of Brinell hardness test

Brinell, Johan August. 1849–1925. The developer of the Brinell hardness test, Brinell began his career at the Lesjörfors Ironworks. From 1903 to 1914 he was chief engineer of the Fagersta Ironworks. He is best known for his invention of the Brinell hardness testing machine in 1910, but he made many contributions to the study of phase transformations in steel.

Britannia metal. A tin-base alloy, named for the early name of the English island, containing 5–10% Sb and 1–3% Cu. It is commonly used for decorative purposes and is said to be the material of the gold-plated Oscar awards presented in Hollywood.

brittle crack propagation. A very sudden propagation of a crack with the absorption of no energy except that stored elastically in the body. Microscopic examination may reveal some deformation even though it is not noticeable to the unaided eye.

brittle fracture. Separation of a solid accompanied by little or no macroscopic plastic deformation. Typically, brittle fracture occurs by rapid crack propagation with less expenditure of energy than for ductile fracture.

brittleness. The quality of a material that leads to crack propagation without appreciable plastic deformation.

broach. A bar-shaped cutting tool with a series of cutting edges or teeth that increase in size or change in shape from the starting point to the finishing end. The tool cuts in the axial direction when pushed or pulled and is used to shape either holes or outside surfaces.

bronze. A copper-rich, copper-tin alloy with or without small proportions of other elements such as zinc and phosphorus. By extension, certain copper-base alloys containing considerably less tin than other alloying elements, such as manganese bronze (copper-zinc plus manganese, tin, and iron), leaded tin bronze (copper-aluminum), and silicon bronze (copper-beryllium). Also, trade designations for certain specific copper-base alloys that actually are brasses, such as architectural bronze (57Cu-40Zn-3Pb) and commercial bronze (90Cu-10 Zn).

Bronze Age. The period in history that followed the Stone Age and preceded the Iron Age. In the Middle East it occurred as early as 3500 B.C., based on the discovery of copper-tin or copper-arsenic artifacts.

bronzing. (1) Applying a chemical finish to copper or copper-alloy surfaces to alter the color. (2) Plating a copper-tin alloy on various materials.

brush anodizing. An *anodizing* process similar to *brush plating*.

brush plating. Plating with a concentrated solution or gel held in or fed to an absorbing medium, pad, or brush carrying the anode (usually insoluble). The brush is moved back and forth over the area of the cathode to be plated.

brush polishing (electrolytic). A method of *electropolishing* in which the electrolyte is applied with a pad or brush in contact with the part to be polished.

buckle. (1) A local waviness in metal bar or sheet, usually transverse to the direction of rolling. (2) An indentation in a casting resulting from expansion of molding sand into the mold cavity.

buckling. Producing a bulge bend, bow kink, or other wavy condition by compressively stressing a beam, column, plate, bar, or sheet.

Bucky diaphragm. An x-ray scatter-reducing device originally intended for medical radiography but also applicable to industrial radiography in some circumstances. Thin strips of lead, with their widths held parallel to the primary radiation, are used to absorb scattered radiation preferentially; the array of strips is in motion during exposure, to prevent formation of a pattern on the film.

buffer. A substance whose purpose is to maintain a constant hydrogen-ion concentration in water solutions, even where acids or alkalis are added. Each buffer has a characteristic limited range of pH over which it is effective.

buffing. Developing a lustrous surface by contacting the work with a rotating *buffing wheel.*

buffing wheel. Buff sections assembled to the required face width for use on a rotating shaft between flanges. Sometimes called a buff.

buff section. A number of fabric, paper, or leather disks with concentric center holes held together by various types of sewing to provide degrees of flexibility or hardness. These sections are assembled to make wheels for polishing.

builder. A material, such as an alkali, a buffer, or a water softener, added to a soap or synthetic surface-active agent to produce a mixture having enhanced detergency. Examples: (1) alkalis—caustic soda, soda ash, and trisodium phosphate; (2) *buffers*—sodium metasilicate and borax; and (3) water softeners—sodium tripolyphosphate, sodium tetraphosphate, sodium hexametaphosphate, and ethylene diamine tetraacetic acid.

buildup. Excessive electrodeposition that occurs on high-current-density areas, such as corners or edges.

buildup sequence. The order in which weld beads are deposited, generally designated in cross section, as shown in Fig. 11.

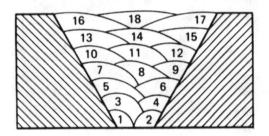

Fig. 11 Buildup weld sequence

built-up edge. Chip material that adheres to the tool face adjacent to the cutting edge during cutting.

bulging. Expanding the walls of a cup, shell, or tube with an internally expanded segmented punch or a punch composed of air, liquids, or semiliquids such as waxes, rubber, and other elastomers.

bull block. A machine with a power-driven revolving drum for cold drawing wire through a drawing die as the wire winds around the drum.

bulldozer. A horizontal machine, usually mechanical, having two bull gears with eccentric pins, two connecting links to a ram, and dies to perform bending, forming, and punching of narrow plate and bars. Railroad car sills are formed with a bulldozer.

bullion. (1) A semirefined alloy containing sufficient precious metal to make recovery profitable. (2) Refined gold or silver, uncoined.

bull's-eye structure. The microstructure of malleable or ductile cast iron in which graphite nodules are surrounded by a ferrite layer in a pearlitic matrix (Fig. 12).

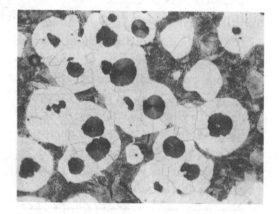

Fig. 12 Ductile iron, with the bull's-eye structure of graphite nodules surrounded by ferrite. Original magnification: 100×

bumper. A machine used for packing molding sand in a flask by repeated jarring or jolting.

bumping. (1) Forming a dish in metal by means of many repeated blows. (2) Forming a head. (3) Setting the seams on sheet metal parts. (4) Ramming sand in a flask by repeated jarring and jolting.

burned deposit. A dull, nodular electrodeposit resulting from excessive current density.

burned-on sand. A mixture of sand and cast metal adhering to the surface of the casting. In some instances, may resemble *metal penetration*.

burning. (1) Permanently damaging a metal or alloy by heating to cause either incipient melting or intergranular oxidation. See *overheating*. (2) In grinding, getting the work hot enough to cause discoloration or to change the microstructure by tempering or hardening.

burnishing. Smoothing surfaces through frictional contact between the work and some hard pieces of material such as hardened metal balls.

burnoff. (1) Unintentional removal of an autocatalytic deposit from a nonconducting substrate, during subsequent electroplating operations, owing to the application of excessive current or a poor contact area. (2) Removal of volatile lubricants such as metallic stearates from metal powder compacts by heating immediately prior to sintering. (3) See *melting rate*.

burr. (1) A turned-over edge on work resulting from cutting, punching, or grinding. (2) A rotary tool having teeth similar to those on hand files.

burring. Removing burrs, sharp edges, or fins from metal parts by filing, grinding, or rolling the work in a barrel containing abrasives suspended in a suitable liquid medium. Same as *deburring*.

bushing. A bearing or guide.

buster. A pair of shaped dies used to combine preliminary forging operations such as edging and blocking, or to loosen the scale.

butler finish. A semilustrous metal finish composed of fine, uniformly distributed parallel lines, usually produced with a soft abrasive wheel; similar in appearance to the traditional hand-rubbed finish on silver.

buttering. A form of surfacing in which one or more layers of weld metal is deposited on the groove face of one member (e.g., a high-alloy weld deposit on steel base metal that is to be welded to a dissimilar base metal). The buttering provides a suitable transition weld deposit for subsequent completion of the butt weld.

butt joint. A joint between two abutting members lying approximately in the same plane. A welded butt joint may contain a variety of grooves. See *groove weld*.

button. (1) A globule of metal remaining in an assaying crucible or cupel after fusion has been completed. (2) That part of a weld that tears out in destructive testing of a spot, seam, or projection welded specimen.

butt seam welding. See *seam welding*.

butt welding. Welding a *butt joint*.

Dictionary of Metals
H.M. Cobb, editor

cadmium. A chemical element having atomic number 48, atomic weight 112, and the symbol Cd, from the Greek *Kadmeia,* a name for calamine (zinc carbonate) where cadmium was discovered. Kadmeia was the name of a Greek fortress, so-named in honor of Cadmus, the son of Agenor, a Phoenician king. Cadmium was discovered simultaneously by Friedrich Strohmeyer and Karl Samuel Leberecht Hermann, German chemists, in 1817. Cadmium is a by-product of electrolytic zinc production and from zinc ores containing 0.1–0.5% Cd. Cadmium was formerly used as an electrocoating for steel. It is currently used in applications requiring low-melting-point alloys and bearing alloys, and in copper-base alloys, particularly for electrical equipment. Cadmium-copper alloys containing 1% Cd increase the tensile strength of copper by 50%, with only a small reduction in electrical conductivity. The principal use of cadmium is in nickel-cadmium batteries.

caesium. Also known as cesium. A chemical element having atomic number 55, atomic weight 133, and symbol Cs, from the Latin *caesius,* meaning sky blue, which refers to the color of the element's emission spectrum. The element was identified with a spectroscope by German physicist Gustav Kirchhoff and German chemist Robert Bunsen. Caesium occurs in nature as pollux or pollucite, a caesium-aluminum-sodium silicate. It was the first metal to be discovered by spectrographical analysis. It is the most strongly basic and electropositive metal known, reacting violently with air or water; therefore, it usually is stored under oil. The principal application has been in photoelectric cells.

cake. (1) A copper or copper alloy casting, rectangular in cross section, used for rolling into sheet or strip. (2) A coalesced mass of unpressed metal powder.

calamine brass. Brass made by the calamine process that was used first in Asia Minor and then by the Romans from the first millennium. Calamine, a zinc carbonate mineral, is a zinc ore also known as smithsonite. Powdered calamine, charcoal, and pieces of copper are heated in a closed crucible to 1300 °C (2400 °F). The ore reduces to zinc vapor, which diffuses into the copper to form brass. The Romans, however, were unaware of the zinc and believed they had developed a copper-coloring process that made it look like gold. The calamine process was used until the 18th century to make brass.

calcination. Heating ores, concentrates, precipitates, or residues to decompose carbonates, hydrates, or other compounds.

calcium. A chemical element having atomic number 20, atomic weight 40, and symbol Ca, from the Latin *calx,* for the mineral lime in which calcium occurs. Calcium was isolated by Sir Humphry Davy in 1808. Calcium is the fifth most abundant metal in the earth's crust, making up approximately 3.6% of the total. Occurring as the carbonate in limestone, marble, and chalk, it is widespread throughout the world. Calcium is used as a deoxidizing element for a wide range of ferrous and nonferrous alloys. It is used in lead-acid batteries and in gray iron castings.

californium. A chemical element having atomic number 98, atomic weight 251, and symbol Cf, named for the state in which it was discovered. Small amounts of the element were synthesized in 1950 by Glenn T. Seaborg, Albert Ghiorso, and co-workers.

calomel electrode (calomel half cell). A secondary reference electrode of the composition: $Pt/Hg-Hg_2Cl_2/KCl$ solution. For $1.0\,N$ KCl solution, its potential versus a hydrogen electrode at 25 °C (77 °F) and one atmosphere is +0.281 V.

calorizing. Imparting resistance to oxidation to an iron or steel surface by heating in aluminum powder at 800 to 1000 °C (1470 to 1830 °F).

camber. (1) Deviation from edge straightness, usually referring to the greatest deviation of side edge from a straight line. (2) Sometimes used to denote crown in rolls where the center diameter has been increased to compensate for deflection caused by the rolling pressure.

canning. (1) A dished distortion in a flat or nearly flat surface; sometimes referred to as oil canning. (2) Enclosing a highly reactive metal within a relatively inert one for the purpose of hot working without undue oxidation of the active metal.

capillary attraction. The combined force of adhesion and cohesion that causes liquids, including molten metals, to flow between very closely spaced solid surfaces, even against gravity.

capped steel. A type of steel similar to rimmed steel, usually cast in a bottle-top ingot mold, in which the application of a mechanical or a chemical cap renders the rimming action incomplete by causing the top metal to solidify. The surface condition of capped steel is much like that of *rimmed steel,* but certain other characteristics are intermediate between those of rimmed steel and those of *semi-killed steel.*

capping. Partial or complete separation of a powder metallurgy compact into two or more portions by cracks that originate near the edges of the punch faces and that proceed diagonally into the compact.

carbide. A compound of carbon with one or more metallic elements.

carbide tools. Cutting or forming tools, usually made from tungsten, titanium, tantalum, or niobium carbides, or a combination of them, in a matrix of cobalt, nickel, or other metals. Carbide tools are characterized by high hardness and compressive strength and may be coated to improve wear resistance.

carbon dioxide welding. *Gas metal arc welding* using carbon dioxide as the shielding gas.

carbon edges. Carbonaceous deposits in a wavy pattern along the edges of a sheet or strip. Also known as *snaky edges*.

carbon electrode. A carbon or graphite rod used in carbon arc welding or cutting torches.

carbon equivalent (CE). (1) For cast iron, the equivalent carbon content concept is used to understand how alloying elements will affect the heat treatment and casting behavior. An empirical relationship of the total carbon, silicon, and phosphorus contents is expressed by the formula:

$$CE = \%C + 0.3(\%Si) + 0.33(\%P) - 0.027(\%Mn) + 0.4(\%S)$$

(2) For rating of weldability, this formula has been found suitable for predicting hardenability in a large range of commonly used plain carbon and carbon-manganese steels, but not in microalloyed high-strength low-alloy steels or low-alloy chromium-molybdenum steels:

$$CE = \%C + \frac{\%Mn}{6} + \frac{\%Cr + \%Mo + \%V}{5} + \frac{\%Ni + \%Cu}{15}$$

carbonitriding. A *case-hardening* process in which a suitable ferrous material is heated above the lower transformation temperature in a gaseous atmosphere of such composition as to cause simultaneous absorption of carbon and nitrogen by the surface and, by diffusion, create a concentration gradient. The process is completed by cooling at a rate that produces the desired properties in the workpiece.

carbonization. Conversion of an organic substance into elemental carbon. (Not the same as car*buri*zation.)

carbon potential. A measure of the ability of an environment containing active carbon to alter or maintain, under prescribed conditions, the carbon level of the steel. In any particular environment, the carbon level attained will depend on such factors as temperature, time, and steel composition.

carbon restoration. Replacing the carbon lost in the surface layer from previous processing by carburizing this layer to substantially restore the original carbon level. Sometimes called *recarburizing*.

carbon steel. Steel having no specified minimum quantity for any alloying element (other than the commonly accepted amounts of manganese, silicon, and copper) and containing only an incidental amount of any element other than carbon, silicon, manganese, copper, sulfur, and phosphorus.

carbonyl powder. A metal powder prepared by the thermal decomposition of a metal carbonyl.

carburizing. Absorption and diffusion of carbon into solid ferrous alloys by heating, to a temperature usually above Ac_3, in contact with a suitable carbonaceous material. A form of *case hardening* that produces a carbon gradient extending inward from the surface, enabling the surface layer to be hardened either by quenching directly from the carburizing temperature or by cooling to room temperature, then reaustenitizing and quenching.

carburizing flame. A gas flame that introduces carbon into some heated metals, as during a gas welding operation. A carburizing flame is a *reducing flame*, but a reducing flame is not necessarily a carburizing flame.

cartridge brass. An alloy with 68.5–71.5% Cu, 0.05% maximum iron, 0.05% maximum lead, and the remainder zinc.

cascade sequence. A welding sequence in which a continuous multiple-pass weld is built up by depositing weld beads in overlapping layers, usually laid in a *backstep sequence* (Fig. 13). Compare with *block sequence*.

Fig. 13 Cascade welding sequence of reducing stress

case. The portion of a ferrous alloy, extending inward from the surface, whose composition has been altered so it can be *case hardened*. Typically considered to be the portion of the alloy (a) whose composition has been measurably altered from the original composition, (b) that appears light on an etched cross section, or (c) that has a hardness, after hardening, equal to or greater than a specified value. Contrast with *core* (2).

case hardening. A generic term covering several processes applicable to steel that change the chemical composition of the

surface layer by absorption of carbon, nitrogen, or a mixture of the two and, by diffusion, create a concentration gradient. The processes commonly used are *carburizing* and *quench hardening*; *cyaniding*; *nitriding*; and *carbonitriding*. The use of the applicable specific process name is preferred.

cassette. A light-tight holder used to contain radiographic films during exposure to x-rays or gamma rays, that may or may not contain intensifying or filter screens, or both. A distinction is often made between a cassette, which has positive means for ensuring contact between screens and film and is usually rigid, and an exposure holder, which is rather flexible.

CASS test. Abbreviation for *copper-accelerated salt-spray test*. An accelerated corrosion test for some electrodeposits and for anodic coatings on aluminum.

cast. A casting of the die impression made to confirm the exactness of the impression. Also called *die proof*.

cast-alloy tool. A cutting tool made by casting a cobalt-base alloy and used at machining speeds between those for high-speed steels and sintered carbides.

casting. (1) An object at or near finished shape obtained by solidification of a substance in a mold. (2) Pouring molten metal into a mold to produce an object of desired shape.

casting copper. Fire-refined tough pitch copper usually cast from melted secondary metal into ingot bars only, and used for making foundry castings but not wrought products.

casting shrinkage. (1) Liquid shrinkage: the reduction in volume of liquid metal as it cools to the liquidus. (2) Solidification shrinkage: the reduction in volume of metal from beginning to end of solidification. (3) Solid shrinkage: the reduction in volume of metal from the solidus to room temperature.

casting strains. Strains in a casting caused by *casting stresses* that develop as the casting cools.

casting stresses. Residual stresses set up when the shape of a casting impedes contraction of the solidified casting during cooling.

cast iron. A generic term for a large family of cast ferrous alloys in which the carbon content exceeds the solubility of carbon in austenite at the eutectic temperature. Most cast irons contain at least 2% C, plus silicon and sulfur, and may or may not contain other alloying elements. For the various forms— *gray cast iron, white cast iron, malleable cast iron,* and *ductile cast iron*—the word *cast* often is left out, resulting in gray iron, white iron, malleable iron, and ductile iron, respectively. See *Technical Note 2.*

TECHNICAL NOTE 2

Cast Irons

CAST IRON is a generic term that identifies a large family of cast ferrous alloys that solidify with a eutectic and in which the carbon content exceeds the solubility of carbon in austenite at the eutectic temperature. Cast irons primarily are alloys of iron that contain more than 2% C and 1–3% Si. Wide variations in properties can be achieved by varying the balance between carbon and silicon, by alloying with various metallic or nonmetallic elements, and by varying melting, casting, and heat-treating practices.

- Cast irons can be classified according to their graphite shape, matrix microstructure (austenitic, ferritic, etc.), or fracture type:

Range of compositions for typical unalloyed common cast irons

Type of iron	Composition, %				
	C	Si	Mn	P	S
Gray (FG)	2.5–4.0	1.0–3.0	0.2–1.0	0.002–1.0	0.02–0.25
Compacted graphite (CG)	2.5–4.0	1.0–3.0	0.2–1.0	0.01–0.1	0.01–0.03
Ductile (SG)	3.0–4.0	1.8–2.8	0.1–1.0	0.01–0.1	0.01–0.03
White	1.8–3.6	0.5–1.9	0.25–0.8	0.06–0.2	0.06–0.2
Malleable (TG)	2.2–2.9	0.9–1.9	0.15–1.2	0.02–0.2	0.02–0.2

- *White iron* is essentially free of graphite, and most of the carbon content is present as separate grains of hard Fe_3C. White iron exhibits a white, crystalline fracture surface, because fracture occurs along the iron carbide plates. It is usually not heat treated, but is stress relieved.

TECHNICAL NOTE 2 (*continued*)

- *Malleable iron* contains compact nodules of graphite flakes called "temper carbon," because they form during an extended annealing of white iron of a suitable composition.
- *Gray iron* contains carbon in the form of graphite flakes. Gray iron exhibits a gray fracture surface, because fracture occurs along the graphite plates (flakes).

Classification of cast iron by commercial designation, microstructure, and fracture

Commercial designation	Carbon-rich phase	Matrix(a)	Fracture	Final structure after
Gray iron	Lamellar graphite	P	Gray	Solidification
Ductile iron	Spheroidal graphite	F, P, A	Silver-gray	Solidification or heat treatment
Compacted graphite iron	Compacted vermicular graphite	F, P	Gray	Solidification
White iron	Fe_3C	P, M	White	Solidification and heat treatment(b)
Mottled iron	Lamellar Gr + Fe_3C	P	Mottled	Solidification
Malleable iron	Temper graphite	F, P	Silver-gray	Heat treatment
Austempered ductile iron	Spheroidal graphite	At	Silver-gray	Heat treatment

(a) F, ferrite; P, pearlite; A, austenite; M, martensite; At, austempered (bainite). (b) White irons are not usually heat treated, except for stress relief and to continue austenite transformation.

- *Mottled iron* falls between gray and white iron, with the fracture showing both gray and white zones.
- *Ductile iron*, also known as spheroidal graphite iron, contains spherulitic graphite, in which the graphite flakes form into balls as do cabbage leaves. Ductile iron is so named because in the as-cast form it exhibits measurable ductility.
- *Austempered ductile iron* is a moderately alloyed ductile iron that is austempered for high strength with appreciable ductility. Its microstructure is different from austempered steel, and its heat treatment is a specialty.
- *Compacted graphite iron* contains graphite in the form of thick, stubby flakes. Its mechanical properties are between those of gray and ductile iron.
- *High-alloy iron* contains more than 3% alloy content and is commercially classified separately. High-alloy irons may be a type of white iron, gray iron, or ductile iron. The matrix may be ferritic or austenitic.

Selected References

- *ASM Handbook*, Vol 4, *Heat Treating*, ASM International, 1991, p 667–708
- *Metals Handbook*, 10th ed., Vol 1, *Properties and Selection: Irons, Steels, and High-Performance Alloys*, ASM International, 1990, p 3–104
- *Metals Handbook*, 9th ed., Vol 15, *Casting*, ASM International, 1988, p 627–710

cast steel. Steel in the form of a *casting*.

cast structure. The metallographic structure of a *casting* evidenced by shape and orientation of grains and by segregation of impurities.

catalyst. A substance capable of changing the rate of a reaction without itself undergoing any net change.

catastrophic failure. Sudden failure of a component or assembly that frequently results in extensive secondary damage to adjacent components or assemblies.

cathode. The electrode where electrons enter an operating system such as a battery, an electrolytic cell, an x-ray tube, or a vacuum tube. In the first of these, it is positive; in the other three, negative. In a battery or electrolytic cell, it is the electrode where reduction occurs. Contrast with *anode*.

cathode compartment. In an electrolytic cell, the enclosure formed by a diaphragm around the cathode.

cathode copper. Copper deposited at the cathode in electrolytic refining.

cathode efficiency. *Current efficiency* at the *cathode*.

cathode film. The portion of solution in immediate contact with the *cathode* during *electrolysis*.

cathodic cleaning. *Electrolytic cleaning* in which the work is the *cathode*.

cathodic pickling. *Electrolytic pickling* in which the work is the *cathode*.

cathodic protection. Partial or complete protection of a metal from corrosion by making it a *cathode,* using either a galvanic or an impressed current. Contrast with *anodic protection*.

catholyte. The *electrolyte* adjacent to the cathode in an electrolytic cell; in a divided cell, the portion on the cathode side of the diaphragm.

cation. A positively charged ion; it flows to the *cathode* in *electrolysis*.

cationic detergent. A detergent in which the *cation* is the active part.

caustic cracking. A form of *stress-corrosion cracking* most frequently encountered in carbon steels or Fe-Cr-Ni alloys that are exposed to concentrated hydroxide solutions at temperatures of 200 to 250 °C (400 to 480 °F).

caustic dip. A strongly alkaline solution into which metal is immersed for etching, for neutralizing acid, or for removing organic materials such as greases or paints.

cavitation. The formation and instantaneous collapse of innumerable tiny voids or cavities within a liquid subjected to rapid and intense pressure changes. Cavitation produced by ultrasonic radiation is sometimes used to create violent localized agitation. Cavitation caused by severe turbulent flow often leads to *cavitation damage*.

cavitation damage. Erosion of a solid surface through the formation and collapse of cavities in an adjacent liquid.

cavitation erosion. See preferred term *cavitation damage*.

cell feed. The material supplied to the cell in the electrolytic production of metals.

cementation. Introduction of one or more elements into the outer portion of a metal object by means of diffusion at high temperature.

cement copper. Impure copper recovered by *chemical deposition* when iron (most often shredded steel scrap) is brought into prolonged contact with a dilute copper sulfate solution.

cemented carbide. A solid and coherent mass made by pressing and sintering a mixture of powders of one or more metallic carbides and a much smaller amount of a metal, such as cobalt, to serve as a binder.

cementite. A compound of iron and carbon, known chemically as iron carbide and having the approximate chemical formula Fe_3C. It is characterized by an orthorhombic crystal structure. When it occurs as a phase in steel, the chemical composition will be altered by the presence of manganese and other carbide-forming elements.

centerless grinding. Grinding the outside or inside of a workpiece mounted on rollers rather than on centers. The workpiece may be in the form of a cylinder or the frustum of a cone.

centrifugal casting. A casting made by pouring metal into a mold that is rotated or revolved.

ceramic tools. Cutting tools made from fused, sintered, or cemented metallic oxides.

cereal. An organic *binder,* usually corn flour.

cerium. A chemical element having atomic number 58, atomic weight 140, and the symbol Ce, named for the asteroid, Ceres, that was discovered in 1801 and named for the Roman goddess of earth, agriculture, fertility, and grain. The element was identified in 1803 by Jöns Jakob Berzelius and Wilhelm Hisinger, Swedish chemist and Swedish geologist at Uppsala University. Cerium is the most valuable and most abundant of the rare earth metals. It is a soft metal that oxidizes in the atmosphere. Commercial applications include catalysts, additives to fuel to decrease emissions, and additives to glass and enamels to change their color. Cerium oxide is an important component of glass polishing compounds and phosphors used in screens and fluorescent lamps. Cerium improves the high-temperature strength and ductility of magnesium alloys and is used in the production of spheroidal graphite cast iron. A major application of cerium oxide is as a catalytic converter for the reduction of carbon monoxide emissions in the exhaust gases of motor vehicles.

cermet. A powder metallurgy product consisting of ceramic particles bonded with a metal.

C-frame press. Same as *gap-frame press.*

CG iron. Same as *compacted graphite cast iron.*

chafing fatigue. Fatigue initiated in a surface damaged by rubbing against another body. See *fretting.*

chain-intermittent fillet welding. Depositing a line of intermittent fillet welds on each side of a member at a joint so that the increments on one side are essentially opposite those on the other. Contrast with *staggered-intermittent fillet welding.*

chamfer. (1) A beveled surface to eliminate an otherwise sharp corner. (2) A relieved angular cutting edge at a tooth corner.

chamfer angle. (1) The angle between a reference surface and the bevel. (2) On a milling cutter, the angle between a beveled surface and the axis of the cutter.

chamfering. Making a sloping surface on the edge of a member. Also called beveling.

chaplet. Metal support that holds a core in place within a mold; molten metal solidifies around a chaplet and fuses it into the finished casting (Fig. 14).

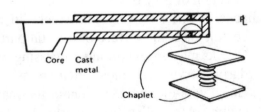

Fig. 14 Chaplet in casting core

characteristic radiation. High-intensity single-wavelength x-rays, characteristic of the element emitting the rays, that appear in addition to continuous "white" radiation whenever the element is bombarded with electrons whose energy exceeds a specific critical value that is different for each element.

charge. (1) The materials fed into a furnace. (2) Weights of various liquid and solid materials put into a furnace during one feeding cycle.

charging. (1) For a lap, impregnating the surface with fine abrasive. (2) Placing materials into a furnace.

Charpy test. A pendulum-type single-blow impact test in which the specimen, usually notched, is supported at both ends as a simple beam and broken by a falling pendulum. The energy absorbed, as determined by the subsequent rise of the pendulum, is a measure of impact strength or notch toughness. Contrast with *Izod test*.

chatter. In machining or grinding, (1) a vibration of the tool, wheel, or workpiece producing a wavy surface on the work and (2) the finish produced by such vibration.

checked edges. Sawtooth edges seen after hot rolling and/or cold rolling.

checkers. In a chamber associated with a metallurgical furnace, bricks stacked openly so that heat may be absorbed from the combustion products and later transferred to incoming air when the direction of flow is reversed.

checks. Numerous, very fine cracks in a coating or at the surface of a metal part. Checks may appear during processing or during service and most often are associated with thermal treatment or thermal cycling. Also called check marks, checking, and heat checks.

cheek. The intermediate section of a flask that is used between the *cope* and the *drag* when molding a shape that requires more than one parting plane.

chelating agent. A substance used in metal finishing to control or eliminate certain metallic ions present in undesirable quantities.

chemical deposition. The precipitation or plating-out of a metal from solutions of its salts through the introduction of another metal or reagent to the solution.

chemically precipitated powder. Metal powder produced as a precipitate by chemical displacement.

chemical machining. Removing metal stock by controlled selective chemical dissolution.

chemical metallurgy. See *process metallurgy*.

chemical polishing. Improving the surface luster of a metal by chemical treatment.

chevron pattern. A fractographic pattern of radial marks (shear ledges) that look like nested Vs; sometimes called a herringbone pattern. Chevron patterns typically are found on brittle fracture surfaces in parts whose widths are considerably greater than their thicknesses. The points of the chevrons can be traced back to the fracture origin.

chill. (1) A metal or graphite insert embedded in the surface of a sand mold or core, or placed in a mold cavity to increase the cooling rate at that point. (2) White iron occurring on a gray or ductile iron casting, such as the chill in the wedge test. Compare with *inverse chill*.

chill time. In resistance welding, the time from the finish of the welding operation to the beginning of tempering. Also called *quench time*.

Chinese script. The angular microstructural form suggestive of Chinese writing and characteristic of the constituents α(Al-Fe-Si) and α(Al-Fe-Mn-Si) in cast aluminum alloys. A similar microstructure is found in cast magnesium alloys containing silicon as Mg_2Si.

chip breaker. (1) Notch or groove in the face of a tool parallel to the cutting edge, designed to break the continuity of the chip. (2) A step formed by an adjustable component clamped to the face of the cutting tool.

chipping. (1) Removing seams and other surface imperfections in metals manually with a chisel or gouge, or by a continuous machine, before further processing. (2) Similarly, removing excessive metal.

chips. Pieces of material removed from a workpiece by cutting tools or by an abrasive medium.

chlorination. (1) Roasting ore in contact with chlorine or a chloride salt to produce chlorides. (2) Removing dissolved gases and entrapped oxides by passing chlorine gas through molten metal such as aluminum and magnesium.

chromadizing. Improving paint adhesion on aluminum or aluminum alloys, mainly aircraft skins, by treatment with a solution of chromic acid. Also called chromodizing or chromatizing. Not the same as *chromating* or *chromizing*.

chromate treatment. A treatment of metal in a solution of a hexavalent chromium compound to produce a *conversion coating* consisting of trivalent and hexavalent chromium compounds.

chromating. Performing a *chromate treatment*.

Chromel. (1) A 90Ni-10Cr alloy used in thermocouples. (2) A series of nickel-chromium alloys, some with iron, used for heat-resistant applications.

chrome plating. (1) Producing a chromate *conversion coating* on magnesium for temporary protection or for a paint base. (2) The solution that produces the conversion coating.

chromium. A chemical element having atomic number 24, atomic weight 52, and symbol Cr, named for the Greek *chroma,* meaning color, because of the colors of its compounds. Metallic chromium was first obtained by Nicolas-Louis Vaquelin in 1797. Hans Goldschmidt discovered a pure form in 1895. In the early part of the 19th century, chromium was found in France and America, with Pennsylvania and Maryland producing almost the entire world's requirements. Chromium is a steel-gray, hard lustrous metal. In the pure form it is extremely brittle and can be used only for decorative plating. It is used extensively to make high-strength low-alloy steels containing 0.5–3% Cr as well as chromium-molybdenum steels. Chromium is alloyed with copper and with cobalt to make heat-resistant alloys such as the Stellites. The largest use of chromium is in stainless steels that contain 10.5–30% Cr. See Table 8.

chromizing. A surface treatment at elevated temperature, generally carried out in pack, vapor, or salt bath, in which an alloy is formed by the inward diffusion of chromium into the base metal.

circle grinding. Either *cylindrical grinding* or *internal grinding*, the preferred terms.

circle shear. A shearing machine with two rotary disk cutters mounted on parallel shafts driven in unison and equipped with an attachment for cutting circles where the desired piece of material is inside the circle.

Table 8 Nominal chemical compositions for heat-resistant chromium-molybdenum steels

Type	UNS designation	Composition, %(a)						
		C	Mn	S	P	Si	Cr	Mo
1/2Cr–1/2Mo	K12122	0.10–0.20	0.30–0.80	0.040	0.040	0.10–0.60	0.50–0.80	0.45–0.65
1C–1/2Mo	K11562	0.15	0.30–0.60	0.045	0.045	0.50	0.80–1.25	0.45–0.65
1 1/4Cr–1/2Mo	K11597	0.15	0.30–0.60	0.030	0.030	0.50–1.00	1.00–1.50	0.45–0.65
1 1/4Cr–1/2Mo	K11592	0.10–0.20	0.30–0.80	0.040	0.040	0.50–1.00	1.00–1.50	0.45–0.65
2 1/4Cr–1Mo	K21590	0.15	0.30–0.60	0.040	0.040	0.50	2.00–2.50	0.87–1.13
3Cr–1Mo	K31545	0.15	0.30–0.60	0.030	0.030	0.50	2.65–3.35	0.80–1.06
3Cr–1MoV(b)	K31830	0.18	0.30–0.60	0.020	0.020	0.10	2.75–3.25	0.90–1.10
5Cr–1/2Mo	K41545	0.15	0.30–0.60	0.030	0.030	0.50	4.00–6.00	0.45–0.65
7Cr–1/2Mo	K61595	0.15	0.30–0.60	0.030	0.030	0.50–1.00	6.00–8.00	0.45–0.65
9Cr–1Mo	K90941	0.15	0.30–0.60	0.030	0.030	0.50–1.00	8.00–10.00	0.90–1.10
9Cr–1MoV(c)	...	0.08–0.12	0.30–0.60	0.010	0.020	0.20–0.50	8.00–9.00	0.85–1.05

(a) Single values are maximums. (b) Also contains 0.02–0.030% V, 0.001–0.003% B, and 0.015–0.035% Ti. (c) Also contains 0.40% Ni, 0.18-0.25% V, 0.06–0.10% Nb, 0.03–0.07% N, and 0.04% Al

It cannot be employed to cut circles where the desired material is outside the circle.

circular field. The magnetic field that (a) surrounds a nonmagnetic conductor of electricity, (b) is completely contained within a magnetic conductor of electricity, or (c) both exists within and surrounds a magnetic conductor. Generally applied to the magnetic field within any magnetic conductor resulting from a current being passed through the part or through a section of the part. Compare with *bipolar field.*

clad metal. A composite metal containing two or three layers that have been bonded together. The bonding may have been accomplished by co-rolling, welding, casting, heavy chemical deposition, or heavy electroplating.

clamshell marks. Same as *beach marks.*

classification. (1) The separation of ores into fractions according to size and specific gravity, generally in accordance with Stokes' law of sedimentation. (2) Separation of a metal powder into fractions according to particle size.

clay. An earthy or stony mineral aggregate consisting essentially of hydrous silicates of alumina. It is plastic when sufficiently pulverized and wetted, rigid when dry, and vitreous when fired at a sufficiently high temperature. Clay minerals most commonly used in the foundry are montmorillonites and kaolinites.

cleanup allowance. The amount of excess metal surrounding the intended final configuration of a formed part; also called *finish allowance*, *machining allowance*, or *forging envelope.*

clearance. (1) The gap or space between two mating parts. (2) The space provided between the relief of a cutting tool and the surface that has been cut.

clearance angle. The angle between a plane containing the flank of the tool and a plane passing through the cutting edge in the direction of relative motion between the cutting edge and the work. See also the figures accompanying the terms *face mill* and *single-point tool.*

clearance fit. Any of various classes of fit between mating parts where there is a positive allowance (gap) between the parts, even when they are made to the respective extremes of individual tolerances that ensure the tightest fit between the parts. Contrast with *interference fit*.

cleavage. Splitting (fracture) of a crystallographic plane of low index.

cleavage fracture. A fracture, usually of a polycrystalline metal, in which most of the grains have failed by cleavage, resulting in bright reflecting facets. It is one type of *crystalline fracture* and is associated with low-energy brittle fracture. Contrast with *shear fracture*.

cleavage plane. A characteristic crystallographic plane or set of planes on which *cleavage fracture* occurs easily.

climb cutting. Analogous to *climb milling*.

climb milling. Milling in which the cutter moves in the direction of feed at the point of contact.

clip and shave. In forging, a dual operation in which one cutting surface in the clipping die removes the *flash* and then another shaves and sizes the piece.

close annealing. Same as *box annealing*.

closed-die forging. See *impression-die forging*.

closed dies. Forging or forming impression dies designed to restrict the flow of metal to the cavity within the die set, as opposed to open dies, in which there is little or no restriction to lateral flow.

closed pass. A pass of metal through rolls where the bottom roll has a groove deeper than the bar being rolled and the top roll has a collar fitting into the groove, thus producing the desired shape free from *flash* or fin.

close-tolerance forging. A forging held to unusually close dimensional tolerances. Often, little or no machining is required after forging.

cloudburst treatment. A form of *shot peening*.

cluster mill. A rolling mill in which each of the two working rolls of small diameter is supported by two or more backup rolls (Fig. 15).

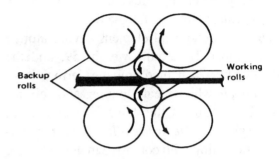

Fig. 15 Cluster mill

CO_2 welding. See *carbon dioxide welding*.

coalesced copper. Massive oxygen-free copper made by briquetting ground, brittle cathode copper, and then sintering the briquettes in a pressurized reducing atmosphere, followed by hot working.

coalescence. (1) The union of particles of a dispersed phase into larger units, usually effected at temperatures below the fusion point. (2) In welding, brazing, or soldering, the union of two or more components into a single body, which usually involves the melting of a filler metal or of the base metal.

coarse grain practice. A steelmaking practice for steels other than stainless steel that is intended to produce a *killed steel* in which aluminum, niobium (columbium), titanium, and vanadium are residual elements.

coarsening. An increase in grain size, usually, but not necessarily, by *grain growth*.

coated abrasive. An abrasive product (sandpaper, for example) in which a layer of abrasive particles is firmly attached to a paper, cloth, or fiber backing by means of glue or synthetic-resin adhesive.

coated electrode. See preferred term *lightly coated electrode.*

coaxing. Improvement of the fatigue strength of a specimen by the application of a gradually increasing stress amplitude, usually starting below the fatigue limit.

cobalt. A chemical element having atomic number 27, atomic weight 59, and the symbol Co, named from the German word *kobelt,* which means gnomes and goblins. The origin of the name is explained in Agricola's *De Re Metallica.* German miners, toiling with corrosive materials, were tricked into thinking that the shiny mineral, now known as cobaltite, would yield a precious metal; instead, it exuded an offensive arsenic odor. They despised it and gave it the epithet kobelt, not realizing it contained the valuable cobalt metal. In 1735, the cobalt metal was identified by Georg Brandt, a Swedish professor and mineralogist. It is believed that Brandt, when naming his metal, was engaged in playing a practical joke on the establishment. It is especially interesting to note that one of his students, Axel Friedrich Cronstedt, continued the game 15 years later when he discovered a metal that he named nickel, which, in German, was associated with "imp" or "little devil." See *Technical Note 3.*

coefficient of elasticity. Same as *modulus of elasticity.*

coercive force. The magnetizing force that must be applied in the direction opposite that of the previous magnetizing force to reduce magnetic flux density to zero; thus, a measure of the magnetic retentivity of magnetic materials.

cogging mill. A *blooming mill.*

coherency. The continuity of lattice of precipitate and parent phase (solvent) maintained by mutual strain and separated by a phase boundary.

coherent precipitate. A crystalline precipitate that forms from solid solution with an orientation that maintains continuity between the crystal lattice of the precipitate and the lattice of the matrix, usually accompanied by some strain in both lattices. Because the lattices fit at the interface between precipitate and matrix, there is no discernible phase boundary.

cohesion. Force of attraction between the molecules (or atoms) within a single phase. Contrast with adhesion.

cohesive strength. (1) The hypothetical stress causing tensile fracture without plastic deformation. (2) The stress corresponding to the forces between atoms. (3) Same as *technical cohesive strength.* (4) Same as *disruptive strength.*

coil breaks. Creases or ridges in sheet or strip that appear as parallel lines across the direction of rolling and that generally extend across the full width of the sheet or strip.

coining. (1) A closed-die squeezing operation, usually performed cold, in which all surfaces of the work are confined or restrained, resulting in a well-defined imprint of the die upon the work. (2) A *restriking* operation used to sharpen or change an existing radius or profile. (3) The final pressing of a sintered powder metallurgy compact to obtain a definite surface configuration (not to be confused with *repressing* or *sizing*).

coin silver. An alloy containing 90% Ag, with copper being the usual alloying element.

cold-chamber machine. A *die casting* machine in which the metal chamber and plunger are not heated.

cold extrusion. See *extrusion.*

TECHNICAL NOTE 3

Cobalt and Cobalt Alloys

COBALT is a tough silver-gray magnetic metal that resembles iron and nickel in appearance and in some properties. Cobalt is useful in applications that utilize its magnetic properties, corrosion resistance, wear resistance, and/or strength at elevated temperatures. Some cobalt-base alloys are biocompatible, which has prompted their use as orthopedic implants.

Much of cobalt today derives from copper and copper-nickel-rich sulfide deposits in Zaire and Zambia in Africa. The ore is subjected to crushing, grinding, and flotation, prior to a magnetic concentrating process. This concentrate is then leached in sulfuric acid and the cobalt and copper extracted by electrolysis.

With an atomic number of 27, cobalt falls between iron and nickel on the periodic table. The density of cobalt is 8.8 g/cm^3, similar to that of nickel. At temperatures below 417 °C (783 °F), cobalt exhibits an hcp structure. Between 417 °C (783 °F) and its melting point of 1494 °C (2719 °F), cobalt has an fcc structure.

The single largest application area for cobalt-base alloys is for wear resistance. These alloys are available as castings and weld overlays, with some alloys available in wrought (plate, sheet, and bar) form. In heat-resistant applications, cobalt is more widely used as an alloying element in nickel-base alloys, with cobalt tonnages in excess of those used in cobalt-base heat-resistant alloys. Cobalt is also an important ingredient in:

- Paint pigments
- Nickel-base alloys. See also *superalloys* (Technical Note 13).
- Cemented carbides. See also *cemented carbides*.
- Tool steels. See also *tool steels* (Technical Note 17).
- Magnetic materials. See also *permanent magnet materials* and *soft magnetic materials*.
- Artificial γ-ray sources

Selected References

- P. Crook, Cobalt and Cobalt Alloys, *Metals Handbook*, 10th ed., Vol 2, ASM International, 1990, p 446–454
- J.R. Davis, Cast Cobalt Alloys, *Metals Handbook*, 9th ed., Vol 16, ASM International, 1989, p 69–70
- A.I. Asphahani et al., Corrosion of Cobalt-Base Alloys, *Metals Handbook*, 9th ed., Vol 13, ASM International, 1987, p 658–668

TECHNICAL NOTE 3 (continued)

Nominal compositions of various cobalt-base alloys

Alloy trade name	Co	Cr	W	Mo	C	Fe	Ni	Si	Mn	Others
					Nominal composition, %					
Cobalt-base wear-resistant alloys										
Stellite 1	bal	31	12.5	1 (max)	2.4	3 (max)	3 (max)	2 (max)	1 (max)	...
Stellite 6	bal	28	4.5	1 (max)	1.2	3 (max)	3 (max)	2 (max)	1 (max)	...
Stellite 12	bal	30	8.3	1 (max)	1.4	3 (max)	3 (max)	2 (max)	1 (max)	...
Stellite 21	bal	28	...	5.5	0.25	2 (max)	2.5	2 (max)	1 (max)	...
Haynes alloy 6B	bal	30	4	1	1.1	3 (max)	2.5	0.7	1.5	...
Tribaloy T-800	bal	17.5	...	29	0.08 (max)	...	...	3.5	...	...
Stellite F	bal	25	12.3	1 (max)	1.75	3 (max)	22	2 (max)	1 (max)	...
Stellite 4	bal	30	14.0	1 (max)	0.57	3 (max)	3 (max)	2 (max)	1 (max)	...
Stellite 190	bal	26	14.5	1 (max)	3.3	3 (max)	3 (max)	2 (max)	1 (max)	...
Stellite 306	bal	25	2.0	...	0.4	...	5	...	...	6 Nb
Stellite 6K	bal	31	4.5	1.5 (max)	1.6	3 (max)	3 (max)	2 (max)	2 (max)	...
Cobalt-base high-temperature alloys										
Haynes alloy 25 (L605)	bal	20	15	...	0.10	3 (max)	10	1 (max)	1.5	...
Haynes alloy 188	bal	22	14	...	0.10	3 (max)	22	0.35	1.25	0.05 La
MAR-M alloy 509	bal	22.5	7	...	0.60	1.5 (max)	10	0.4 (max)	0.1 (max)	3.5 Ta, 0.2 Ti, 0.5 Zr
Cobalt-base corrosion-resistant alloys										
MP35N, Multiphase alloy	bal	20	...	10	...	...	35	...	...	...
Haynes alloy 1233	bal	25.5	2	5	0.08 (max)	3	9	...	...	0.1N (max)

bal, balance

cold heading. Working metal at room temperature in such a manner that the cross-sectional area of a portion or all of the stock is increased.

cold inspection. A visual (usually final) inspection of forgings for visible imperfections, dimensions, weight, and surface condition at room temperature. The term also may be used to describe certain nondestructive tests such as magnetic-particle, dye-penetrant, and sonic inspection.

cold lap. Wrinkled markings on the surface of an ingot, caused by incipient freezing of the surface while the liquid is still in motion, resulting from insufficient pouring temperature. See also *cold shut* (1).

cold mill. A mill for cold rolling of sheet or strip.

cold pressing. Forming a powder metallurgy *compact* at a temperature low enough to avoid *sintering*, usually room temperature. Contrast with *hot pressing*.

cold rolled sheets. A mill product produced from a hot rolled pickled coil that has been subjected to substantial cold reduction at room temperature. The resulting product usually requires further processing to make it suitable for most common applications. The usual end product is characterized by improved surface, greater uniformity in thickness, and improved mechanical properties compared with hot rolled sheet.

cold shortness. Brittleness that exists in some metals at temperatures below the recrystallization temperature.

cold shot. A portion of the surface of an ingot or casting showing premature solidification; caused by splashing of molten metal onto a cold mold wall during pouring.

cold shut. (1) A discontinuity that appears on the surface of cast metal as a result of two streams of liquid meeting and failing to unite. (2) A lap on the surface of a forging or billet that was closed without fusion during deformation. (3) Freezing of the top surface of an ingot before the mold is full.

cold treatment. Exposing steel to suitable subzero temperatures for the purpose of obtaining desired conditions or properties such as dimensional or microstructural stability. When the treatment involves the transformation of retained austenite, it usually is followed by tempering.

cold trimming. Removing flash or excess metal from a forging in a trimming press when the forging is at room temperature.

cold welding. A solid-state welding process in which pressure is used at room temperature to produce coalescence of metals with substantial deformation at the weld. Compare to *hot pressure welding*, *diffusion welding*, and *forge welding*.

cold work. Permanent strain in a metal accompanied by strain hardening.

cold working. Deforming metal plastically under conditions of temperature and strain rate that induce *strain hardening*. Usually, but not necessarily, conducted at room temperature. Contrast with *hot working*.

collapsibility. The tendency of a sand mold or core to break down under the pressure and temperature of casting in order to avoid *hot tears* or to facilitate the separation of sand and casting.

color buffing. Producing a final high luster by buffing. Sometimes called *coloring*.

coloring. Producing desired colors on metal by a chemical or electrochemical reaction. See also *color buffing*.

columbium. Also known as *niobium*. A chemical element having atomic number 41, atomic weight 93, and the symbol Nb. In approximately 1734, John Winthrop the Younger, the first governor of Connecticut, discovered a new mineral that he called columbite (Columbia is a synonym for America), a sample of which was sent to the British Museum. The sample finally was examined in 1801 by the British chemist and manufacturer, Charles Hatchett, who discovered in it a new element that he called columbium. Columbium was rediscovered and renamed niobium in approximately 1844 by Heinrich Rose. The principal ore is columbite, which often is associated with tantalite. The metal is exceptionally corrosion resistant, being attacked only by aqua regia. Small additions of columbium to the 18%Cr-8%Ni alloy are made to create the stable type 347 alloy that is not subject to carbide precipitation and intergranular corrosion. The addition of up to 5% Nb improves the creep strength of nickel-base high-temperature alloys. The metal is used as a cladding (canning) metal for fuel elements in nuclear reactors and is a principal element in *superconductor* alloys.

columnar structure. A coarse structure of parallel elongated grains formed by unidirectional growth, most often observed in castings, but sometimes seen in structures resulting from diffusional growth accompanied by a solid-state transformation.

combination die. (1) A die-casting die having two or more different cavities for different castings. (2) For forming, see *compound die*.

combination mill. An arrangement of a continuous mill for roughing and a *guide mill* or *looping mill* for shaping.

combined carbon. The part of the total carbon in steel or cast iron that is combined chemically with other elements; distinguished from *free carbon*.

combined cyanide. The cyanide of a metal-cyanide complex ion.

combined stresses. Any state of stress that cannot be represented by a single component of stress; that is, one that is more complicated than simple tension, compression, or shear.

commercial bronze. An alloy with 89–91% Cu, 0.05% maximum iron, 0.05% maximum lead, and the balance zinc.

commercial steel (CS). The designation established by ASTM in 2000 for cold rolled carbon steel sheet intended for exposed or unexposed parts where bending, moderate forming, and welding may be involved. The steel, which is produced in coils and cut lengths, is produced in three types: A, B, and C, which have different composition requirements, particularly with regard to the carbon contents. The typical yield strengths of the three types are 140 to 275 MPa (20 to 40 ksi) and typical elongation in two inches is 15 to 30% or more (ASTM A1008).

comminution. (1) Breaking up or grinding an ore into small fragments. (2) Reducing metal to powder by mechanical means.

common brass. A 37% brass that is inexpensive and standard for cold working; also known as rivet brass.

commutator-controlled welding. Spot or projection welding in which several electrodes, in simultaneous contact with the work, function progressively under the control of an electrical commutating device.

compact. An object produced by the compression of metal powder, generally while confined in a die, with or without the inclusion of nonmetallic constituents. See also *compound compact* and *composite compact*.

compacted graphite cast iron. Cast iron having a graphite shape intermediate between the flake form typical of gray cast iron and the spherical form of fully spherulitic ductile cast iron. Compacted graphite cast iron is produced in a manner similar to that for ductile cast iron but using a technique that inhibits the formation of fully spherulitic graphite nodules. The same as *CG iron* or *vermicular iron.*

complete fusion. Fusion that has occurred over the entire base metal surfaces exposed for welding.

complexing agent. A substance that is an electron donor and that will combine with a metal ion to form a soluble complex ion.

complex ion. An ion that may be formed by the addition reaction of two of more other ions.

component. (1) One of the elements or compounds used to define a chemical (or alloy) system, including all phases, in terms of the fewest substances possible. (2) One of the individual parts of a vector as referred to a system of coordinates.

composite compact. A powder metallurgy *compact* consisting of two or more adhering layers of different metals or alloys with each layer retaining its original identity.

composite electrode. A welding electrode made from two or more distinct components, at least one of which is filler metal. A composite electrode may exist in any of various physical forms, such as stranded wires, filled tubes, or covered wire.

composite joint. A joint in which welding is used in conjunction with mechanical joining.

composite material. A heterogeneous, solid structural material consisting of two or more distinct components that are mechanically or metallurgically bonded together (such as a *cermet*, or boron wire embedded in a matrix of epoxy resin).

composite plate. An electrodeposit consisting of layers of at least two different compositions.

composite structure. A structural member (such as a panel, plate, pipe, or other shape) that is built up by bonding together two or more distinct components, each of which may be made of a metal, alloy, nonmetal, or *composite material.* Examples of composite structures include: honeycomb panels, clad plate, electrical contacts, sleeve bearings, carbide-tipped drills or lathe tools, and elements constructed of two or more different alloys.

compound compact. A powder metallurgy *compact* consisting of mixed metals, the particles of which are joined by pressing or sintering, or both, with each metal particle retaining substantially its original composition.

compound die. Any die so designed that it performs more than one operation on a part with one stroke of the press, such as blanking and piercing, where all functions are performed simultaneously within the confines of the particular blank size being worked.

compressibility. In powder metallurgy, the reciprocal of the *compression ratio,* where a compact is made following a procedure in which the die, the pressure, and the pressing speed are specified.

compression ratio. In powder metallurgy, the ratio of the volume of the loose powder to the volume of the compact made from it.

compressive strength. The maximum compressive stress that a material is capable of developing based on original area of cross section. If a material fails in compression by a shattering fracture, the compressive strength has a very definite value. If a material does not fail in compression by a shattering fracture, the value obtained for compressive strength is an arbitrary value depending on the degree of distortion that is regarded as indicating complete failure of the material.

concentration. A process for enrichment of an ore in valuable mineral content by separation and removal of waste material, or *gangue*.

concentration polarization. That part of the total *polarization* that is caused by changes in the activity of the potential-determining components of the electrolyte.

concurrent heating. Using a second source of heat to supplement the primary heat in cutting or welding.

conditioning heat treatment. A preliminary heat treatment used to prepare a material for a desired reaction to a subsequent heat treatment. For the term to be meaningful, the exact heat treatment must be specified.

cone angle. The angle that the cutter axis makes with the direction along which the blades are moved for adjustment, as in adjustable-blade reamers where the base of the blade slides on a conical surface.

congruent melting. An isothermal or isobaric melting in which both the solid and liquid phases have the same composition throughout the transformation.

congruent transformation. An isothermal or isobaric phase change in which both of the phases concerned have the same composition throughout the process.

constantan. A group of copper-nickel alloys containing 45–60% Cu with minor amounts of iron and manganese, characterized by relatively constant electrical resistivity irrespective of temperature; used in resistors and thermocouples.

constituent. (1) One of the ingredients that make up a chemical system. (2) A phase or a combination of phases that occurs in a characteristic configuration in an alloy microstructure.

constitution diagram. A graphical representation of the temperature and composition limits of phase fields in an alloy system as they actually exist under the specific conditions of heating or cooling (synonymous with phase diagram). A constitution diagram may be an equilibrium diagram, an approximation to an equilibrium diagram, or a representation of metastable conditions or phases. Compare with *equilibrium diagram*.

constraint. Any restriction that limits the transverse contraction normally associated with a longitudinal tension, and that hence causes a secondary tension in the transverse direction; usually used in connection with welding. Contrast with *restraint*.

consumable electrode. A general term for any arc welding electrode made chiefly of filler metal. Use of specific names such as *covered electrode*, *bare electrode*, flux-cored electrode, and *lightly coated electrode* is preferred.

consumable-electrode remelting. A process for refining metals in which an electric current passes between an electrode made of the metal to be refined and an ingot of the refined metal, which is contained in a water-cooled mold. As a result of the passage of electric current, droplets of molten metal form on the electrode and fall to the ingot. The refining action occurs from contact with the atmosphere, vacuum, or slag through which the drop falls. See *electroslag remelting* and *vacuum arc remelting*.

contact fatigue. Cracking and subsequent pitting of a surface subjected to alternating Hertzian stresses such as those produced under rolling contact or combined rolling and sliding. The phenomenon of contact fatigue is encountered most often in rolling-element bearings or in gears, where the surface stresses are high due to the concentrated loads and are repeated many times during normal operation.

contact plating. A metal plating process wherein the plating current is provided by galvanic action between the work metal and a second metal, without the use of an external source of current.

contact potential. The potential difference at the junction of two dissimilar substances.

contact scanning. In ultrasonic inspection, a planned systematic movement of the beam relative to the object being inspected, the search unit being in contact with and coupled to this object by a thin film of coupling material.

container. The chamber into which an ingot or billet is inserted prior to extrusion. The container for backward extrusion of cups or cans is sometimes called a *die*.

continuous casting. A casting technique in which a cast shape is continuously withdrawn through the bottom of the mold as it solidifies, so that its length is not determined by mold dimensions. Used chiefly to produce semifinished mill products such as billets, blooms, ingots, slabs, and tubes. See also *strand casting*.

continuous mill. A rolling mill consisting of a number of strands of synchronized rolls (in tandem) in which metal undergoes successive reductions as it passes through the various stands.

continuous phase. In an alloy or portion of an alloy containing more than one phase, the phase that forms the matrix in which the other phase or phases are present as isolated units.

continuous precipitation. Precipitation from a supersaturated solid solution in which the precipitate particles grow by long-range diffusion without recrystallization of the matrix. Continuous precipitates grow from nuclei distributed more or less uniformly throughout the matrix. They usually are randomly oriented, but may form a *Widmanstätten structure*. Also called general precipitation. Compare with *discontinuous precipitation* and *localized precipitation*.

continuous weld. A weld extending continuously from one end of a joint to the other or, where the joint essentially is circular, completely around the joint. Contrast with *intermittent weld*.

contour forming. See *stretch forming, tangent bending,* and *wiper forming*.

contour machining. Machining of irregular surfaces, such as those generated in tracer turning, tracer boring, and *tracer milling*.

contour milling. Milling of irregular surfaces. See *tracer milling*.

controlled cooling. Cooling from an elevated temperature in a predetermined manner to avoid hardening, cracking, or internal damage, or to produce desired microstructure or mechanical properties.

controlled-pressure cycle. A forming cycle during which the hydraulic pressure in the forming cavity is controlled by an adjustable cam that is coordinated with the punch travel.

conventional forging. A forging characterized by design complexity and tolerances that fall within the broad range of general forging practice.

conventional milling. Milling in which the cutter moves in the direction opposite the feed at the point of contact.

conventional strain. See *strain*.

conventional stress. See *stress*.

conversion coating. A coating consisting of a compound of the surface metal, produced by chemical or electrochemical treatments of the metal. Examples are chromate coatings on zinc, cadmium, magnesium, and aluminum, and oxides and phosphate coatings on steel.

converter. A furnace in which air is blown through a bath of molten metal or matte, oxidizing the impurities and maintaining the temperature through the heat produced by the oxidation reaction.

convex fillet weld. A fillet weld having a convex face (Fig. 16).

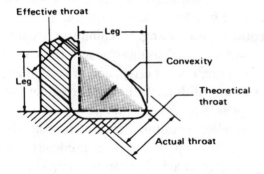

Fig. 16 Convex fillet weld

coolant. In metal cutting, the preferred term is *cutting fluid*.

cooling curve. A curve showing the relation between time and temperature during the cooling of a material.

cooling stresses. Residual stresses resulting from nonuniform distribution of temperature during cooling.

cooling table. Same as *hot bed*.

coordination number. (1) Number of atoms or radicals coordinated with the central atom in a complex covalent compound. (2) Number of nearest neighboring atoms to a selected atom in crystal structure.

cope. The upper or topmost section of a *flask, mold,* or *pattern.*

copper. A chemical element having atomic number 29, atomic weight 63.5, and symbol Cu, from the Latin *cuprum,* for Cyprus, the island where copper was first found. Copper is one of the two colored metals and was one of the seven metals of antiquity. Copper may be found naturally. It is a soft metal but one that can be hardened to make bronze with the addition of small amounts of tin or arsenic. See *Technical Note 4.*

copper-accelerated salt-spray test. An accelerated corrosion test for some electrodeposits and for anodic coatings on aluminum. Often referred to as CASS test.

copper brazing. A term improperly used to denote joining with a copper-base filler metal. See preferred terms *brazing* and *braze welding.*

copperhead. A reddish spot in a porcelain enamel coating caused by iron pickup during enameling, iron oxide left on poorly cleaned basis metal, or burrs on iron or steel basis metal that protrude through the coating and are oxidized during firing.

core. (1) A specially formed material inserted in a mold to shape the interior or other part of a casting that cannot be shaped as easily by the pattern. (2) In a ferrous alloy prepared for *case hardening*, that portion of the alloy that is not part of the *case*. Typically considered

TECHNICAL NOTE 4

Copper and Copper Alloys

COPPER and copper alloys constitute one of the major groups of commercial metals. They are widely used because of their excellent electrical and thermal conductivities, outstanding resistance to corrosion, ease of fabrication, and good strength and fatigue resistance. They are generally nonmagnetic. They can be readily soldered and brazed, and many coppers and copper alloys can be welded by various gas, arc, and resistance methods. For decorative parts, standard alloys having specific colors are readily available. They can be plated, coated with organic substances, or chemically colored to further extend the variety of available finishes.

Generic classification of copper alloys

Generic name	UNS numbers	Composition
Wrought alloys		
Coppers	C10100–C15760	>99% Cu
High-copper alloys	C16200–C19600	>96% Cu
Brasses	C20500–C28580	Cu·Zn
Leaded brasses	C31200–C38590	Cu·Zn·Pb
Tin brasses	C40400–C49080	Cu·Zn·Sn·Pb
Phosphor bronzes	C50100–C52400	Cu·Sn·P
Leaded phosphor bronzes	C53200–C54800	Cu·Sn·Pb·P
Copper-phosphorus and copper-silver-phosphorus alloys	C55180–C55284	Cu·P·Ag
Aluminum bronzes	C60600–C64400	Cu·Al·Ni·Fe·Si·Sn
Silicon bronzes	C64700–C66100	Cu·Si·Sn
Other copper-zinc alloys	C66400–C69900	. . .
Copper-nickels	C70000–C79900	Cu·Ni·Fe
Nickel silvers	C73200–C79900	Cu·Ni·Zn
Cast alloys		
Coppers	C80100–C81100	>99% Cu
High-copper alloys	C81300–C82800	>94% Cu
Red and leaded red brasses	C83300–C85800	Cu·Zn·Sn·Pb (75–89% Cu)
Yellow and leaded yellow brasses	C85200–C85800	Cu·Zn·Sn·Pb (57–74% Cu)
Manganese and leaded manganese bronzes	C86100–C86800	Cu·Zn·Mn·Fe·Pb
Silicon bronzes, silicon brasses	C87300–C87900	Cu·Zn·Si
Tin bronzes and leaded tin bronzes	C90200–C94500	Cu·Sn·Zn·Pb
Nickel-tin bronzes	C94700–C94900	Cu·Ni·Sn·Zn·Pb
Aluminum bronzes	C95200–C95810	Cu·Al·Fe·Ni
Copper-nickels	C96200–C96800	Cu·Ni·Fe
Nickel silvers	C97300–C97800	Cu·Ni·Zn·Pb·Sn
Leaded coppers	C98200–C98800	Cu·Pb
Miscellaneous alloys	C99300–C99750	. . .

Pure copper is used extensively for cables and wires, electrical contacts, and a wide variety of other parts that are required to pass electrical current. Coppers,

---

TECHNICAL NOTE 4 (*continued*)

and certain *brasses*, *bronzes*, and cupronickels are used extensively for automobile radiators, heat exchangers, home heating systems, panels for absorbing solar energy, and various other applications requiring rapid conduction of heat. Because of their outstanding ability to resist corrosion, coppers, brasses, some bronzes, and cupronickels are used for pipes, valves, and fittings in systems carrying potable water, process water, or other aqueous fluids.

In all classes of copper alloys, certain alloy compositions for wrought products have counterparts among the cast alloys. Most wrought alloys are available in various cold-worked conditions, and the room-temperature strengths and fatigue resistances of these alloys depend on the amount of cold work as well as the alloy content. Typical applications of cold worked wrought alloys include springs, fasteners, hardware, small gears, cams, and electrical components.

Copper powder metallurgy (P/M) products based on pressed and sintered atomized or hydrometallurgical copper powders are also produced. Applications for copper P/M parts include self-lubricated sintered bearings, structural parts, friction materials, and porous bronze filters. Dispersion-strengthened copper alloys are also produced.

Selected References

- D.E. Tyler and W.T. Black, Introduction to Copper and Copper Alloys, *Metals Handbook*, 10th ed., Vol 2, ASM International, 1990, p 217–240
- N.W. Polan et al. Corrosion of Copper and Copper Alloys, *Metals Handbook*, 9th ed., Vol 13, ASM International, 1987, p 610–640
- R.F. Schmidt, D.G. Schmidt, and M. Sahoo, Cast Copper and Copper Alloys, *Metals Handbook*, 9th ed., Vol 15, ASM International, 1988, p 771–785
- E. Klar and D.F. Berry, Copper P/M Products, *Metals Handbook*, 10th ed., Vol 2, ASM International, 1990, p 392–402

---

to be the portion that (a) appears dark on an etched cross section, (b) has an essentially unaltered chemical composition, or (c) has a hardness, after hardening, less than a specified value.

core blower. A machine for making foundry cores using compressed air to blow and pack the sand into the core box.

cored bar. A powder metallurgy *compact* of bar shape, the interior of which has been melted by passage of electricity.

core forging. (1) Displacing metal with a punch to fill a die cavity. (2) The product of such an operation.

core rod. The part of a die used to produce a hole in a powder metallurgy *compact*.

coring. (1) A condition of variable composition between the center and surface of a unit of microstructure (such as a dendrite, grain, or carbide particle); results from nonequilibrium solidification, which occurs over a range of temperature. (2) A central cavity at the butt end of a rod extrusion, sometimes called extrusion pipe.

corner angle. On face milling cutters, the angle between an angular cutting edge of a cutter tooth and the axis of the cutter, measured by rotation into an axial plane. See the figure accompanying the term *face mill*.

corner joint. A joint between two members located approximately at right angles to each other in the form of an L (Fig. 17).

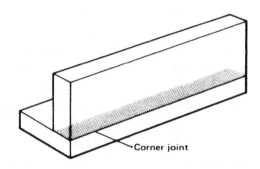
Corner joint

Fig. 17 Corner weld joint

corona. In spot welding, an area sometimes surrounding the nugget at the faying surfaces, where solid-state welding occurs. The corona contributes variably to overall bond strength, depending on the size of the corona and the degree of solid-state bonding achieved.

corrodkote test. An *accelerated corrosion test* for electrodeposits.

corrosion. The deterioration of a metal by chemical or electrochemical reaction with its environment.

corrosion embrittlement. The severe loss of ductility of a metal resulting from corrosive attack, usually intergranular and often not visually apparent.

corrosion fatigue. Cracking produced by the combined action of repeated or fluctuating stress and a corrosive environment.

corrugating. Forming sheet metal into a series of straight, parallel ridges and grooves by using a rolling mill equipped with matched roller dies or by using a *press brake* equipped with a specially shaped punch and die.

corrugations. Transverse ripples caused by a variation in strip shape during hot or cold reduction.

corundum. Natural abrasive of the aluminum oxide type that has higher purity than emery.

Cottrell process. Removal of solid particulates from gases with electrostatic precipitation.

coulometer. An electrolytic cell arranged to measure the quantity of electricity by the chemical action produced in accordance with Faraday's law.

counterblow hammer. A forging hammer in which both the *ram* and the anvil are driven simultaneously toward each other by air or steam pistons.

counterboring. Drilling or boring a flat-bottomed hole, often concentric with other holes.

counterlock. A jog in the mating surfaces of dies to prevent lateral die shifting from side thrusts developed in forging irregularly shaped pieces.

countersinking. Forming a flaring depression around the top of a hole for deburring, for receiving the head of a fastener, or for receiving a center.

coupling. The degree of mutual interaction between two or more elements resulting from mechanical, acoustical, or electrical linkage.

coupon. A piece of metal from which a test specimen is to be prepared—often an extra piece (as on a casting or forging) or a separate piece made for test purposes (such as a test weldment).

covalent bond. A bond between two or more atoms resulting from the completion of shells by the sharing of electrons.

covered electrode. A composite filler-metal welding electrode consisting of a bare wire or a metal-cored electrode plus a covering sufficient to provide a layer of slag on deposited weld metal. The covering often contains materials that provide shielding during welding, deoxidizers for weld metal, and arc stabilization; it also may contain alloying elements or other additives for the weld metal.

cover half. The stationary half of a die-casting die.

covering power. The ability of a solution to give satisfactory plating at very low current densities, a condition that exists in recesses and pits. This term suggests an ability to cover, but not necessarily to build up, a uniform coating, whereas *throwing power* suggests the ability to obtain a coating of uniform thickness on an irregularly shaped object.

"C" process. See *Croning process*.

crank press. A mechanical press whose slides are actuated by a crankshaft.

crater. (1) In machining, a depression in a cutting tool face eroded by chip contact. (2) In arc welding, a depression at the termination of a bead or in the weld pool beneath the electrode.

crater crack. A crack, often star-shaped, that forms in the crater of a weld bead, usually during cooling after welding.

creep. Time-dependent strain occurring under stress. The creep strain occurring at a diminishing rate is called primary creep; that occurring at a minimum and almost constant rate, secondary creep; and that occurring at an accelerating rate, tertiary creep.

creep limit. (1) The maximum stress that will cause less than a specified quantity of creep in a given time. (2) The maximum nominal stress under which the creep strain rate decreases continuously with time under constant load and at constant temperature. Sometimes used synonymously with *creep strength*.

creep recovery. Time-dependent strain after release of load in a creep test.

creep-rupture test. A method of evaluating elevated-temperature durability in which a tension-test specimen is stressed under constant load until it breaks. Data recorded commonly include: initial stress, time to rupture, initial extension, creep extension, and reduction of area at fracture. The same as *stress-rupture test*.

creep strength. (1) The constant nominal stress that will cause a specified quantity of creep in a given time at a constant temperature. (2) The constant nominal stress that will cause a specified rate of secondary creep in a given time at a constant temperature.

crevice corrosion. A type of concentration-cell corrosion; corrosion caused by the concentration or depletion of dissolved salts, metal ions, oxygen, or other gases, and such, in crevices or pockets remote from the principal fluid stream, with a resultant building up of differential cells that ultimately cause deep pitting.

crimping. Forming relatively small corrugations in order to (a) set down and lock a seam, (b) create an arc in a strip of metal, or (c) reduce an existing arc or diameter.

critical cooling rate. The rate of continuous cooling required to prevent undesirable transformation. For steel, it is the minimum rate at which austenite must be continuously cooled to suppress transformations above the M_s temperature.

critical current density. In an electrolytic process, a current density at which an abrupt change occurs in an operating variable or in the nature of an electrodeposit or electrode film.

critical point. (1) The temperature or pressure at which a change in crystal structure, phase, or physical properties occurs. Same as *transformation temperature*. (2) In an equilibrium diagram, that specific value of composition, temperature, or pressure, or combinations thereof, at which the phases of a heterogeneous system are in equilibrium.

critical shear stress. The shear stress required to cause slip in a designated slip direction on a given slip plane. It is called the critical resolve shear stress if the shear stress is induced by tensile or compressive forces acting on the crystal.

critical strain. The strain just sufficient to cause *recrystallization*; because the strain is small, usually only a few percent, recrystallization takes place from only a few nuclei, which produces a recrystallized structure consisting of very large grains.

critical temperature. (1) Synonymous with *critical point* if the pressure is constant. (2) The temperature above which the vapor phase cannot be condensed to liquid by an increase in pressure.

critical temperature ranges. Synonymous with *transformation ranges*, which is the preferred term.

Croning process. A *shell molding* process using a phenolic resin binder. Sometimes referred to as "C" process.

crop. (1) An end portion of an ingot that is cut off as scrap. (2) To shear a bar or billet.

cross breaks. Same as *coil breaks*.

cross-country mill. A rolling mill in which the mill stands are so arranged that their tables are parallel with a transfer (or crossover) table connecting them. Such a mill is used for rolling structural shapes, rails, and any special form of bar stock not rolled in the ordinary bar mill.

crossed joint. In soldering, an unintended solder connection between two or more conductors, either securely or by mere contact. Also called a *solder short* or *bridging* (5).

cross forging. Preliminary working of forging stock in flat dies to develop mechanical properties, particularly in the center portions of heavy sections.

cross rolling. Rolling of sheet or plate so that the direction of rolling is approximately 90° from the direction of a previous rolling.

cross-roll straightener. A machine having paired rolls of special design for straightening round bars or tubes, the pass being made with the work parallel to the axes of the rolls.

cross-wire weld. A weld made at the junction between crossed wires or bars.

crown. (1) A contour on a sheet or roll where the thickness or diameter increases from edge to center. (2) The top section of a press structure where the cylinders and other working parts may be mounted. Also called dome, head, or top platen.

crucible. A vessel or pot, made of a refractory substance or of a metal with a high melting point, used for melting metals or other substances.

crucible steelmaking process. It is said that the art of making steel in crucibles was known in ancient times, but the art had long been lost. In 1740, Benjamin Huntsman, a young clockmaker who had a shop near Sheffield, England, was dissatisfied with the springs he had made from blister steel. This led him to reinvent what came to be known as crucible steel. He first obtained wrought iron that he broke into pieces and packed in a clay container, covering the iron with powdered charcoal, and the container with a tight-fitting lid. He baked the container for approximately two days to carburize the iron, creating blister steel by a process also known as cementation. Huntsman then built a furnace that would have had an air draft sufficient to reach a temperature high enough to melt the carburized iron. There were no pyrometers in those days, and the color would have been the only way to gage temperature. The iron pieces were loaded into a clay crucible fitted with a cover. The making of the crucibles, which also were called pots, was a craft in itself, and there is no record of how much time was spent to create pots that would be strong enough to withstand heating for three hours at a red heat, as well as having the strength necessary to hold 60 or 70 pounds of steel and slag.

The tall furnace was built like a chimney to maximize the draft through the charcoal surrounding the pot so that the very high temperature necessary to melt wrought iron could be reached. Huntsman would have tested for melting with a rod or a stick. When the metal was molten, the pot would have been lifted from its low position to the floor where a mold would be ready. The teemer would pour the metal slowly into a mold while a helper would hold back the slag. The steel would have been of fairly good quality with few blow holes and no surface blisters. Later on it was discovered how to "kill" the steel to minimize internal gas holes. Crucible steel (Fig. 18) remained the best quality steel, even after the invention of the *bessemer process* and the open hearth furnace. It was the process selected for tool steel. Crucible steel, made in 60- to 80-pound lots, however, was very expensive. The process was largely replaced by the electric furnace in the 20th century.

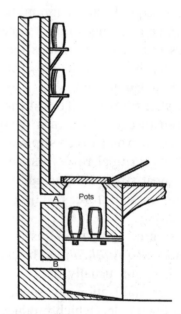

Fig. 18 Crucible steel illustration

crush. (1) Buckling or breaking of a section of a casting mold due to incorrect register when the mold is closed. (2) An indentation in the surface of a casting due to displacement of sand when the mold was closed.

crush forming. Shaping a grinding wheel by forcing a rotating metal roll into its face so as to reproduce the desired contour.

crushing test. (1) A radial compressive test applied to tubing, sintered-metal bearings, or other similar products for determining radial crushing strength (maximum load in compression). (2) An axial compressive test for determining quality of tubing, such as weld soundness in welded tubing.

crystal. A solid composed of atoms, ions, or molecules arranged in a pattern that is repetitive in three dimensions.

crystalline fracture. A pattern of brightly reflecting crystal facets on the fracture surface of a polycrystalline metal, resulting from cleavage fracture of many individual crystals. Contrast with *fibrous fracture*, *silky fracture*.

crystallization. (1) The separation, usually from a liquid phase on cooling, of a solid crystalline phase. (2) Sometimes erroneously used to explain fracturing that actually has occurred by fatigue.

crystal orientation. See *orientation*.

CS. See *commercial steel*.

cubic plane. A plane perpendicular to any one of the three crystallographic axes of the cubic (isometric) system; the *Miller indices* are {100}.

cup. (1) Sheet metal part, the product of the first step in deep drawing. (2) Any cylindrical part or shell closed at one end.

cupellation. Oxidation of molten lead containing gold and silver to produce lead oxide, thereby separating the precious metals from the base metal.

cup fracture (cup-and-cone fracture). A mixed-mode fracture, often seen in tensile-test specimens of a ductile material, where the central portion undergoes *plane-strain* fracture and the surrounding region undergoes *plane-stress* fracture. It is called a cup fracture (or cup-and-cone fracture) because one of the mating fracture surfaces looks like a miniature cup—that is, it has a central depressed flat-face region surrounded by a shear lip; the other fracture surface looks like a miniature truncated cone.

cupola. A cylindrical vertical furnace for melting metal, especially cast iron, by having the charge come in contact with the hot fuel, usually metallurgical coke.

cupping. (1) The first step in deep drawing. (2) A fracture of severely worked rods or wire in which one end has the appearance of a cup and the other that of a cone.

Curie temperature. The temperature of magnetic transformation below which a metal or alloy is ferromagnetic and above which it is paramagnetic.

curium. A chemical element having atomic number 96, atomic weight 247, and symbol Cm, named after Marie Sklodowska-Curie and her husband Pierre Curie. Curium was intentionally produced and identified in the summer of 1944 by the Glenn T. Seaborg group at the University of California, Berkeley. The discovery was released to the public in 1945. Curium usually is produced by bombarding uranium or plutonium with neutrons.

curling. Rounding the edge of sheet metal into a closed or partly closed loop.

current decay. In spot, seam, or projection welding, the controlled reduction of the welding current from its peak amplitude to a lower value to prevent excessively rapid cooling of the weld nugget.

current efficiency. The proportion of current used in a given process to accomplish a desired result; in electroplating, the proportion used in depositing or dissolving metal.

cushion. Same as *die cushion*.

cut. (1) In castings, a rough spot or area of excess metal caused by erosion of the mold or core surface by metal flow. (2) In powder metallurgy, same as *fraction*.

cut and carry method. Stamping method wherein the part remains attached to the strip or is forced back into the strip to be fed through the succeeding stations of a progressive die.

cut edge. A mechanically sheared edge obtained by slitting, shearing, or blanking.

cutoff wheel. A thin abrasive wheel for severing or slotting any material or part.

cutting down. Removing roughness or irregularities of a metal surface by abrasive action.

cutting edge. The leading edge of a cutting tool (such as a lathe tool, drill, or milling cutter) where a line of contact is made with the work during machining. See the figure accompanying the term *single-point tool*.

cutting fluid. A fluid used in metal cutting to improve finish, tool life, or dimensional accuracy. On being flowed over the tool and work, the fluid reduces friction, the heat generated, and tool wear, and prevents galling. It conducts the heat away from the point of generation and also serves to wash away the *chips*.

cutting speed. The linear or peripheral speed of relative motion between the tool and workpiece in the principal direction of cutting.

cyanide copper. Copper electrodeposited from an alkali-cyanide solution containing a complex ion made up of univalent copper and the cyanide radical; also, the solution itself.

cyanide slimes. Finely divided metallic precipitates that are formed when precious metals are extracted from their ores using cyanide solutions.

cyaniding. A case-hardening process in which a ferrous material is heated above the lower transformation range in a molten salt containing cyanide to cause simultaneous absorption of carbon and nitrogen at the surface and, by diffusion, create a concentration gradient. *Quench hardening* completes the process.

cycle annealing. An annealing process employing a predetermined and closely controlled time-temperature cycle to produce specific properties or microstructures.

cylindrical grinding. Grinding the outer cylindrical surface of a rotating part.

cylindrical land. *Land* having zero relief.

damping capacity. The ability of a material to absorb vibration (cyclical stresses) by internal friction, converting the mechanical energy into heat.

dangler. The flexible electrode used in *barrel plating* to conduct current to the work.

Davy, Humphry. 1778–1829. An outstanding English scientist who contributed important discoveries in the fields of metallurgy and chemistry. At age 23, he was assistant lecturer in chemistry, a director of the Royal Institution, and assistant editor of its journals. He was a pioneer in the field of electrolysis using a voltaic cell to electrolyze molten salts. Davy discovered the alkali metals sodium and potassium, as well as magnesium, boron, and calcium, making him one of the major discoverers of metals.

daylight. The maximum clear distance between the pressing surfaces of a hydraulic press when the surfaces are in their usable open position. Where a bolster is supplied, it is considered the pressing surface. See also *shut height*.

dc casting. Same as *direct chill casting*. A continuous method of making ingots for rolling or extrusion by pouring the metal into a short mold. The base of the mold is a platform that is gradually lowered while the metal solidifies, the frozen shell of metal acting as a retainer for the liquid metal below the wall of the mold. The ingot usually is cooled by the impingement of water directly on the mold or on the walls of the solid metal as it is lowered. The length of the ingot is limited by the depth to which the platform can be lowered; therefore, it often is called semi-continuous casting.

dead roast. A *roasting* process for the complete elimination of sulfur. The same as *sweet roast*.

dead soft. A *temper* of nonferrous alloys and some ferrous alloys corresponding to the condition of minimum hardness and tensile strength produced by *full annealing*.

deburring. Removing burrs, sharp edges, or fins from metal parts by filing, grinding, or rolling the work in a barrel containing abrasives suspended in a suitable liquid medium. Sometimes called *burring*.

decalescence. A phenomenon, associated with the transformation of α iron to γ iron on the heating (superheating) of iron or steel, revealed by the darkening of the metal surface owing to the sudden decrease in temperature caused by the fast absorption of the latent heat of transformation. Contrast with *recalescence*.

decarburization. The loss of carbon from the surface layer of a carbon-containing alloy due to reaction with one or more chemical substances in a medium that contacts the surface.

decomposition potential. The minimum potential difference necessary to decompose the electrolyte of a cell.

deep drawing. Forming deeply recessed parts by forcing sheet metal to undergo plastic flow between dies, usually without substantial thinning of the sheet.

deep etching. Severe *macroetching*.

defect. A departure of any *quality characteristic* from its intended (usually specified) level that is severe enough to cause the product or service not to fulfill its anticipated function. According to ANSI standards, defects are classified according to severity:

- *Serious* defects lead directly to significant injury or significant economic loss.
- *Major* defects are related to major problems with respect to anticipated use.
- *Minor* defects are related to minor problems with respect to anticipated use.

defective. A quality control term describing a unit of product or service containing at least one *defect*, or having several lesser imperfections that, in combination, cause the unit to not perform its anticipated function. The term *defective* is not synonymous with

nonconforming (or rejectable), and should be applied only to those units incapable of performing their anticipated functions.

deformation bands. Parts of a crystal that have rotated differently during deformation to produce bands of varied orientation within individual grains.

degasifier. A substance that can be added to molten metal to remove soluble gases that might otherwise be occluded or entrapped in the metal during solidification.

degassing. Removing gases from liquids or solids.

degreasing. Removing oil or grease from a surface. See also *vapor degreasing*.

degrees of freedom. The number of independent variables (such as temperature, pressure, or concentration within the phases present) that may be altered at will without causing a phase change in an alloy system at equilibrium, or the number of such variables that must be fixed arbitrarily to define the system completely.

delayed yield. A phenomenon involving a delay in time between the application of a stress and the occurrence of the corresponding yield-point strain.

delta (δ) ferrite. See *ferrite*.

dendrite. A crystal that has a treelike branching pattern, being most evident in cast metals slowly cooled through the solidification range (Fig. 19).

Fig. 19 Illustration of a dendrite

dendritic powder. Particles of metal powder, usually of electrolytic origin, having typical pine tree structure.

denickelification. Corrosion in which nickel is selectively leached from nickel-containing alloys after extended service in fresh water.

density ratio. The ratio of the determined density of a powder metallurgy compact to the absolute density of metal of the same composition, usually expressed as a percentage.

deoxidized copper. Copper from which cuprous oxide has been removed by adding a *deoxidizer,* such as phosphorous, to the molten bath.

deoxidizer. A substance that can be added to molten metal to remove either free or combined oxygen.

deoxidizing. (1) The removal of oxygen from molten metals by use of suitable *deoxidizers.* (2) Sometimes refers to the removal of the undesirable elements other than oxygen by the introduction of elements or compounds that readily react with them. (3) In metal finishing, the removal of oxide films from metal surfaces by chemical or electrochemical reaction.

depolarization. A decrease in the *polarization* of an electrode.

depolarizer. A substance that produces *depolarization.*

deposition efficiency. In welding, the ratio of the weight of deposited weld metal to the net weight of electrodes consumed, exclusive of stubs.

deposition sequence. The order in which increments of weld metal are deposited.

depth of cut. The thickness of material removed from a workpiece in a single machining pass.

depth of fusion. In welding, the distance that fusion extends into the base metal or into the previous pass (Fig. 20).

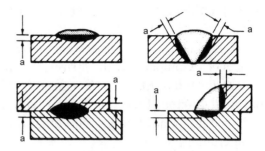

Fig. 20 Illustration of depth of fusion

depth of penetration. See *joint penetration* and *root penetration.*

derby. A massive piece (intermediate in size, extending to more than 45 kg, or 100 lb, and usually cylindrical) of primary metal made by bomb reduction (such as uranium from uranium tetrafluoride reduced with magnesium). Compare with *biscuit* and *dingot.*

descaling. Removing the thick layer of oxides formed on some metals at elevated temperatures.

deseaming. Analogous to *chipping,* the surface imperfections being removed by gas cutting.

detergent. A chemical substance, generally used in aqueous solution, that removes *soil.*

detritus. Wear debris. Particles that become detached in a wear process.

developed blank. A blank that requires little or no trimming when formed.

dewaxing. Removing the expendable wax pattern from an investment mold by heat or solvent.

dezincification. Corrosion in which zinc is selectively leached from zinc-containing alloys. Most commonly found in copper-zinc alloys containing less than 85% Cu after extended service in water containing dissolved oxygen.

diamagnetic material. A material whose specific permeability is less than unity and is therefore repelled weakly by a magnet. Compare with *ferromagnetic material* and *paramagnetic material*.

diamond boring. Precision boring with a shaped diamond (but not with other tool materials).

diamond pyramid hardness test. A microindentation hardness test using a 136° diamond pyramid indenter (Vickers) and variable loads, enabling the use of one hardness scale for all ranges of hardness—from very soft lead to tungsten carbide. See *Vickers hardness test.*

diamond tool. (1) A diamond, shaped or formed to the contour of a single-point cutting tool, for use in precision machining of nonferrous or nonmetallic materials. (2) Sometimes, an insert made from multicrystalline diamond compacts.

diamond wheel. A grinding wheel in which crushed and sized industrial diamonds are held in a resinoid, metal, or vitrified bond.

diaphragm. (1) A porous or permeable membrane separating anode and cathode compartments of an electrolytic cell from each other or from an intermediate compartment. (2) Universal die member made of rubber or similar material used to contain hydraulic fluid within the forming cavity and to transmit pressure to the part being formed.

dichromate treatment. A chromate *conversion coating* produced on magnesium alloys in a boiling solution of sodium dichromate.

didymium. A natural mixture of the rare-earth elements praseodymium and neodymium, often given the quasi-chemical symbol Di.

die. A tool, usually containing a cavity, that imparts shape to solid, molten, or powdered metal primarily because of the shape of the tool itself. Used in many press operations (including blanking, drawing, forging, and forming), in die casting, and in forming green powder metallurgy compacts. Die-casting and powder metallurgy dies are sometimes referred to as *molds.*

die block. A block, usually of tool steel, into which the desired impressions are sunk, formed, or machined and from which forgings or die castings are made.

die body. The stationary or fixed part of a powder pressing die.

die casting. (1) A casting made in a die. (2) A casting process wherein molten metal is forced under high pressure into the cavity of a metal mold.

die clearance. The clearance between a mated punch and die; commonly expressed as clearance per side. Also called *clearance* and *punch-to-die clearance.*

die cushion. A press accessory located beneath or within a *bolster* or *die block* to provide additional motion or pressure for stamping operations; actuated by air, oil, rubber, or springs, or by a combination thereof.

die forging. A forging whose shape is determined by impressions in specially prepared dies.

die forming. Shaping of solid or powdered metal by forcing it into or through the cavity in a die.

die holder. A plate or block, on which the die block is mounted, having holes or slots for fastening to the *bolster* or the bed of the press.

die insert. A removable liner or part of a *die body* or punch.

die layout. The transfer of drawing or sketch dimensions to templates or die surfaces for use in sinking dies.

die life. The productive life of a die impression, usually expressed as the number of units produced before the impression has worn beyond permitted tolerances.

die lines. Lines or markings on formed, drawn, or extruded metal parts caused by imperfections in the surface of the die. See also *draw marks*.

die lubricant. A lubricant applied to working surfaces of dies and punches to facilitate drawing, pressing, stamping, and/or ejection. In powder metallurgy, the die lubricant sometimes is mixed into the powder before pressing into a compact.

die match. The condition where dies, after having been set up in a press or other equipment, are in proper alignment relative to each other.

die opening. In flash or upset welding, the distance between the electrodes, usually measured with the parts in contact before welding has commenced or immediately upon completion of the cycle but before upsetting.

die proof. A casting of the die impression made to confirm the exactness of the impression. Also called *cast*.

die radius. The radius on the exposed edge of a drawing die, over which the sheet flows in forming drawn shells.

die scalping. Removing surface layers from bar, rod, wire, or tube by drawing through a sharp-edged die to eliminate minor surface defects.

die set. A tool or tool holder consisting of a die base and punch plate for the attachment of a die and punch, respectively.

die shift. A condition requiring correction where, after dies have been set up in the forging equipment, displacement of a point in one die from the corresponding point in the opposite die occurs in a direction parallel to the fundamental parting line of the dies.

die sinking. Forming or machining a depressed pattern in a die.

die welding. Forge welding between dies.

differential coating. A coated product having a specified coating on one surface and a significantly lighter coating on the other surface (such as a hot dip galvanized product or electrolytic tin plate).

differential flotation. Separating a complex ore into two or more valuable minerals and *gangue* by *flotation*. Also called *selective flotation*.

differential heating. Heating that intentionally produces a temperature gradient within an object such that, after cooling, a desired stress distribution or variation in properties is present within the object.

diffusion. (1) The spreading of a constituent in a gas, liquid, or solid, tending to make the composition of all parts uniform. (2) The spontaneous movement of atoms or molecules to new sites within a material.

diffusion aid. A solid filler metal sometimes used in *diffusion welding*.

diffusion bonding. See preferred terms *diffusion welding* and *diffusion brazing*.

diffusion brazing. A brazing process that joins two or more components by heating them to suitable temperatures and by using a filler metal or an in situ liquid phase. The filler metal may be distributed by capillary attraction or may be placed or formed at the faying surfaces. The filler metal is diffused with the base metal to the extent that the joint properties are changed to approach those of the base metal.

diffusion coating. Any process whereby a basis metal or alloy is either (1) coated with another metal or alloy and heated to a sufficient temperature in a suitable environment or (2) exposed to a gaseous or liquid medium containing the other metal or alloy, thus causing *diffusion* of the coating or of the other metal or alloy into the basis metal with resultant changes in the composition and properties of its surface.

diffusion coefficient. A factor of proportionality representing the amount of substance diffusing across a unit area through a unit concentration gradient in unit time.

diffusion welding. A high-temperature solid-state welding process that permanently joins faying surfaces by simultaneous application of pressure and heat. The process does not involve macroscopic deformation, melting, or relative motion of parts. A solid filler metal (*diffusion aid*) may or may not be inserted between the faying surfaces.

digging. A sudden erratic increase in cutting depth, or in the load on a cutting tool, caused by unstable conditions in the machine setup. Usually the machine is stalled, or either the tool or the workpiece is destroyed.

dilatometer. An instrument for measuring the linear expansion or contraction in a metal resulting from changes in such factors as temperature and allotropy.

dimple rupture. A fractographic term describing *ductile fracture* that occurs through formation and coalescence of microvoids along the fracture path. The fracture surface of such a ductile fracture appears dimpled when observed at high magnification and usually is most clearly resolved when viewed in a scanning electron microscope.

dimpling. (1) Stretching a relatively small, shallow indentation into sheet metal. (2) In aircraft, stretching thin metal into a conical flange for use with a countersunk head rivet.

DIN Designation Systems for Nonferrous Metals. According to DIN 17007, there is a numbering system that includes most commercial nonferrous metals and their alloys. There are two main groups, including No. 2 for heavy metals excluding iron and No. 3 for light (nonheavy metals). The initial number 2 or 3 is followed by four digits. The number assignments are:
- Copper and copper alloys: 2.0000 to 2.1799
- Zinc and cadmium: 2.2000 to 2.2499
- Lead: 2.3000 to 2.3499
- Tin: 2.3500 to 2.3999
- Nickel and cobalt: 2.4000 to 2.4999
- Magnesium: 3.5000 to 3.5999
- Aluminum: 3.0000 to 3.4999

dingot. An oversized *derby* (possibly a ton or more) of a metal produced in a bomb reaction (such as uranium from uranium tetrafluoride reduced with magnesium). For these metals, the term *ingot* is reserved for massive units produced in vacuum melting and casting. See *biscuit* and *derby*.

dinking. Cutting of nonmetallic materials or light-gage soft metals by using a hollow punch with a knifelike edge acting against a wooden fiber or resiliently mounted metal plate.

dip brazing. Brazing by immersing the assembly to be joined in a bath of hot molten chemicals or metal. A molten chemical bath may provide brazing flux; molten metal bath may provide the filler metal.

diphase cleaning. Removing *soil* by an emulsion that produces two phases in the cleaning tank: a solvent phase and an aqueous phase. Cleaning is effected by both solvent action and emulsification.

dip plating. Depositing a metallic coating on a metal immersed in a liquid solution, without the aid of an external electric current. The same as *immersion plating*.

direct arc furnace. An electric arc furnace in which the metallic charge is one of the poles of the arc.

direct chill casting. A continuous method of making ingots for rolling or extrusion by pouring the metal into a short mold. The base of the mold is a platform that is gradually lowered while the metal solidifies, the frozen shell of metal acting as a retainer for the liquid metal below the wall of the mold. The ingot is usually cooled by the impingement of water directly on the mold or on the walls of the solid metal as it is lowered. The length of the ingot is limited by the depth to which the platform can be lowered; therefore, it often is called semicontinuous casting. Same as *dc casting*.

direct-current cleaning. *Electrolytic cleaning* in which the work is the cathode. Same as *cathodic cleaning*.

direct extrusion. See *extrusion*.

directional property. Property whose magnitude varies depending on the relation of the test axis to a specific direction within the metal. The variation results from preferred orientation or from the fibering of constituents or inclusions.

directional solidification. Solidification of molten metal in such a manner that feed metal is always available for that portion that is just solidifying.

direct quenching. (1) Quenching carburized parts directly from the carburizing operation. (2) Quenching pearlitic malleable parts directly from the malleabilizing operation. See *quenching*.

discontinuity. Any interruption in the normal physical structure or configuration of a part, such as cracks, laps, seams, inclusions, or porosity. A discontinuity may or may not affect the usefulness of the part.

discontinuous precipitation. Precipitation from a supersaturated solid solution in which the precipitate particles grow by short-range diffusion, accompanied by recrystallization of the matrix in the region of precipitation. Discontinuous precipitates grow into the matrix from nuclei near grain boundaries, forming cells of alternate lamellae of precipitate and depleted (and recrystallized) matrix. Often referred to as cellular or nodular precipitation. Compare with *continuous precipitation* and *localized precipitation*.

discontinuous yielding. Nonuniform plastic flow of a metal exhibiting a yield point in which plastic deformation is inhomogeneously distributed along the *gage length*. Under some circumstances, it may occur in metals not exhibiting a distinct yield point, either at the onset of or during plastic flow.

dishing. Forming a shallow concave surface, the area being large compared to the depth.

disk grinding. Grinding with the flat side of an abrasive disk or segmented wheel.

dislocation. A linear imperfection in a crystalline array of atoms. Two basic types are recognized: an edge dislocation corresponds to the row of mismatched atoms along the edge formed by an extra, partial plane of atoms within the body of a crystal; a screw dislocation corresponds to the axis of a spiral structure in a crystal, characterized by a distortion that joins together normally parallel planes to form a continuous helical ramp (with a pitch of one interplanar distance) winding about the dislocation. Most prevalent is the so-called mixed dislocation, which is the name given to any combination of an edge dislocation and a screw dislocation.

disordering. Forming a lattice arrangement in which the solute and solvent atoms of a solid solution occupy lattice sites at random. Contrast with *ordering* and *superlattice*.

dispersing agent. A material that increases the stability of a suspension of powder particles in a liquid medium by deflocculation of the primary particles.

disruptive strength. The stress at which a metal fractures under hydrostatic tension.

distortion. Any deviation from an original size, shape, or contour that occurs because of the application of stress or the release of residual stress.

disturbed metal. The cold worked metal layer formed at a polished surface during the process of mechanical grinding and polishing.

divided cell. A cell containing a diaphragm or other means for physically separating the *anolyte* from the *catholyte*.

divorced eutectic. A metallographic appearance in which two constituents of a eutectic structure appear as massive phases rather than the finely divided mixture characteristic of normal eutectics. Often, one of the constituents of the eutectic is continuous and indistinguishable from an accompanying proeutectic constituent.

domain. A substructure in a ferromagnetic material within which all of the elementary magnets (electron spins) are held aligned in one direction by interatomic forces; if isolated, a domain would be a saturated permanent magnet.

doré silver. Crude silver containing a small amount of gold, obtained after removing lead in a cupellation furnace. Same as doré bullion and doré metal.

double-acting hammer. A forging hammer in which the ram is raised by admitting steam or air into a cylinder below the piston, and the blow is intensified by admitting steam or air above the piston on the downward stroke.

double-action die. A die designed to perform more than one operation in a single stroke of the press.

double-action forming. Forming or drawing in which more than one action is achieved in a single stroke of the press.

double-action mechanical press. A press having two independent parallel movements by means of two slides, one moving within the other. The inner slide or plunger is usually operated by a crankshaft, whereas the outer or blankholder slide, which dwells during the drawing operation, is usually operated by a toggle mechanism or by cams.

double aging. Employment of two different aging treatments to control the type of precipitate formed from a supersaturated matrix in order to obtain the desired properties. The first aging treatment, sometimes referred to as intermediate or stabilizing, is usually carried out at a higher temperature than the second.

double-bevel groove weld. A groove weld in which the joint edge of one member is beveled from both sides (Fig. 21).

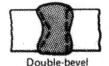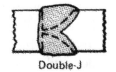

Fig. 21 Double-bevel and double-J groove welds

double-J groove weld. A groove weld in which the joint edge of one member is in the form of two Js, one on either side of the member (Fig. 21).

double salt. A compound of two salts that crystallize together in a definite proportion.

double tempering. A treatment in which a quench-hardened ferrous metal is subjected to two complete tempering cycles, usually at substantially the same temperature, for the purpose of ensuring completion of the tempering reaction and promoting stability of the resulting microstructure.

double-U groove weld. A groove weld in which the joint edge is in the form of two Js or two half-Us, one on either side of the member (Fig. 22).

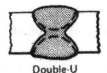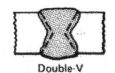

Fig. 22 Double-U and double-V groove welds

double-V groove weld. A groove weld in which the joint edge is beveled on both sides (Fig. 22).

double-welded joint. A butt, edge, tee, corner, or lap joint in which welding has been performed from both sides.

down cutting. See preferred term, *climb cutting,* which is analogous to *climb milling.* Milling in which the cutter moves in the direction of the feed at the point of contact.

downgate. Same as *sprue.*

downhand welding. See *flat-position welding.*

down milling. See preferred term, *climb milling.*

down slope time. In resistance welding, time associated with current decrease using *slope control.*

downsprue. Same as *sprue.*

Dow process. A process for the production of magnesium by electrolysis of molten magnesium chloride.

draft. (1) An angle or taper on the surface of a pattern, core box, punch, or die (or of the parts made with them) that makes it easier to remove the parts from a mold or die cavity, or to remove a core from a casting. (2) The change in cross section that occurs during rolling or cold drawing.

drag. The bottom section of a *flask, mold,* or *pattern.*

drag angle. In welding, the angle between the axis of the electrode or torch and a line normal to the plane of the weld when welding is being done with the torch positioned ahead of the weld puddle (Fig. 23).

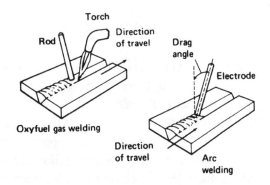

Fig. 23 Backhand welding technique

drag-in. Water or solution carried into another solution by the work and its associated handling equipment.

drag-out. Solution carried out of a bath by the work and its associated handling equipment.

drag technique. A method used in manual arc welding wherein the electrode is in contact with the assembly being welded without being in short circuit. The electrode is usually used without oscillation.

drawability. A measure of the workability of a metal subject to a drawing process. A term usually used to indicate the ability of a metal to be deep drawn.

draw bead. (1) A bead or offset used for controlling metal flow. (2) Riblike projections on draw rings or hold-down surfaces for controlling metal flow (Fig. 24).

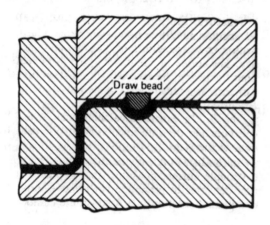

Fig. 24 Use of draw beads

drawbench. The stand that holds the die and draw head used in the drawing of wire, rod, and tubing.

draw forging. A process using two or more moving anvils or dies for producing shafts with constant or varying diameters along their length, or tubes with internal or external variations in diameter; also known as draw forging or rotary swaging. The same as *radial forging*.

draw forming. A method of curving bars, tubes, or rolled or extruded sections in which the stock is bent around a rotating *form block*. Stock is bent by clamping it to the form block, then rotating the form block and a pressure die held against the periphery of the form block. Contrast with *wiper forming*.

draw head. A set of rolls or dies mounted on a drawbench for forming a section from strip, tubing, or solid stock. See *Turk's-head rolls*.

drawing. (1) Forming recessed parts by forcing the plastic flow of metal in dies. (2) Reducing the cross section of bar stock, wire, or tubing by pulling through a die. (3) A misnomer for tempering (see *temper*).

drawing compound. A substance applied to prevent *pickup* and *scoring* during drawing or pressing operations by preventing metal-to-metal contact of the work and die. Also known as *die lubricant*.

drawing out. A stretching operation resulting from forging a series of upsets along the length of the workpiece.

drawing steel (DS). Cold rolled steel normally furnished in coils and cut to lengths for use in fabricating an identified part requiring deformation too severe for the fabrication properties of commercial steel (CS). Drawing steel is produced in two types: A and B, which have different chemical composition requirements. The typical yield strength for types A and B is in the range of 150 to 240 MPa (22 to 35 ksi) and typical elongation in two inches is 36% or more (ASTM A1008).

draw marks. Lines or markings on formed, drawn, or extruded metal parts caused by imperfections in the surface of the die. See *die lines, scoring, galling,* and *pickup.*

drawn shell. An article formed by drawing sheet metal into a hollow structure having a predetermined geometrical configuration.

draw plate. A circular plate with a hole in the center contoured to fit a forming punch, used to support the *blank* during the forming cycle.

draw radius. The radius at the edge of a die or punch over which the work is drawn.

draw ring. A ring-shaped die part over the inner edge of which the metal is drawn by the punch.

dresser. A tool used for *truing* and *dressing* a grinding wheel.

dressing. Cutting, breaking down, or crushing the surface of a grinding wheel to improve its cutting ability and accuracy.

drift. (1) A flat piece of steel of tapering width used to remove taper shank drills and other tools from their holders. (2) A tapered rod used to force mismated holes into line for riveting or bolting. Sometimes called a drift pin.

drive fit. A type of *force fit,* which is any of various interference fits between parts assembled under various amounts of force.

drop. A casting imperfection due to a portion of the sand dropping from the cope or other overhanging section of the mold.

drop forging. A shallow forging made in impression dies, usually with a drop hammer.

drop hammer. A forging hammer that depends on gravity for its force.

dross. The scum that forms on the surface of molten metal largely because of oxidation but sometimes because of the rising of impurities to the surface.

dry cyaniding. (obsolete) Same as *carbonitriding.*

dry sand mold. A casting mold made of sand and dried at 100 °C (212 °F) or above before being used. Contrast with *green sand mold.*

ductile cast iron. A *cast iron* that has been treated while molten with an element such as magnesium or cerium to induce the formation of free graphite as nodules of spherulites, which imparts a measurable degree of ductility to the cast metal. Also known as nodular cast iron, spherulitic graphite cast iron, and SG iron.

ductile crack propagation. Slow crack propagation that is accompanied by noticeable and that requires energy to be supplied from outside the body.

ductile fracture. Fracture characterized by the tearing of metal accompanied by appreciable gross plastic deformation and expenditure of considerable energy.

ductility. The ability of a material to deform plastically without fracturing, measured by elongation or reduction of area in a tensile test, by height of cupping in an Erichsen test, or by other means.

dummy block. In *extrusion,* a thick unattached disk placed between the ram and billet to prevent overheating of the ram.

dummy cathode. (1) A *cathode,* usually corrugated to give variable current densities, that is plated at low current densities to preferentially remove impurities from a plating solution. (2) A substitute cathode that is used during adjustment of operating conditions.

dummying. Plating with *dummy cathodes.*

duplex coating. See *composite plate.* An electrodeposit consisting of layers of at least two different compositions.

duplexing. Any two-furnace melting or refining process. Also called duplex melting or duplex processing.

duplicating. In machining and grinding, reproducing a form from a master with an appropriate type of machine tool, using a suitable tracer or program-controlled mechanism.

duralumin. (obsolete) The trade name of one of the earliest types of age-hardenable aluminum-copper alloys containing manganese, magnesium, or silicon.

Duranel. The Alcoa (Aluminum Company of America) trade name of stainless steel clad aluminum alloy used for hollowware and saucepans.

Durville process. A casting process that involves rigid attachment of the mold in an inverted position above the crucible. The melt is poured by tilting the entire assembly, causing the metal to flow along a connecting *launder* and down the side of the mold.

dusting. Applying a powder, such as sulfur to molten magnesium or graphite to a mold surface.

duty cycle. For electric welding equipment, the percentage of time that current flows during a specified period. In arc welding, the specified period is 10 min.

dynamic creep. Creep that occurs under conditions of fluctuating load or fluctuating temperature.

dysprosium. A chemical element having atomic number 66, atomic weight 163, and symbol Dy, named for the Greek *dysprositos,* meaning hard to get at. The metal was identified by Paul Émile (François) Lecoq de Boisbaudran in 1886. The element was isolated in 1906 and was finally obtained in pure form by Frank H. Spedding and co-workers in 1950.

Dictionary of Metals
H.M. Cobb, editor

earing. Formation of scallops (ears) around the top edge of a drawn part, caused by directional differences in the properties of the sheet metal used.

eccentric press. A *mechanical press* in which the eccentric and strap are used to move the *slide*, rather than a crankshaft and connection.

ECM. An abbreviation for *electrochemical machining*. Removal of stock from an electrochemically conductive material by anodic dissolution in an electrolyte flowing rapidly through a gap between the workpiece and a shaped electrode. Variations of the process include electrochemical deburring and electrochemical grinding.

eddy-current testing. An electromagnetic nondestructive testing method in which eddy-current flow is induced in the test object. Changes in flow caused by variations in the object are reflected into a nearby coil or coils where they are detected and measured by suitable instrumentation.

edge dislocation. See *dislocation.* A linear imperfection in a crystalline array of atoms. Two basic types are recognized: an edge dislocation corresponds to the row of mismatched atoms along the edge formed by an extra, partial plane of atoms within the body of a crystal; a screw dislocation corresponds to the axis of a spiral structure in a crystal, characterized by a distortion that joins together normally parallel planes to form a continuous helical ramp (with a pitch of one inter-planar distance) winding about the dislocation. Most prevalent is the so-called mixed dislocation, which refers to any combination of an edge dislocation and a screw dislocation.

edge joint. A joint between the edges of two or more parallel or nearly parallel members (Fig. 25).

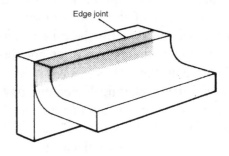

Fig. 25 Edge joint

edger. In forging, the portion of the die that generally distributes the metal in amounts required for the shape to be forged, usually a gathering operation. A rolling edger shapes the stock into various solids of revolution; a ball edger forms a ball.

edge strain. Transverse strain lines or Lüders lines located 25 to 300 mm (1 to 12 in.) in from the edges of cold rolled steel sheet or strip.

edging. (1) In forming, reducing the flange radius by retracting the forming punch a small amount after the stroke but prior to releasing the pressure. (2) In forging, removing flash that is directed upward between dies, usually accomplished in a lathe. (3) In rolling, the working of metal where the axis of the roll is parallel to the thickness dimension. Also called edge rolling.

EDM. An abbreviation for *electrical discharge machining*. Removal of stock from an electrically conductive material by rapid, repetitive spark discharge through a dielectric fluid flowing between the workpiece and a shaped electrode. Variations of the process include electrical discharge grinding and electrical discharge wire cutting.

effective rake. The angle between a plane containing a tooth face and the axial plane through the tooth point as measured in the direction of chip flow through the tooth point. Thus, it is the rake resulting from both cutter configuration and direction of chip flow.

einsteinium. A chemical element having atomic number 92, atomic weight 254, and symbol Es, named for Albert Einstein. It is ironic because Einstein was a pacifist and the element was found in the debris of the first thermonuclear explosion (November 1952) by Albert Ghiorso and co-workers.

ejector. A device mounted in such a way that it removes or assists in removing a formed part from a die.

ejector half. The moveable half of a die-casting die containing the ejector pins.

ejector rod. A rod used to push out a formed piece.

elastic after effect. Time-dependent recovery, toward original dimensions, after the load has been reduced or removed from an elastically or plastically strained body. See *anelasticity*.

elastic constants. Factors of proportionality that describe elastic responses of a material to applied forces, including *modulus of elasticity* (either in tension, compression, or shear), *Poisson's ratio, compressibility,* and bulk modulus.

elastic deformation. A change in dimensions directly proportional to and in phase with an increase or decrease in applied force.

elastic hysteresis. A misnomer for an anelastic strain that lags a change in applied stress, thereby creating energy loss during cyclic loading. More properly termed *mechanical hysteresis.*

elasticity. Ability of a solid to perform in direct proportion to and in phase with increases or decreases in applied force.

elastic limit. The maximum stress to which a material may be subjected without any permanent strain remaining upon complete release of stress.

elastic modulus. Same as *modulus of elasticity*. A measure of the rigidity of metal. Ratio of stress, below the proportional limit, to corresponding strain. Specifically, the modulus obtained in tension or compression is Young's modulus, stretch modulus, or modulus of extensibility; the modulus obtained in torsion or shear is modulus of rigidity, shear modulus, or modulus of torsion; the modulus covering the ratio of the mean normal stress to the change in volume per unit volume is the bulk modulus. The tangent modulus and secant modulus are not restricted within the proportional limit; the former is the slope of the stress-strain curve at a specified point; the latter is the slope of a line from the origin to a specified point on the stress-strain curve. Also called coefficient of elasticity.

elastic ratio. *Yield point* divided by *tensile strength.*

elastic strain. Same as *elastic deformation.*

elastic strain energy. The work done in deforming a body within the elastic limit of the material. It can be recovered as work rather than heat. See *strain energy.*

elastic waves. Mechanical vibrations in an elastic medium.

electrical discharge machining (EDM). Removal of stock from an electrically conductive material by rapid, repetitive spark discharge through a dielectric fluid flowing between the workpiece and a shaped electrode. Variations of the process include electrical discharge grinding and electrical discharge wire cutting.

electrical disintegration. Metal removed by an electrical spark acting in air. It is not subject to precise control, the most common application being the removal of broken tools such as taps and drills; hence the shop name "tap buster."

electrical resistance alloys. See *resistance alloys.*

electrical steels. A group of sheet steels used for generators, motors, and transformers.

The steels have a very low carbon content and contain 0.5–5.0% Si.

electrochemical corrosion. Corrosion that is accompanied by a flow of electrons between cathodic and anodic areas on metallic surfaces.

electrochemical equivalent. The weight of an element, compound, radical, or ion involved in a specified electrochemical reaction during the passage of a unit quantity of electricity.

electrochemical machining. Removal of stock from an electrochemically conductive material by anodic dissolution in an electrolyte flowing rapidly through a gap between the workpiece and a shaped electrode. Often abbreviated ECM. Variations of the process include electrochemical deburring and electrochemical grinding.

electrochemical series. Same as *electromotive force series.*

electrode. (1) In arc welding, a current-carrying rod that supports the arc between the rod and the work, or between two rods as in twin carbon-arc welding. It may or may not furnish filler metal. See *bare electrode, covered electrode,* and *lightly coated electrode.* (2) In resistance welding, a part of a resistance welding machine through which current and, in most instances, pressure are applied directly to the work. The electrode may be in the form of a rotating wheel, rotating roll, bar, cylinder, plate, clamp, chuck, or modification thereof. (3) An electrical conductor for leading current into or out of a medium.

electrode cable. Same as *electrode lead.*

electrode deposition. The weight of weld-metal deposit obtained from a unit length of electrode.

electrode force. The force between electrodes in spot, seam, and projection welding.

electrode lead. The electrical conductor between the source of arc welding current and the electrode holder. Same as *electrode cable*.

electrodeposition. The deposition of a substance on an electrode by passing electric current through an electrolyte. *Electroplating (plating), electroforming, electrorefining,* and *electrowinning* result from electrodeposition.

electrode potential. The potential of a *half cell* as measured against a standard reference half cell.

electrode skid. In spot, seam, or projection welding, the sliding of an electrode along the surface of the work.

electroforming. Making parts by electrodeposition on a removable form.

electrogalvanizing. The *electroplating* of zinc upon iron or steel.

electrogas welding. A process for *vertical position welding* in which molding shoes confine the molten weld metal. Welding may be done by either *gas metal arc welding* or *flux cored arc welding*.

electroless plating. A process in which metal ions in a dilute aqueous solution are plated out on a substrate by means of autocatalytic chemical reduction.

electrolysis. Chemical change resulting from the passage of an electric current through an *electrolyte*.

electrolyte. (1) An ionic conductor. (2) A liquid, most often a solution, that will conduct an electric current.

electrolytic brightening. Same as *electropolishing*.

electrolytic cell. An assembly, consisting of a vessel, electrodes, and an electrolyte, in which electrolysis can be carried out.

electrolytic cleaning. Removing soil from work by *electrolysis*, the work being one of the electrodes.

electrolytic copper. Copper that has been refined by electrolytic deposition, including cathodes that are the direct product of the refining operation, refinery shapes cast from melted cathodes, and, by extension, fabricators' products made therefrom. Usually when this term is used alone, it refers to electrolytic tough pitch copper without elements other than oxygen being present in significant amounts.

electrolytic deposition. Same as *electrodeposition*.

electrolytic grinding. A combination of grinding and machining wherein a metal-bonded abrasive wheel, usually diamond, is the cathode in physical contact with the anodic workpiece, the contact being made beneath the surface of a suitable electrolyte. The abrasive particles produce grinding and act as nonconducting spacers permitting simultaneous machining through electrolysis.

electrolytic machining. Controlled removal of metal by use of an applied potential and a suitable electrolyte to produce the shapes and dimensions desired.

electrolytic pickling. *Pickling* in which electric current is used, the work being one of the electrodes.

electrolytic powder. Metal powder produced by electrolytic deposition or by pulverization of an electrodeposit, or from metal made from *electrodeposition*.

electrolytic protection. See preferred term *cathodic protection*. Partial or complete protection of a metal from corrosion by making it a cathode, using either a galvanic or an impressed current. Contrast with *anodic protection*.

electrometallurgy. Industrial recovery or processing of metals and alloys by electric or electrolytic methods.

electromotive force. Electrical potential; voltage.

electromotive force series. A series of elements arranged according to their *standard electrode potentials*. In corrosion studies, the analogous but more practical *galvanic series* of metals is generally used. The relative positions of a given metal are not necessarily the same in the two series.

electron bands. Energy states for the free electrons in a metal, as described by use of the band theory (zone theory) of electron structure. Also called Brillouin zones.

electron beam cutting. A cutting process that uses the heat obtained from a concentrated beam composed primarily of high-velocity electrons, which impinge upon the workpieces to be cut; it may or may not use an externally supplied gas.

electron beam machining. Removing material by melting and vaporizing the workpiece at the point of impingement of a focused high-velocity beam of electrons. The machining is done in high vacuum to eliminate scattering of the electrons due to interaction with gas molecules.

electron beam microprobe analyzer. An instrument for selective analysis of a microscopic component or feature in which an electron beam bombards the point of interest in a vacuum at a given energy level. Scanning of a larger area permits determination of the distribution of selected elements. The analysis is made by measuring the wavelengths and intensities of secondary electromagnetic radiation resulting from the bombardment.

electron beam welding. A welding process that produces coalescence of metals with the heat obtained from a concentrated beam composed primarily of high-velocity electrons impinging upon the surfaces to be joined.

electron compound. An intermediate phase on a *constitution diagram*, usually a binary phase, that has the same crystal structure and the same ratio of valence electrons to atoms as those of intermediate phases in several other systems. An electron compound is often a solid solution of variable composition and good metallic properties. Occasionally, an ordered arrangement of atoms is characteristic of the compound, in which case the range of composition is usually small. Phase stability depends essentially on electron concentration and crystal structure and has been observed at valence-electron-to-atom ratios of $^3/_2$, $^{21}/_{13}$, and $^7/_4$.

electrophoresis. The transport of charged colloidal or macromolecular materials in an electric field.

electroplating. Electrodepositing a metal or alloy in an adherent form on an object serving as a cathode.

electropolishing. (1) A technique commonly used to prepare metallographic specimens, in which a high polish is produced making the specimen the anode in an electrolytic cell, where preferential dissolution at high points smooths the surface. (2) A variation of *chemical machining* wherein electrolytic deplating promotes chemical cutting, especially at surface irregularities.

electrorefining. Using electric or electrolytic methods to convert impure metal to purer metal, or to produce an alloy from impure or partly purified raw materials.

electroslag remelting. A *consumable-electrode remelting* process in which heat is generated by the passage of electric current through a conductive slag. The droplets of metal are refined by contact with the slag. Sometimes abbreviated ESR.

electroslag welding. A fusion welding process in which the welding heat is provided by passing an electric current through a layer of molten conductive slag contained in a pocket formed by molding shoes that bridge the gap between the members being welded. The resistance heated slag not only melts filler-metal electrodes as they are fed into the slag layer, but also provides shielding for the massive weld puddle characteristic of the process.

electrostrictive effect. The reversible interaction, exhibited by some crystalline materials, between an elastic strain and an electric field. The direction of the strain is independent of the polarity of the field. Compare with *piezoelectric effect.*

electrotinning. *Electroplating* tin on an object.

electrotyping. The production of printing plates by *electroforming.*

electrowinning. The recovery of a metal from an ore by means of an electrochemical process.

elongation. In tensile testing, the increase in the gage length, measured after fracture of the specimen within the gage length, usually expressed as a percentage of the original gage length.

elutriation. Separation of metal powder into particle-size fractions by means of a rising stream of gas or liquid.

embossing. Raising a design in relief against a surface.

embossing die. A die used for producing embossed designs.

embrittlement. The reduction in the normal ductility of a metal due to a physical or chemical change. Examples include *blue brittleness, hydrogen embrittlement,* and *temper brittleness.*

embrittlement, 475 °C (885 °F). An embrittlement caused by the exposure of ferritic stainless steel in the 400 to 565 °C (750 to 1050 °F) temperature range or by slow cooling through this range, which results in a pronounced increase in hardness with a corresponding decrease in ductility. This embrittlement is with the higher chromium contents. The 13% Cr steels are rarely susceptible. For steels over 18% Cr, the onset of embrittlement is fast enough to require rapid cooling from the annealing temperature in order to ensure optimum ductility.

emery. An impure mineral of the corundum or aluminum oxide type used extensively as an abrasive before the development of electric furnace products.

emf. An abbreviation for *electromotive force.*

emissivity. The ratio of the amount of energy or of energetic particles radiated from a unit area of a surface to the amount radiated from a unit area of an ideal emitter under the same conditions.

emulsion. A dispersion of one liquid phase in another.

emulsion cleaner. A cleaner consisting of organic solvents dispersed in an aqueous medium with the aid of an emulsifying agent.

enameling steels. Steel sheets used for vitreous (porcelain) enameling with very low carbon, phosphorus, sulfur, and silicon contents. When the steel is to be used for deep drawing prior to enameling, the carbon content is limited to 0.008–0.05% maximum, depending on the application (ASTM A424). The same as *porcelain enameling steels.*

enantiotropy. The relation of crystal forms of the same substance in which one form is stable above a certain temperature and the other form is stable below that temperature. Ferrite and austenite are enantiotropic in ferrous alloys, for example.

end mark. A roll mark caused by the end of a sheet marking the roll during hot or cold rolling.

end milling. A method of machining with a rotating peripheral and end cutting tool. See also *face milling*.

end-quench hardenability test. A laboratory procedure for determining the hardenability of a steel or other ferrous alloy. Hardenability is determined by heating a standard specimen above the upper critical temperature, placing the hot specimen in a fixture so that a stream of cold water impinges on one end, and, after cooling to room temperature is completed, measuring the hardness near the surface of the specimen at regularly spaced intervals along its length. The data are normally plotted as hardness versus distance from the quenched end. Widely referred to as the Jominy test.

end relief. See sketch accompanying *single-point tool*.

endurance limit. The maximum stress below which a material presumably can endure an infinite number of stress cycles. If the stress is not completely reversed, the value of the mean stress, the minimum stress, or the stress ratio also should be stated. Compare with *fatigue limit*.

endurance ratio. The ratio of the *endurance limit* for completely reversed flexural stress to the tensile strength of a given material.

entry mark (exit mark). A slight corrugation caused by the entry or exit rolls of a *roller leveling* unit.

epitaxy. Growth of an electrodeposit or vapor deposit in which the orientations of the crystals in the deposit are directly related to crystal orientations in the underlying crystalline substrate.

epsilon (ε) structure. A Hume-Rothery designation for structurally analogous close-packed phases or electron compounds, such as $CuZn_3$, that have ratios of seven valence electrons to four atoms. Not to be confused with the ε phase on a *constitution diagram.*

equiaxed grain structure. A structure in which the grains have approximately the same dimensions in all directions.

equilibrium. A dynamic condition of physical, chemical, mechanical, or atomic balance that appears to be a condition of rest rather than one of change.

equilibrium diagram. A graphical representation of the temperature, pressure, and composition limits of phase fields in an alloy system as they exist under conditions of complete equilibrium. In metal systems, pressure usually is considered constant.

erbium. A chemical element having atomic number 68, atomic weight 167, and the symbol Er. It is one of the 16 rare earth elements. Erbium was named for Ytterby, a Swedish village having a large concentration of erbium, by Carl Gustaf Mosander, the mineralogist who discovered it in 1843. The metal is made more malleable by the addition of vanadium. Erbium is used as a doping agent in fiber optics for signal amplification. It is also used in the nuclear industry.

Erichsen test. A cupping test in which a piece of sheet metal, restrained except at the center, is deformed by a cone-shaped, spherical-end plunger until fracture occurs. The height of the cup in millimeters at the fracture is a measure of the ductility of the metal.

erosion. The destruction of metals or other materials by the abrasive action of moving fluids, usually accelerated by the presence of solid particles or matter in suspension. When *corrosion* occurs simultaneously, the term *erosion-corrosion* often is used.

erosion-corrosion. The simultaneous occurrence of erosion and corrosion.

etchant. A chemical substance or mixture used for etching.

etch cleaning. Removing soil by dissolving away some of the underlying metal.

etch cracks. Shallow cracks in hardened steel containing high residual surface stresses, produced by etching in an embrittling acid.

etch figures. Characteristic markings produced on crystal surfaces by chemical attack, usually having facets that are parallel to low-index crystallographic planes.

etching. (1) Subjecting the surface of a metal to preferential chemical or electrolytic attack in order to reveal structural details for metallographic examination. (2) Chemically or electrochemically removing tenacious films from a metal surface to condition the surface for a subsequent treatment, such as painting or electroplating.

eutectic. (1) An isothermal reversible reaction in which a liquid solution is converted into two or more intimately mixed solids on cooling, the number of solids formed being the same as the number of components in the system. (2) An alloy having the composition indicated by the eutectic point on an equilibrium diagram. (3) An alloy structure of intermixed solid constituents formed by a eutectic reaction.

eutectic carbide. Carbide formed during freezing as one of the mutually insoluble phases participating in the eutectic reaction of ferrous alloys.

eutectic melting. Melting of localized microscopic areas whose composition corresponds to that of the eutectic in the system.

eutectoid. (1) An isothermal reversible reaction in which a solid solution is converted into two or more intimately mixed solids on cooling, the number of solids formed being the same as the number of components in the system. (2) An alloy having the composition indicated by the eutectoid point on an equilibrium diagram. (3) An alloy structure of intermixed solid constituents formed by a eutectoid reaction.

exfoliation. A type of corrosion that progresses approximately parallel to the outer surface of the metal, causing layers of the metal to be elevated by the formation of corrosion product.

expanding. A process used to increase the diameter of a cup, shell, or tube. See *bulging*.

expansion fit. An *interference* or *force fit* made by placing a cold (subzero) inside member into a warmer outside member and allowing an equalization of temperature.

explosion welding. A solid-state welding process effected by a controlled detonation, which causes the parts to move together at high velocity.

explosive forming. Shaping of metal parts wherein the forming pressure is generated by an explosive charge.

extensometer. An instrument for measuring changes in length caused by application or removal of a force. Commonly used in tension testing of metal specimens.

extractive metallurgy. The branch of *process metallurgy* dealing with the *winning* of metals from their ores. Compare with *refining*.

extra hard. A *temper* of nonferrous alloys and some ferrous alloys characterized by values of tensile strength and hardness approximately one-third of the way from those of *full hard* to those of *extra spring* temper.

Extra Low Carbon Steel 1002. SAE 1002 (UNS G10020) is the number for an extra-low carbon steel added to the former list of AISI carbon steels in 2009. The steel appears in SAE J403 with the chemical composition: 0.02–0.04% C, 0.35% maximum manganese, 0.030% maximum phosphorus, and 0.050% maximum sulfur.

Extra Low Carbon Steel 1003. SAE 1003 (UNS G10030) is the number for an extra-low carbon steel added to the former list of AISI carbon steels in 2009. The steel appears in SAE J403 with the chemical composition: 0.03–0.06% C, 0.35% maximum manganese, 0.030% maximum phosphorus, and 0.050% maximum sulfur.

Extra Low Carbon Steel 1004. SAE 1004 (UNS G10040) is the number for an extra-low carbon steel added to the former list of AISI carbon steels in 2009. The steel appears in SAE J403 with the chemical composition: 0.02–0.08% C, 0.35% maximum manganese, 0.030% maximum phosphorus, and 0.050% maximum sulfur.

Extra Low Carbon Steel 1007. SAE 1007 (UNS G10070) is the number for an extra-low carbon steel added to the former list of AISI carbon steels in 2009. The steel appears in SAE J403 with the chemical composition: 0.02–0.10% C, 0.35% maximum manganese, 0.030% maximum phosphorus, and 0.050% maximum sulfur.

extra spring. A *temper* of nonferrous alloys and some ferrous alloys corresponding approximately to a cold worked state above *full hard* beyond which further cold work will not measurably increase strength or hardness.

extruded hole. A hole formed by a punch that first cleanly cuts a hole and then is pushed farther through to form a flange with an enlargement of the original hole.

extrusion. The conversion of an ingot or billet into lengths of uniform cross section by forcing metal to flow plastically through a die orifice. In direct extrusion (forward extrusion), the die and ram are at opposite ends of the extrusion stock, and the product and ram travel in the same direction. There is also relative motion between the extrusion stock and the *container*. In indirect extrusion (backward extrusion), the die is at the ram end of the stock and the product travels in the direction opposite that of the ram, either around the ram (as in impact extrusion of cylinders such as cases for dry cell batteries) or up through the center of a hollow ram.

Impact extrusion is the process (or resultant product) in which a punch strikes a slug (usually unheated) in a confining die. The metal flow may be either between punch and die or through another opening. Impact extrusion of unheated slugs often is called cold extrusion. Also see *Hooker process*, in which a pierced slug is used.

Stepped extrusion is the process whereby a single product has one or more abrupt changes in cross section, produced by stopping extrusion to change dies. Often, such an extrusion is made in a complex die having a die section that can be freed from the main die and allowed to ride out with the product when extrusion is resumed.

face. In a lathe tool, the surface against which the chips bear as they are formed. See Fig. 50 accompanying *single-point tool*.

face mill. A milling cutter that cuts metal with its face. See illustration of nomenclature in Fig. 26.

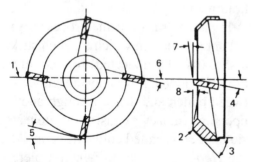

1. Reference plane
2. Tooth point
3. Corner angle (bevel)
4. Axial rake (positive)
5. Peripheral clearance angle
6. Radial rake (negative)
7. End clearance angle
8. End cutting edge angle

Fig. 26 Face mill nomenclature

face milling. Milling a surface that is perpendicular to the cutter axis.

face of weld. The exposed surface of an arc or gas weld on the side from which the welding was done. See Fig. 27 accompanying *fillet weld*.

face-type cutters. Cutters that can be mounted directly on and driven from the machine spindle nose.

facing. (1) In machining, generating a surface on a rotating workpiece by the traverse of a tool perpendicular to the axis of rotation. (2) In founding, special sand placed against a pattern to improve the surface quality of the casting. (3) For abrasion resistance, see preferred term *hardfacing*.

fagot. In forging work, a bundle of iron bars that will be heated and then hammered and welded to form a single bar.

failure. A general term used to imply that a part in service (a) has become completely inoperable, (b) is still operable but is incapable of satisfactorily performing its intended function, or (c) has deteriorated seriously, to the point that it has become unreliable or unsafe for continued use.

false bottom. An *insert* put in either member of a die set to increase the strength and improve the life of the die.

false brinelling. Evenly spaced depressions in a raceway of a rolling-element bearing caused by fretting that occurs when the bearing is subjected to vibration while it is not rotating. Compare with *brinelling*.

false indication. In nondestructive inspection, an *indication* that may be interpreted erroneously as an *imperfection*. See also *artifact*.

false wiring. Rounding the edge of sheet metal into a closed or partly closed loop. Same as *curling*.

fatigue. The phenomenon leading to fracture under repeated or fluctuating stresses having a maximum value less than the tensile strength of the material. Fatigue fractures are progressive, beginning as minute cracks that grow under the action of the fluctuating stress.

fatigue life. The number of cycles of stress that can be sustained prior to failure under a stated test condition.

fatigue limit. The maximum stress that presumably leads to fatigue fracture in a specified number of stress cycles. If the stress is not completely reversed, the value of the *mean stress,* the minimum stress, or the *stress ratio* also should be stated. Compare with *endurance limit*.

fatigue notch factor (K_f). The ratio of the *fatigue strength* of an unnotched specimen to the fatigue strength of a notched specimen of the same material and condition; both strengths are determined at the same number of stress cycles.

fatigue notch sensitivity (q). An estimate of the effect of a notch or hole on the fatigue properties of a material; measured by $q = (K_f - 1)/(K_t - 1)$. K_f is the *fatigue notch factor,* and K_t is the *stress-concentration factor,* for a specimen of the material containing a notch or hole of a given size and shape. A material is said to be fully notch sensitive if q approaches a value of 1.0; it is not notch sensitive if the ratio approaches 0.

fatigue ratio. The *fatigue limit* under completely reversed flexural stress divided by the tensile strength for the same alloy and condition.

fatigue strength. The maximum stress that can be sustained for a specified number of cycles without failure, the stress being completely reversed within each cycle unless otherwise stated.

fatigue-strength reduction factor. The ratio of the *fatigue strength* of a member or specimen with no stress concentration to the fatigue strength with stress concentration. This factor has no meaning unless the stress range and the shape, size, and material of the member or specimen are stated.

fatigue striations. Parallel lines frequently observed in electron microscope fractographs of fatigue fracture surfaces. The lines are transverse to the direction of local crack propagation; the distance between successive lines represents the advance of the crack front during one cycle of stress variation.

faying surface. The surface of a piece of metal (or a member) in contact with another to which it is joined or is to be joined.

feed. The rate at which a cutting tool or grinding wheel advances along or into the surface of a workpiece, the direction of advance depending on the type of operation involved.

feeder (feeder head, feedhead). A reservoir of molten metal connected to a casting to provide additional metal to the casting, required as the result of shrinkage before and during solidification Also known as a *riser.*

feeding. (1) Conveying metal stock or workpieces to a location for use or processing, such as wire to a consumable electrode, strip to a die, or workpieces to an assembler. (2) In casting, providing molten metal to a region undergoing solidification, usually at a rate sufficient to fill the mold cavity ahead of the solidification front and to make up for any shrinkage accompanying solidification.

feed lines. Linear marks on a machined or ground surface that are spaced at intervals equal to the *feed* per revolution or per stroke.

Feltmetal™. A porous felted fiber metal product with sinter-bonded fibers that produce a stiff plate with interconnecting pores. The product, which was developed by the Huyck Felt Company in 1960, was especially adapted to the use of stainless steel fibers. The product is made by Technetics Group (United States) and used primarily for abradable seals and sound suppression of aircraft jet engines.

fermium. A chemical element having atomic number 100, atomic weight 257, and the symbol Fm, named for nuclear physics pioneer Enrico Fermi. Fermium was discovered by Albert Ghiorso and colleagues in 1952 while studying debris produced by the detonation of the first hydrogen bomb.

ferrimagnetic material. A material that macroscopically has properties similar to those of a *ferromagnetic material* but that microscopically also resembles an antiferromagnetic material in that some of the elementary magnetic moments are aligned antiparallel. If the moments are of different magnitudes, the material may still have a large resultant magnetization.

ferrite. (1) A solid solution of one or more elements in body-centered cubic iron. Unless otherwise designated (for instance, as chromium ferrite), the solute is generally assumed to be carbon. On some equilibrium diagrams, there are two ferrite regions separated by an austenite area. The lower area is α ferrite; the upper, δ ferrite. If there is no designation, α ferrite is assumed. (2) In the field of magnetics, substances having the general formula: $M^{++}O \cdot M_2^{+++}O_3$, with the trivalent metal often being iron.

ferrite banding. Parallel bands of free *ferrite* aligned in the direction of working. Sometimes referred to as *ferrite streaks*.

ferrite number. An arbitrary, standardized value designating the ferrite content of an austenitic stainless steel weld metal. This value directly replaces percent ferrite or volume percent ferrite and is determined by the magnetic test described in AWS A4.2.

ferrite streaks. Parallel bands of free ferrite aligned in the direction of working. Same as *ferrite banding*.

ferritic malleable. A cast iron made by prolonged annealing of *white cast iron* in which decarburization or graphitization, or both, take place to eliminate some or all of the *cementite*. The graphite is in the form of temper carbon. If decarburization is the predominant reaction, the product will exhibit a light fracture surface, hence the name "whiteheart malleable," otherwise, the fracture surface will be dark, hence the name "blackheart malleable." Ferritic malleable has a predominantly ferritic matrix; pearlitic malleable may contain pearlite, spheroidite, or tempered martensite, depending on heat treatment and desired hardness. See *malleable cast iron.*

ferritizing anneal. A treatment given as-cast gray or ductile (nodular) iron to produce an essentially ferritic matrix. For the term to be meaningful, the final microstructure desired or the time-temperature cycle used must be specified.

ferroalloy. An alloy of iron that contains a sufficient amount of one or more other chemical elements to be useful as an agent for introducing these elements into molten metal, especially into steel or cast iron.

ferrograph. An instrument used to determine the size distribution of wear particles in lubricating oils of mechanical systems.

ferromagnetic material. A material that in general exhibits the phenomena of hysteresis and saturation, and whose permeability is dependent on the magnetizing force. Microscopically, the elementary magnets are aligned parallel in volumes called domains. The unmagnetized condition of a ferromagnetic material results from the overall neutralization of the magnetization of the domains to produce zero external magnetization. Compare with *paramagnetic material, diamagnetic material,* and *ferrimagnetic material.*

fiber. (1) The characteristic of wrought metal that indicates *directional properties* and is revealed by etching a longitudinal section or is manifested by the fibrous or woody appearance of a fracture. It is caused chiefly by extension of the constituents of the metal, both metallic and nonmetallic, in the direction of working. (2) The pattern of preferred orientation of metal crystals after a given deformation process, usually wiredrawing. See *preferred orientation.*

fiber stress. Local stress through a small area (a point or line) on a section where the stress is not uniform, as in a beam under a bending load.

fibrous fracture. A fracture whose surface is characterized by a dull gray or silky appearance. Contrast with *crystalline fracture.*

fibrous structure. (1) In forgings, a structure revealed as laminations, not necessarily detrimental, on an etched section or having a ropy appearance on a fracture. It is not the same as silky or ductile fracture of a clean metal. (2) In wrought iron, a structure consisting of slag fibers embedded in *ferrite.* (3) In rolled steel plate stock, a uniform, fine-grained structure on a fractured surface, free of laminations or shale-type discontinuities. As contrasted with (1), it is virtually synonymous with silky or ductile fracture.

filamentary shrinkage. A fine network of shrinkage cavities, occasionally found in steel castings, that produces a radiographic image resembling lace.

file hardness. Hardness as determined by the use of a steel file of standardized hardness on the assumption that a material that cannot be cut with the file is as hard as, or harder than, the file. Files covering a range of hardnesses may be employed.

filler. A material used to increase the bulk of a product without improving its functional performance.

filler metal. Metal added in making a brazed, soldered, or welded joint.

fillet. (1) A radius (curvature) imparted to inside meeting surfaces. (2) A concave corner piece used on foundry patterns.

fillet weld. A weld, approximately triangular in cross section, joining two surfaces, essentially at right angles to each other in a lap, tee, or corner joint (Fig. 27).

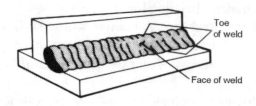

Fig. 27 Example of a fillet weld

final annealing. An imprecise term used to denote the last anneal given to a nonferrous alloy prior to shipment.

fine grain practice. A steelmaking practice for steel other than stainless steel that is intended to produce a *killed steel* capable of meeting the requirements specified for austenitic grain size.

fineness. A measure of the purity of gold or silver expressed in parts per thousand.

fines. (1) The product that passes through the finest screen in sorting crushed or ground material. (2) Sand grains that are substantially smaller than the predominating size in a batch or lot of foundry sand. (3) The portion of a metal powder composed of particles smaller than a specified size, usually 44 μm. See also *superfines*.

fine silver. Silver with a fineness of 999; equivalent to a minimum content of 99.9% Ag with the remaining content unrestricted.

finish. (1) Surface condition, quality, or appearance of a metal. (2) Stock on a forging or casting to be removed in finish machining.

finish allowance. The amount of excess metal surrounding the intended final configuration of a formed part; sometimes called forging envelope, machining allowance, or cleanup allowance.

finished steel. Steel that is ready for the market and has been processed beyond the stages of billets, blooms, sheet bars, slabs, and wire rods.

finish grinding. The final grinding action on a workpiece, of which the objectives are surface finish and dimensional accuracy.

finishing die. The die used to make the final impression on a forging. Sometimes called finisher.

finishing temperature. The temperature at which *hot working* is completed.

finish machining. A machining process analogous to *finish grinding*.

fire-refined copper. Copper that has been refined by the use of a furnace process only, including refinery shapes and, by extension, fabricators' products made therefrom. Usually, when this term is used alone it refers to fire-refined tough pitch copper without elements other than oxygen being present in significant amounts.

fire scale. Intergranular copper oxide remaining below the surface of silver-copper alloys that have been annealed and pickled.

fir-tree crystal. A type of *dendrite*.

fisheyes. Areas on a steel fracture surface having a characteristic white crystalline appearance.

fishmouthing. The longitudinal splitting of flat slabs in a plane parallel to the rolled surface. Same as *alligatoring*.

fishscale. A scaly appearance in a porcelain enamel coating in which the evolution of hydrogen from the basis metal (iron or steel) causes loss of adhesion between the enamel and the basis metal. Individual scales are usually small, but have been observed in sizes up to 25 mm (1 in.) or more in diameter. The scales are somewhat like blisters that have cracked partway around the perimeter but remain attached to the coating around the rest of the perimeter; if detached completely, it is one form of *pop-off*.

fishtail. (1) In roll forging, the excess trailing end of a forging. It is often used, before being trimmed off, as a tong hold for a subsequent forging operation. (2) In hot rolling or extrusion, the imperfectly shaped trailing end of a bar or special section that must be cut off and discarded as mill scrap.

fit. The amount of clearance or interference between mating parts is called actual fit. Fit is the preferable term for the range of clearance or interference that may result from the specified limits on dimensions (limits of size). Refer to ANSI standards.

fixed-feed grinding. Grinding in which the wheel is fed into the work, or vice versa, by given increments or at a given rate.

fixed position welding. Welding in which the work is held in a stationary position.

fixture. A positioning device used to hold the workpiece only.

flake powder. Metal powder in the form of flat or scalelike particles that are relatively thin.

flakes. Short, discontinuous internal fissures in ferrous metals attributed to stresses produced by localized transformation and decreased solubility of hydrogen during cooling after hot working. In a fracture surface, flakes appear as bright silvery areas; on an etched surface, they appear as short discontinuous cracks. Also called *shatter cracks* or snowflakes.

flame annealing. Annealing in which the heat is applied directly by a flame.

flame cleaning. Cleaning metal surfaces of scale, rust, dirt, and moisture by use of a gas flame.

flame hardening. A process for hardening the surfaces of hardenable ferrous alloys in which an intense flame is used to heat the surface layers above the upper transformation temperature, whereupon the workpiece is immediately quenched.

flame spraying. *Thermal spraying* in which a coating material is fed into an oxyfuel gas flame, where it is melted. Compressed gas may or may not be used to atomize the coating material and propel it onto the substrate.

flame straightening. Correction distortion in metal structures by localized heating with a gas flame.

flank. The end surface of a tool that is adjacent to the cutting edge and below it when the tool is in a horizontal position, as for turning. See the figure accompanying the term *single-point tool*.

flank wear. The loss of relief on the flank of the tool behind the cutting edge due to rubbing contact between the work and the tool during cutting; measured in terms of linear dimension behind the original cutting edge.

flapping. In copper refining, hastening oxidation of molten copper by striking through the slag-covered surface of the melt with a *rabble* just before the bath is poled.

flare test. A test applied to tubing, involving tapered expansion over a cone. Similar to *pin expansion test*.

flaring. (1) Forming an outward acute-angle flange on a tubular part. (2) Forming a flange by using the head of a hydraulic press.

flash. (1) In forging, excess metal forced out between the upper and lower dies. (2) In casting, a fin of metal that results from leakage between mating mold surfaces. (3) In resistance butt welding, a fin formed perpendicular to the direction of applied pressure.

flashback. The recession of a flame into or in back of the interior of a torch.

flash butt welding. A resistance welding process that joins metals by first heating abutting surfaces by passage of an electric current across the joint, then forcing the surfaces together by the application of pressure. Flashing and upsetting are accompanied by the expulsion of metal from the joint. See preferred term *flash welding*.

flash extension. The portion of *flash* remaining after trimming, measured from the intersection of the draft and flash at the body of the forging to the trimmed edge of the stock.

flashing. In *flash welding*, the heating portion of the cycle, consisting of a series of rapidly recurring localized short circuits followed by molten metal expulsions, during which time the surfaces to be welded are moved one toward the other at a predetermined speed.

flash land. Relief at the parting line of a set of closed-die forging dies that is designed either to restrict or to encourage the growth of flash, whichever is required to ensure complete filling of the finishing impression.

flash line. The line left on a forging or casting after the flash has been trimmed off.

flash plate. A very thin final electrodeposited film of metal.

flash welding. A resistance welding process that joins metals by first heating abutting surfaces by passage of an electric current across the joint, then forcing the surfaces together by the application of pressure. Flashing and upsetting are accompanied by the expulsion of metal from the joint.

flask. A metal or wood frame used for making and holding a sand mold (Fig. 28). The upper part is called the *cope;* the lower, the *drag.*

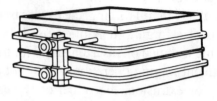

Fig. 28 Flask used to cast metals in sand molds

flat-die forging. Forging metal between flat or simple-contour dies by repeated strokes and manipulation of the workpiece. Also known as open-die forging, hand forging, or *smith forging.*

flat drill. A rotary end-cutting tool constructed from a flat piece of material, provided with suitable cutting lips at the cutting end.

flat edge trimmer. A machine for trimming notched edges on shells. The slide is cam driven so as to obtain a brief dwell at the bottom of the stroke, at which time the die, sometimes called a *shimmy die,* oscillates to trim the part.

flat-position welding. Welding from the upper side, the face of the weld being horizontal. Also called downhand welding.

flattening. (1) A preliminary operation performed on forging stock so as to position the metal for a subsequent forging operation. (2) Removing irregularities or distortion in sheets or plates by a method such as *roller leveling* or *stretcher leveling.*

flattening test. A quality test for tubing in which a specimen is flattened to a specified height between parallel plates.

flat wire. A roughly rectangular or square mill product, narrower than *strip,* in which all surfaces are rolled or drawn without any previous slitting, shearing, or sawing.

flaw. A nonspecific term often used to imply a crack-like discontinuity. See preferred terms *discontinuity, imperfection,* and *defect.*

flexible cam. An adjustable pressure-control cam of spring steel strips used to obtain varying pressure during a forming cycle.

flex roll. A movable jump roll designed to push up against a sheet as it passes through a roller leveler. The flex roll can be adjusted to deflect the sheet any amount up to the roll diameter.

flex rolling. Passing sheets through a *flex roll* unit to minimize yield-point elongation so as to reduce the tendency for *stretcher strains* to appear during forming.

floating die. (1) A die mounted in a die holder or a punch mounted in its holder, such that a slight amount of motion compensates for tolerance in the die parts, the work, or the press. (2) A die mounted on heavy springs to allow vertical motion in some trimming, shearing, and forming operations.

floating plug. In tube drawing, an unsupported mandrel that locates itself at the die inside the tube, causing a reduction in wall thickness while the die reduces the outside diameter of the tube.

floppers. On metals, lines or ridges that are transverse to the direction of rolling and generally confined to the section midway between the edges of a coil as rolled.

flotation. The concentration of valuable mineral from ores by agitation of the ground material with water, oil, and flotation chemicals. The valuable minerals are generally wetted by the oil, lifted to the surface by clinging air bubbles, and then floated off.

flowability. A characteristic of a foundry sand mixture that enables it to move

under pressure or vibration so that it makes intimate contact with all surfaces of the pattern or core box.

flow brazing. Brazing by pouring hot molten nonferrous filler metal over a joint until the brazing temperature is attained. The filler metal is distributed in the joint by capillary action.

flow brightening. The melting of an electrodeposit, followed by solidification, especially of tin plate. Also called *reflowing.*

flow lines. (1) Texture showing the direction of metal flow during hot or cold working. Flow lines can often be revealed by etching the surface or a section of a metal part (Fig. 29). (2) In mechanical metallurgy, paths followed by minute volumes of metal during deformation.

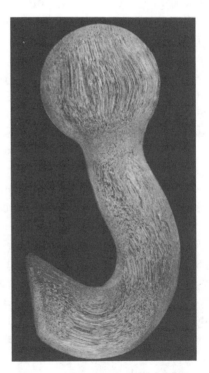

Fig. 29 Hook forged from 4140 steel, showing flow lines in a longitudinal section. Original magnification: 0.5×

flow stress. The uniaxial true stress at the onset of *plastic deformation* in a metal.

fluidity. The ability of liquid metal to run into and fill a mold cavity.

fluorescence. The emission of characteristic electromagnetic radiation by a substance as a result of the absorption of electromagnetic or corpuscular radiation having a greater unit energy than that of the fluorescent radiation. It occurs only so long as the stimulus responsible for it is maintained.

fluorescent magnetic-particle inspection. Inspection with either dry magnetic particles or those in a liquid suspension, the particles being coated with a fluorescent substance to increase the visibility of the indications.

fluorescent penetrant inspection. Inspection using a fluorescent liquid that will penetrate any surface opening; after the surface has been wiped clean, the location of any surface flaws may be detected by the fluorescence, under ultraviolet light, of back-seepage of the fluid.

fluoroscopy. An inspection procedure in which the radiographic image of the subject is viewed on a fluorescent screen, normally limited to low-density materials or thin sections of metals because of the low light output of the fluorescent screen at safe levels of radiation.

flute. (1) As applied to drills, reamers, and taps, the channels or grooves formed in the body of the tool to provide cutting edges and to permit passage of cutting fluid and chips. (2) As applied to milling cutters and hobs, the chip space between the back of one tooth and the face of the following tooth.

fluting. (1) Forming longitudinal recesses in a cylindrical part, or radial recesses in a conical part. (2) A series of sharp parallel kinks or creases occurring in the arc when sheet metal is roll formed into a cylindrical shape.

flux. (1) In metal refining, a material used to remove undesirable substances, such as sand, ash, or dirt, as a molten mixture. It is also used as a protective covering for certain molten metal baths. Lime or limestone is generally used to remove sand, as in iron smelting; sand, to remove iron oxide in copper refining. (2) In brazing, cutting, soldering, or welding, material used to prevent the formation of, or to dissolve and facilitate the removal of, oxides and other undesirable substances.

flux cored arc welding (FCAW). An arc welding process that joins metals by heating them with an arc between a continuous tubular filler-metal electrode and the work. Shielding is provided by a flux contained within the consumable tubular electrode. Additional shielding may or may not be obtained from an externally supplied gas or gas mixture. See also *electrogas welding*.

flux density. In magnetism, the number of *flux lines* per unit area passing through a cross section at right angles. It is given by $B = \mu H$, where μ and H are permeability and magnetic-field intensity, respectively.

flux lines. Imaginary lines used as a means of explaining the behavior of magnetic and other fields. Their concept is based on the pattern of lines produced when magnetic particles are sprinkled over a permanent magnet. Sometimes called magnetic lines of force.

flux oxygen cutting. Oxygen cutting with the aid of a flux.

fly ash. A finely divided siliceous material formed during the combustion of coal, coke, or other solid fuels.

fly cutting. Cutting with a single-tooth milling cutter.

flying shear. A machine for cutting continuous rolled products to length that does not require a halt in rolling, but rather moves along the runout table at the same speed as the product while performing the cutting, then returns to the starting point in time to cut the next piece.

fog quenching. The rapid cooling of a metal (often steel) from a suitable elevated temperature by exposing the metal to a fine vapor or mist.

foil. Metal in sheet form less than 0.15 mm (0.006 in.) thick.

fold. Same as *lap*.

follow board. A board contoured to a pattern to facilitate the making of a sand mold.

follow die. A *progressive die* consisting of two or more parts in a single holder; used with a separate lower die to perform more than one operation (such as piercing and blanking) on a part at two or more stations.

fool's gold. Iron or copper pyrite minerals that resemble gold.

force fit. Any of various interference fits between parts assembled under various amounts of force.

forehand welding. Welding in which the palm of the principal hand (torch or electrode hand) of the welder faces the direction of travel (Fig. 30). It has special significance in oxyfuel gas welding in that the flame is directed ahead of the weld bead, which provides *preheating*.

forgeability. Term used to describe the relative ability of material to flow under a compressive load without rupture.

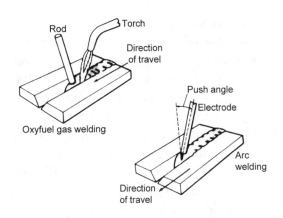

Fig. 30 Forehand welding technique

forge delay time. In spot, seam, or projection welding, the time between the start of the welding and the application of forging pressure.

forge welding. Solid-state welding in which metals are heated in a forge (in air) and then welded together by applying pressure or blows sufficient to cause permanent deformation at the interface. See also *roll welding*.

forging. Plastically deforming metal, usually hot, into desired shapes with compressive force, with or without dies.

forging billet. A wrought metal slug used as *forging stock*.

forging envelope. The amount of excess metal surrounding the intended final configuration of a formed part. Also called *finish allowance*, machining allowance, or cleanup allowance.

forging ingot. A cast metal slug used as *forging stock*.

forging machine. A type of forging equipment, related to the *mechanical press*, in which the main forming energy is applied horizontally to the workpiece, which is held by dies. Commonly called *upsetter* or *header*.

forging plane. In forging, the plane that includes the principal die face and that is perpendicular to the direction of ram travel. When parting surfaces of the dies are flat, the forging plane coincides with the parting line. Contrast with *parting plane.*

forging range. Temperature range in which a metal can be forged successfully.

forging rolls. A machine used in *roll forging.* Also called gap rolls.

forging stock. A rod, bar, or other section used to make forgings.

formability. The relative ease with which a metal can be shaped through plastic deformation. See *drawability.*

form block. Tooling, usually the male part, used for forming sheet metal contours; generally used in *rubber-pad forming.*

form cutter. A cutter, profile sharpened or cam relieved, shaped to produce a specified form on the work.

form die. A die used to change the shape of a blank with minimum plastic flow.

form grinding. Grinding with a wheel having a contour on its cutting face that is a mating fit to the desired form.

forming. Making a change, with the exception of shearing or blanking, in the shape or contour of a metal part without intentionally altering its thickness.

form-relieved cutter. A cutter so relieved that by grinding only the tooth face, the original form is maintained throughout its life.

form rolling. Hot rolling to produce bars having contoured cross sections; not to be confused with *roll forming* of sheet metal or with *roll forging.*

form tool. A single-edge, nonrotating cutting tool, circular or flat, that imparts its inverse or reverse form counterpart upon a workpiece

forward extrusion. The same as *direct extrusion.* Also see *extrusion.* The conversion of an ingot or billet into lengths of uniform cross section by forcing metal to flow plastically through a die orifice. In direct extrusion (forward extrusion), the die and ram are at opposite ends of the extrusion stock, and the product and ram travel in the same direction. Also there is relative motion between the extrusion stock and the *container.*

foundry. A commercial establishment or building where metal castings are produced.

four-high mill. A type of rolling mill, commonly used for flat-rolled mill products, in which two large-diameter backup rolls are used to reinforce two smaller working rolls, which are in contact with the product. Either the working rolls or the backup rolls may be driven. Compare with *two-high mill, cluster mill.*

four-point press. A press whose slide is actuated by four connections and four cranks, eccentrics, or cylinders, the chief merit being to equalize the pressure at the corners of the slides.

fraction. In powder metallurgy, the portion of a powder sample that lies between two stated particle sizes. Same as *cut.*

fractography. Descriptive treatment of fracture, especially in metals, with specific reference to photographs of the fracture surface. Macrofractography involves photographs at low magnification; microfractography, photographs at high magnification.

fracture mechanics. A quantitative analysis for evaluating structural behavior in terms of applied stress, crack length, and specimen or machine component geometry. Same as *linear elastic fracture mechanics.*

fracture stress. (1) The maximum principal true stress at fracture. The term usually refers to unnotched tensile specimens. (2) The (hypothetical) true stress that will cause fracture without further deformation at any given strain.

fracture test. A test in which a specimen is broken and its fracture surface is examined with the unaided eye or with a low-power microscope to determine such factors as composition, grain size, case depth, or discontinuities.

fracture toughness. See *stress-intensity factor.*

fragmentation. The subdivision of a grain into small, discrete crystallites outlined by a heavily deformed network of intersecting slip bands as a result of cold working. These small crystals or fragments differ from one another in orientation and tend to rotate to a stable orientation determined by the slip systems.

francium. A chemical element having atomic number 87, atomic weight 223, and symbol Fr. Named for the country of France, the element was found in 1939 by French chemist Marguerite Catherine Perey.

freckling. A type of segregation revealed as dark spots on a macroetched specimen of a consumable-electrode vacuum-arc-remelted alloy.

free carbon. The part of the total carbon in steel or cast iron that is present in elemental form as graphite or temper carbon. Contrast with *combined carbon.*

free ferrite. Ferrite that is formed directly from the decomposition of hypoeutectoid austenite during cooling, without the simultaneous formation of cementite. Also called proeutectoid ferrite.

free fit. Any of various clearance fits for assembly by hand and free rotation of parts. See *running fit.*

free machining. The machining characteristics of an alloy to which one or more ingredients have been introduced to produce small broken chips, lower power consumption, better surface finish, and longer tool life. Such additions include sulfur or lead to steel, lead to brass, lead and bismuth to aluminum, and sulfur or selenium to stainless steel.

freezing range. The temperature range between *liquidus* and *solidus* temperatures in which molten and solid constituents coexist.

fretting. A type of wear that occurs between tight-fitting surfaces subjected to cyclic relative motion of extremely small amplitude. Usually, fretting is accompanied by corrosion, especially of the very fine wear debris. Also referred to as fretting corrosion, *false brinelling* (in rolling-element bearings), friction oxidation, *chafing fatigue*, molecular attrition, and wear oxidation.

fretting fatigue. Fatigue fracture that initiates at a surface area where fretting has occurred.

friction welding. A solid-state process in which materials are welded by the heat obtained from rubbing together of surfaces that are held against each other under pressure.

full annealing. An imprecise term that denotes an annealing cycle designed to produce minimum strength and hardness. For the term to be meaningful, the composition and starting condition of the material and the time-temperature cycle used must be stated.

full-automatic plating. Electroplating in which the work is automatically conveyed through the complete cycle.

full center. Mild waviness down the center of a sheet or strip.

fuller. In preliminary forging, the portion of a die that reduces the cross-sectional area between the ends of the stock and permits the metal to move outward.

full hard. A *temper* of nonferrous alloys and some ferrous alloys corresponding approximately to a cold worked state beyond which the material can no longer be formed by bending. In specifications, a full hard temper is commonly defined in terms of minimum hardness or minimum tensile strength (or, alternatively, a range of hardness or strength) corresponding to a specific percentage of cold reduction following full annealing. For aluminum, a full hard temper is equivalent to a reduction of 75% from *dead soft;* for austenitic stainless steels, a reduction of approximately 50 to 55%.

furnace brazing. A mass-production *brazing* process in which the filler metal is preplaced on the joint, then the entire assembly is heated to brazing temperature in a furnace. Usually, a protective furnace atmosphere is required, and wetting of the joint surfaces is accomplished without using a brazing flux.

fusible alloys. A group of binary, tertiary, quarternary, and quinary alloys containing bismuth, lead, tin, cadmium, and indium. The term *fusible alloy* refers to any of more than 100 alloys that melt at relatively low temperatures, that is, below the temperature of tin-lead solder (183 °C, or 360 °F). The melting points of these alloys are as low as 47 °C (117 °F), for example, Cerrolow 117, which is a quarternary eutectic alloy composed of 44.7% Bi, 22.6% Pb, 8.3% Sn, and 5.3% Cd. Fusible alloys are used for safety plugs in pressure vessels, fire sprinkler systems, and electrical fuses. They are also used for low-temperature solders and foundry patterns and in tube bending to prevent collapse at the bend.

fusion. A change of state from solid to liquid; melting.

fusion face. The surface of the base metal that will be melted during welding.

fusion welding. A welding process in which filler metal and base metal (substrate), or base metal only, are melted together to complete the weld.

fusion zone. In a weldment, the area of base metal melted as determined on a cross section through the weld (Fig. 31).

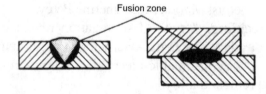

Fig. 31 Cross section of a weldment showing the fusion zone

gadolinium. A chemical element having atomic number 64, atomic weight 157, and the symbol Gd. Gadolinium, one of the rare earth metals, was named after the mineral gadolinite, which in turn was named for Finnish chemist Johan Gadolin, who became the pioneer in the research of the 15 rare earth metals. Gadolinium was first detected spectroscopically in 1880 by Swiss chemist Jean Charles Galissard de Marignac, who is credited with its discovery, and was isolated by Paul Émile (François) Lecoq de Boisbaudran in 1886.

gage. (1) The thickness (or diameter) of sheet or wire. The various standards are arbitrary and differ, ferrous from nonferrous products and sheet from wire. (2) An instrument used to measure thickness or length. (3) An aid for visual inspection that enables the inspector to determine more reliably whether the size or contour of a formed part meets dimensional requirements.

gage length. The original length of that portion of the specimen over which strain, change of length, and other characteristics are measured.

gagger. An irregularly shaped piece of metal used for reinforcement and support in a sand mold.

Galfan. A galvanizing alloy patented in 1981. It is a eutectic composition consisting of 95% Zn, 5% Al, and a small amount of the rare earth mischmetal containing cerium and lanthanum to give better fluidity. The alloy provides considerably better protection than zinc. See ASTM A875.

galling. A condition whereby excessive friction between high spots results in localized welding with subsequent *spalling* and further roughening of the rubbing surface(s) of one or both of two mating parts.

gallium. A chemical element having atomic number 31, atomic weight 70, and the symbol Ga. It was named for Gallia, the Latin word for present-day France, by Paul Émile (François) Lecoq de Boisbaudran, the Frenchman who isolated the metal.

Gallium is historically unique in that its discovery was predicted twice before the metal was isolated. Dmitri Mendeleev, in his work on the correlation of the properties of the elements with their atomic weights, found evidence that led him to the discovery of three elements. He named them eka-boron, eka-silicon, and eka-aluminum because of their similarity to these elements, and predicted the properties of the three eka-elements in 1871. During the same timeframe, de Boisbaudran, in studying the spectral lines of the elements, was led to believe there was an element missing between aluminum and indium. He predicted its discovery and its main spectral lines and undertook the search for it. After considerable effort, de Boisbaudran isolated a small quantity of gallium in 1875 from zinc blende in the Pyrenees. Later, de Boisbaudran recovered 75 grams of the metal and studied its properties. Mendeleev, upon learning of the new element, realized that it was his predicted eka-aluminum: the properties of the new element coincided almost exactly with those he had published before the discovery. Similarly, de Boisbaudran found that the spectral lines of gallium fit into the pattern he had predicted. See *Technical Note 5*.

TECHNICAL NOTE 5

Gallium and Gallium Compounds

GALLIUM-BASE COMPONENTS are found in a variety of products, ranging from compact disc players to advanced military electronic warfare systems. Compared with components made of silicon, a material gallium arsenide (GaAs) has replaced in some of these applications, components made of GaAs can emit light, have greater resistance to radiation, and operate at faster speeds and higher temperatures.

Gallium occurs in very low concentrations in the earth's crust, and virtually all primary gallium is recovered as a by-product, principally from the processing of bauxite to alumina. Most gallium applications require very high purity levels, and the metal must be refined before use. Commercially available gallium metal ranges in purity from 99.5 to 99.9999+%. The most common impurities are mercury, lead, tin, zinc, and copper. If impurity limits of high-purity gallium are exceeded, optoelectric properties are degraded or destroyed.

The principal use of gallium is in the manufacture of semiconducting compounds. More than 90% of the gallium consumed in the United States is used for optoelectronic devices and integrated circuits. Optoelectric devices—light-emitting diodes (LEDs), laser diodes, photodiodes, and solar (photovoltaic) cells—take advantage of the ability of GaAs to convert electrical energy into optical energy and vice versa. An LED, which is a semiconductor that emits light when an electric current is passed through it, consists of layers of epitaxially grown material on a substrate. These epitaxial layers are normally gallium aluminum arsenide (GaAlAs), gallium arsenide phosphide (GaAsP), or indium gallium arsenide phosphide (InGaAsP); the substrate material is either GaAs or gallium phosphide (GaP). Laser diodes operate on the same principle as LEDs, but they convert electrical energy to a coherent light output. Laser diodes principally consist of an epitaxial layer of GaAs, GaAlAs, or InGaAsP on a GaAs substrate. Photodiodes are used to detect a light impulse generated by a source, such as an LED or laser diode, and convert it to an electrical impulse. Photodiodes are fabricated from the same materials as LEDs. Gallium arsenide solar cells have been demonstrated to convert 22% of the available sunlight to electricity, compared with approximately 16% for silicon solar cells.

Although integrated circuits (ICs) currently represent a smaller share of the GaAs market than optoelectronic devices, they are important for military and defense applications. Two types of ICs are produced commercially: analog and digital. Analog ICs are designed to process signals generated by military radar systems, as well as those generated by satellite communications systems. Digital ICs essentially function as memory and logic elements in computers.

Selected physical properties of GaAs

Property	Amount
Molecular weight	144.6
Melting point, K	1511
Density, g/cm^3	
At 300 K (solid)	5.3165 ± 0.0015
At 1511 K (solid)	5.2
At 1511 K (liquid)	5.7
Lattice constant nm	0.5654
Adiabatic bulk modulus, dyne $\cdot$ cm^{-2}	7.55×10^{11}
Thermal expansion, K^{-1}	
At 300 K	6.05×10^{-6}
At 1511 K	7.97×10^{-6}
Specific heat, J $\cdot$ g^{-1} $\cdot$ K^{-1}	
At 300 K	0.325
At 1511 K	0.42
Thermal diffusivity at 300 K, cm^2 $\cdot$ s^{-1}	0.27
Latent heat, J $\cdot$ cm^{-3}	3290
Band gap, eV	1.44
Refractive index at 10 μm	3.309
Dielectric constant	
Static	12.85
Infrared	10.88
Electron mobility, cm^2 $\cdot$ V^{-1} $\cdot$ s^{-1}	
At 77 K	205 000
At 300 K	8500
Hole mobility at 300 K	400
Intrinsic resistivity at 300 K, $\Omega \cdot$ cm	3.7×10^8

Nonsemiconducting applications include the use of gallium oxide for making single-crystal garnets—such as gallium gadolinium garnet (GGG), which is used as the substrate for magnetic domain (bubble) memory devices. Small quantities of metallic gallium are used for low-melting-point alloys, for dental alloys, and as an alloying element in some magnesium, cadmium, and titanium alloys. Gallium is also used in high-temperature thermometers and as a substitute for mercury in switches. Gallium-base superconducting compounds, such as GaV$_3$, also have been developed.

Selected References

- D.A. Kramer, Gallium and Gallium Compounds, *Metals Handbook*, 10th ed., Vol 1, ASM International, 1990, p 739–749
- M.H. Brodsky, Progress in Gallium Arsenide Semiconductors, *Sci. Am.*, Feb 1990, p 68–75
- K. Zwibel, Photovoltaic Cells, *Chem. Eng. News*, Vol 64 (No. 27), 7 July 1986, p 34–48

Galvalume. The trade name of a galvanizing 55%Al–45%Zn alloy-coated sheet steel. The coating provides greater protection than zinc alone. See ASTM A798.

galvanic cell. A cell in which chemical change is the source of electrical energy. It usually consists of two dissimilar conductors in contact with each other and with an electrolyte, or of two similar conductors in contact with each other and with dissimilar electrolytes.

galvanic corrosion. Corrosion associated with the current of a galvanic cell consisting of two dissimilar conductors in an electrolyte or two similar conductors in dissimilar electrolytes. Where the two dissimilar metals are in contact, the resulting reaction is referred to as a couple action.

galvanic series. A series of metals and alloys arranged according to their relative corrosion potentials in a specified environment. Compare with *electromotive force series*. Table 9 lists various metals and alloys as they appear in the galvanic series.

galvanize. To coat a metal surface with zinc using any of various processes.

galvanneal. To produce a zinc-iron alloy coating on iron or steel by keeping the coating molten after hot-dip galvanizing until the zinc alloys completely with the basis metal.

gamma (γ) iron. The allotropic, nonmagnetic, face-centered cubic form of pure iron, stable from 910 to 1400 °C (1670 to 2550 °F). Gamma iron containing carbon or other elements in solution is known as austenite.

gamma (γ) structure. A Hume-Rothery designation for structurally analogous phases or electron compounds that have ratios of 21 valence electrons to

Table 9   Galvanic series in seawater at 25°C (77°F)

Corroded end (anodic, or least noble)

Magnesium
Magnesium alloys
Zinc
Galvanized steel or galvanized wrought iron
Aluminum alloys
5052, 3004, 3003, 1100, 6053, in this order
Cadmium
Aluminum alloys
2117, 2017, 2024, in this order
Low-carbon steel
Wrought iron
Cast iron
Ni-Resist (high-nickel cast iron)
Type 410 stainless steel (active)
50-50 lead-tin solder
Type 304 stainless steel (active)
Type 316 stainless steel (active)
Lead
Tin
Copper alloy C28000 (Muntz metal, 60% Cu)
Copper alloy C67500 (manganese bronze A)
Copper alloys C46400, C46500, C46600, C46700 (naval brass)
Nickel 200 (active)
Inconel alloy 600 (active)
Hastelloy alloy B
Chlorimet 2
Copper alloy C27000 (yellow brass, 65% Cu)
Copper alloys C44300, C44400, C44500 (admiralty brass)
Copper alloys C60800, C61400 (aluminum bronze)
Copper alloy C23000 (red brass, 85% Cu)
Copper C11000 (ETP copper)
Copper alloys C65100, C65500 (silicon bronze)
Copper alloy C71500 (copper nickel, 30% Ni)
Copper alloy C92300, cast (leaded in bronze G)
Copper alloy C92200, cast (leaded in bronze M)
Nickel 200 (passive)
Inconel alloy 600 (passive)
Monel alloy 400
Type 410 stainless steel (passive)
Type 304 stainless steel (passive)
Type 316 stainless steel (passive)
Incoloy alloy 825
Inconel alloy 625
Hastelloy alloy C
Chlorimet 3
Silver
Titanium
Graphite
Gold
Platinum

Protected end (cathodic, or most noble)

13 atoms; generally, a large, complex cubic structure. Not the same as γ phase on a *constitution diagram*.

gang milling. Milling with several cutters mounted on the same arbor or with workpieces similarly positioned for cutting either simultaneously or consecutively during a single setup.

gang slitter. A machine with a number of pairs of rotary cutters spaced on two parallel shafts, used for slitting sheet metal into strips or for trimming the edges of sheets.

gangue. The worthless portion of an ore that is separated from the desired part before smelting is commenced.

gap. The root opening in a weld joint.

gap-frame press. A general classification of presses in which the uprights or housings are made in the form of a letter C, thereby making three sides of the die space accessible.

gas cyaniding. A misnomer for *carbonitriding*.

gas holes. Holes in castings or welds that are formed by gas escaping from molten metal as it solidifies. Gas holes may occur individually, in clusters, or distributed throughout the solidified metal.

gas metal arc welding (GMAW). A process for welding metals together by heating them with an arc between a continuous and consumable filler metal electrode and the work. Shielding is obtained entirely from an externally supplied gas or gas mixture. Some subtypes are metal inert gas (MIG) and metal active gas (MAG) welding. When carbon dioxide shielding gas is used, the process is also known as CO_2 welding. See also *electrogas welding* and *pulsed power welding*.

gas plating. Same as *vapor plating*. Deposition of a metal or compound on a heated surface by reduction or decomposition of a volatile compound at a temperature below the melting points of the deposit and the base material. The reduction usually is accomplished by a gaseous reducing agent such as hydrogen.

gas pocket. A cavity caused by entrapped gas.

gas porosity. Fine holes or pores within a metal that are caused by entrapped gas or by evolution of dissolved gas during solidification.

gas shielded arc welding. Arc welding in which the arc and molten metal are shielded from the atmosphere by a stream of gas, such as argon, helium, argon-hydrogen mixtures, or carbon dioxide.

gassing. (1) The absorption of gas by a metal. (2) The evolution of gas from a metal during melting operations or on solidification. (3) The evolution of gas from an electrode during electrolysis.

gas tungsten arc cutting. An arc-cutting process in which metals are severed by melting them with an arc between a single tungsten (nonconsumable) electrode and the work. Shielding is obtained from a gas or gas mixture.

gas tungsten arc welding (GTAW). A fusion welding process in which metals are joined by heating them with an electric arc between a nonconsumable tungsten electrode and the work. Shielding is obtained from a gas or gas mixture. Pressure may or may not be applied to the joint, and filler metal may or may not be added. Sometimes referred to as tungsten inert gas (TIG) welding.

gas welding. See preferred term *oxyfuel gas welding*. Any of a group of processes used to fuse metals together by heating them with gas flames resulting from combustion of a specific fuel gas such as acetylene, hydrogen, natural gas, or propane. The process may be used with or without the application of pressure to the joint, and with or without adding any filler metal.

gate. The portion of the runner in a mold through which molten metal enters the mold cavity. Sometimes the generic term is applied to the entire network of connecting channels that conduct metal into the mold cavity.

gated pattern. A *pattern* that includes not only the contours of the part to be cast, but also the *gates.*

gathering. A forging operation that increases the cross section of part of the stock; usually a preliminary operation.

gathering stock. Any operation whereby the cross section of a portion of the forging stock is increased beyond its original size.

geared press. A press whose main crank or eccentric shaft is connected by gears to the driving source.

ghost lines. Lines running parallel to the rolling direction that appear in a sheet metal panel when it is stretched. These lines may not be evident unless the panel has been sanded or painted. (Not the same as *leveler lines.*)

gibs. Guides that ensure the proper restrained motion of the slide of a metal forming press, usually being adjustable to compensate for wear.

gilding metal. An alloy with 94–96% Cu, 0.05% Pb, 0.05% maximum iron, and the balance zinc. Other gilding metal alloys may contain as little as 80% Cu. Gilding metal is used in the production of various items, including bullet and artillery shell parts, badges, and jewelry.

glass electrode. A glass membrane *electrode* used to measure pH or hydrogen-ion activity.

glazing. Dulling the abrasive grains in the cutting face of a wheel during grinding.

glide. (1) Same as *slip.* (2) A noncrystallographic shearing movement, such as of one grain over another.

globular transfer. In consumable-electrode arc welding, a type of metal transfer in which molten filler metal passes across the arc as large droplets. Compare with *spray transfer* and *short circuiting transfer.*

gold. A chemical element having atomic number 79, atomic weight 197, and the symbol Au, for the Latin *aurum,* from Aurora, the goddess of the dawn. The name *gold* was derived from *geolu,* an Old English Anglo-Saxon word that means yellow. Gold is one of the seven metals of antiquity, along with silver, copper, tin, lead, mercury, and iron. It was discovered by the Egyptians around 5000 B.C. It has been widely found in nature in metallic form as nuggets and grains with a purity of approximately 85%. It is always contaminated with silver and also may contain copper, mercury, and bismuth. Gold is the most malleable metal known and may be beaten to a thickness of 0.000127 mm (0.000005 in.). The corrosion resistance and beauty of gold have made it the most popular of all metals for jewelry and ornaments since the dawn of history. Gold, one of the two colored metals, became the most sought-after form of wealth, and was widely used for coinage and bullion.

The use of gold in industry is limited because of its artificially high cost and low strength. It is extremely useful, however, in the electrical and electronics industries because of its ability to resist oxidation over a range of temperatures and its extremely high electrical conductivity. The karat, when used with gold, denotes its purity. Pure gold is 24 karats (24 k). An alloy with 50% Au is 12 k.

gold filled. Covered on one or more surfaces with a layer of gold alloy to form a clad metal. By commercial agreement, a quality mark showing the quantity and fineness of gold alloy may be affixed, indicating the actual proportional weight and karat fineness of the gold alloy cladding. For example, "$^1/_{10}$ 12K Gold Filled" means that the article consists of base metal covered on one or more surfaces with a gold alloy of 12 k fineness comprising $^1/_{10}$th part by weight of the entire metal in the article. No article having a gold alloy coating of less than 10 k fineness may have any quality mark affixed. No article having a gold alloy portion of less than $^1/_{20}$th by weight may be marked "Gold Filled," but may be marked "Rolled Gold Plate," provided that the proportional fraction and fineness designation precedes the mark. These standards do not necessarily apply to watch cases.

gooseneck. In die casting, a spout connecting a molten metal holding pot, or chamber, with a nozzle or sprue hole in the die and containing a passage through which molten metal is forced on its way to the die. It is the metal injection mechanism in a *hot chamber machine.*

G-P zone. A *Guinier-Preston zone.* A small precipitation domain in a supersaturated metallic solid solution. A G-P zone has no well-defined crystalline structure of its own, and contains an abnormally high concentration of solute atoms. The formation of G-P zones constitutes the first stage of precipitation, and usually is accompanied by a change in properties of the solid solution in which they occur.

grain. An individual crystal in a polycrystalline metal or alloy; it may or may not contain twinned regions and subgrains.

grain-boundary corrosion. Same as *intergranular corrosion.* Corrosion occurring preferentially at grain boundaries, usually with slight or negligible attack on the adjacent grains. See also *interdendritic corrosion.*

grain fineness number. A numbering system developed by the American Foundry Society (AFS) that expresses a weighted average grain size of a granular material. The grain fineness number is calculated with prescribed weighting factors from the standard screen analysis and approximates the number of meshes per inch of that sieve that would just pass the sample if its grains were of uniform size. It is approximately proportional to the surface area per unit of weight of sand, exclusive of clay.

grain flow. Fiberlike lines appearing on polished and etched sections of forgings, which are caused by orientation of the constituents of the metal in the direction of working during forging. Grain flow produced by proper die design can improve required *mechanical properties* of forgings.

grain growth. An increase in the average size of the grains in polycrystalline metal, usually as a result of heating at elevated temperature.

grain refiner. A material added to a molten metal to induce a finer-than-normal grain size in the final structure.

grain size. (1) For metals, a measure of the areas or volumes of grains in a polycrystalline material, usually expressed as an average when individual sizes are fairly uniform. In metals containing two or more phases, grain size refers to that of the matrix unless otherwise specified. Grain size is reported in terms of number of grains per unit area or volume, in terms of average diameter, or as a grain-size number derived from area measurements. (2) For grinding wheels, see preferred term *grit size*.

granular fracture. A type of irregular surface produced when metal is broken, characterized by a rough, grainlike appearance as differentiated from a smooth, silky, or fibrous type. It can be subclassified into transgranular and intergranular forms. This type of fracture is frequently called *crystalline fracture,* but the inference that the metal broke because it "crystallized" is not justified, because all metals are crystalline when in the solid state. Contrast with *fibrous fracture* and *silky fracture.*

granular powder. Particles of metal powder having approximately equidimensional nonspherical shapes.

granulated metal. Small pellets produced by pouring liquid metal through a screen or by dropping it onto a revolving disk, and, in both instances, chilling with water.

granulation. Production of coarse metal particles by pouring the molten metal through a screen into water or by agitating the molten metal violently during its solidification.

graphitic carbon. Free carbon in steel or cast iron.

graphitic corrosion. Corrosion of gray iron in which the iron matrix is selectively leached away, leaving a porous mass of graphite behind; it occurs in relatively mild aqueous solutions and on buried pipe and fittings.

graphitic steel. Alloy steel made so that part of the carbon is present as graphite.

graphitization. The formation of graphite in iron or steel. Where graphite is formed during solidification, the phenomenon is called primary graphitization; where formed later by heat treatment, secondary graphitization.

graphitizing. Annealing a ferrous alloy in such a way that some or all of the carbon is precipitated as graphite.

gravity hammer. A class of forging hammer in which energy for forging is obtained by the mass and velocity of a freely falling ram and the attached upper die. Examples are *board hammers* and air-lift hammers.

gravity segregation. The settling out of heavy constituents, or the rising of light constituents, before or during solidification, causing variable composition of a casting or ingot.

gray cast iron. A *cast iron* that gives a gray fracture due to the presence of flake graphite. Often called gray iron.

green compact. An unsintered powder metallurgy compact.

green density. The density of a *green compact*. The same as *pressed density*.

green rot. A form of high-temperature attack on stainless steels, nickel-chromium alloys, and nickel-chromium-iron alloys subjected to simultaneous oxidation and carburization. Basically, attack occurs first by precipitation of chromium as chromium carbide, then by oxidation of the carbon particles.

green sand. A naturally bonded sand, or a compounded molding sand mixture, that has been "tempered" with water and that is used while still moist.

green sand core. (1) A *core* made of *green sand* and used as-rammed. (2) A sand core that is used in the unbaked condition.

green sand mold. A casting mold composed of moist prepared molding sand. Contrast with *dry sand mold.*

grindability. Relative ease of grinding, analogous to *machinability.*

grindability index. A measure of the grindability of the material under specified grinding conditions, expressed in terms of volume of material removed per unit volume of wheel wear.

grinding. Removing material from a workpiece with a grinding wheel or abrasive belt.

grinding burn. Getting the work hot enough to cause discoloration or to change the microstructure by tempering or hardening.

grinding cracks. Shallow cracks formed in the surfaces of relatively hard materials because of excessive grinding heat or the high sensitivity of the material. See *grinding sensitivity.*

grinding fluid. A liquid used during grinding to clean, cool, and lubricate the site; *cutting fluid* used in grinding.

grinding oil. An oil-type grinding fluid; it may contain additives, but not water.

grinding relief. A groove or recess located at the boundary of a surface to permit the corner of the wheel to overhang during grinding.

grinding sensitivity. Susceptibility of a material to surface damage such as *grinding cracks;* it can be affected by such factors as hardness, microstructure, hydrogen content, and residual stress.

grinding stress. *Residual stress,* generated by grinding, in the surface layer of work. It may be tensile or compressive, or both.

grinding wheel. A cutting tool of circular shape made of absrasive grains bonded together.

grit blasting. Abrasive blasting with small irregular pieces of steel, malleable cast iron, or hard nonmetallic materials.

grit size. Nominal size of abrasive particles in a grinding wheel, corresponding to the number of openings per linear inch in a screen through which the particles can just pass. Sometimes, but accurately, called *grain size.*

grizzly. A set of parallel bars (or grating) used for coarse separation or screening of ores, rock, or other material.

groove angle. The total included angle of the groove between parts to be joined (Fig. 32). Thus, the sum of two bevel angles, either or both of which may be zero degrees.

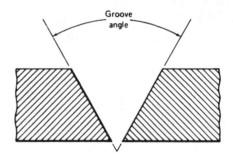

Fig. 32 Weld groove angle

groove face. The portion of a surface or surfaces of a member included in a groove. See the figure accompanying the term *root of joint.*

groove weld. A weld made in the groove between two members. The standard types are square, single-bevel, single flare-bevel, single flare-V, single-J, single-U, single-V, double-bevel, double flare-bevel, double flare-V, double-J, double-U, and double-V.

gross porosity. In a weld metal or a casting, pores, gas holes, or globular voids that are larger and in much greater numbers than those obtained in good practice.

ground connection. In arc welding, a device used for attaching the work lead (ground cable) to the work.

growth. In cast iron, a permanent increase in dimensions resulting from repeated or prolonged heating at temperatures above 480 °C (900 °F) due either to graphitizing of carbides or to oxidation.

Guerin forming. A tradename of a forming operation for shallow parts wherein a pad of rubber or other resilient material is attached to the press slide and becomes the mating die for a punch, or group of punches, that has been placed on the press bed or plate. Also known as *rubber pad forming.*

guided bend test. A test in which the specimen is bent to a definite shape by means of a jig.

guide mill. A small hand mill with several stands in a train and with guides for the work at the entrance to the rolls.

Guinier-Preston (G-P) zone. A small precipitation domain in a supersaturated metallic solid solution. A G-P zone has no well-defined crystalline structure of its own and contains an abnormally high concentration of solute atoms. The formation of G-P zones constitutes the first stage of precipitation and usually is accompanied by a change in properties of the solid solution in which they occur.

gun drill. A drill, usually with one or more flutes and with coolant passages through the drill body, used for deep hole drilling.

gutter. The depression around the periphery of a forging die that provides space for the flash without trapping it in the dies.

habit plane. The plane or system of planes of a crystalline phase along which some phenomenon, such as twinning or transformation, occurs.

Hadfield steel. A ferrous alloy invented by Sir Robert Hadfield in Sheffield, England, in 1882. Also known as Mangalloy and manganese steel, the alloy contains approximately 12% Mg and is especially abrasion resistant because it work hardens rapidly. It is widely used for mining and digging tools and for railroad crossing tracks.

hafnium. A chemical element having atomic number 72, atomic weight 179, and the symbol Hf. Dirk Coster and Georg von Hevesy discovered the metal in 1923 in Copenhagen, and although neither was Danish, they named it hafnium—from the Latin *hafnia,* meaning harbor—in honor of the city.

In 1923, hafnium was one of the last four of the predicted metals to be discovered and it probably was the most difficult. Hafnium was found embedded within the structure of zirconium metal. Its chemical properties were almost identical to those of zirconium so that it was very difficult to separate. When finally separated, the metals were found to have some vastly different properties. Hafnium, with a density of 13.29 g/cc, a little greater than lead, was more than twice the density of zirconium at 6.45 g/cc. The greatest difference, however, was the astounding difference in their capacities to absorb thermal neutrons. For zirconium, the value was 0.180 barn/atom, compared with 105 barns/atom for hafnium. This finding showed that zirconium would be one of the best possible cladding materials for nuclear fuel, while hafnium would be one of the best possible materials for control rods. It was found that zirconium normally contained from 1–3% Hf, about enough to keep up with the requirement for hafnium control rods.

half cell. An *electrode* immersed in a suitable *electrolyte,* designed for measurements of *electrode potential.*

half hard. A *temper* of nonferrous alloys and some ferrous alloys characterized by tensile strength about midway between those of *dead soft* and *full hard* tempers.

Hall process. A commercial process for winning aluminum from alumina by electrolytic reduction of a fused bath of alumina dissolved in cryolite.

hammer forging. Forging in which the work is deformed by repeated blows. Compare with *press forging.*

hammering. Beating metal sheet into a desired shape either over a form or on a high-speed mechanical hammer and a similar anvil to produce the required dishing or thinning.

hammer welding. *Forge welding* by hammering.

hand brake. A small manual folding machine designed to bend sheet metal, similar in design and purpose to a *press brake.*

hand forging. Same as *flat-die forging.* Forging metal between flat or simple-contour dies by repeated strokes and manipulation of the workpiece. Also known as *open-die forging* and *smith forging.*

handling breaks. Irregular *breaks* caused by improper handling of metal sheets during processing. These breaks result from bending or sagging of the sheets during handling.

Hansgirg process. A process for producing magnesium by reduction of magnesium oxide with carbon.

hard chromium. Chromium electrodeposited for engineering purposes (such as to increase the wear resistance of sliding metal surfaces) rather than as a decorative coating. It is usually applied directly to basis metal and is typically thicker than a decorative deposit, but not necessarily harder.

hard drawn. An imprecise term applied to drawn products, such as wire and tubing, that indicates substantial cold reduction without subsequent annealing. Compare with *light drawn.*

hardenability. The relative ability of a ferrous alloy to form martensite when quenched from a temperature above the upper critical temperature. Hardenability is typically measured as the distance below a quenched surface at which the metal exhibits a specific hardness (50 HRC, for example) or a specific percentage of martensite in the microstructure.

hardener. An alloy rich in one or more alloying elements that is added to a melt to permit closer control of composition than is possible by the addition of pure metals, or to introduce refractory elements not readily alloyed with the base metal. Sometimes known as *master alloy* or rich alloy.

hardening. Increasing hardness of metals by suitable treatment, usually involving heating and cooling. When applicable, the following more specific terms should be used: *age hardening, case hardening, flame hardening, induction hardening, precipitation hardening,* and *quench hardening.*

hardfacing. Depositing filler metal on a surface by welding, spraying, or braze welding to increase resistance to abrasion, erosion, wear, galling, impact, or cavitation damage. Also known as *hard surfacing.*

hard head. A hard, brittle, white residue obtained in the refining of tin by liquation containing, among other things, tin, iron, arsenic, and copper. Also, a refractory lump of ore only partly smelted.

hardness. The resistance of metal to plastic deformation, usually by indentation. However, the term also may refer to stiffness, temper, and resistance to scratching, abrasion, or cutting. Indentation hardness may be measured by various hardness tests, such as the *Brinell, Rockwell,* and *Vickers hardness tests.*

hard surfacing. Depositing filler metal on a surface by welding, spraying, or braze welding to increase resistance to abrasion, erosion, wear, galling, impact, or cavitation damage. Also known as *hardfacing.*

hard temper. A temper of nonferrous alloys and some ferrous alloys corresponding approximately to a cold-worked state beyond which the material can no longer be formed by bending. In specifications, a hard temper is commonly defined in terms of minimum hardness or minimum tensile strength (or, alternatively, a range of hardness or strength) corresponding to a specific percentage of cold reduction following full annealing. For aluminum, a hard temper is equivalent to a reduction of 75% from *dead soft*; for austenitic stainless steels, a reduction of about 50 to 55%. Also known as *full hard* temper.

Haring cell. A four-electrode cell that measures electrolyte resistance and electrode polarization during electrolysis.

Hartmann lines. Elongated surface markings or depressions caused by localized plastic deformation that results from discontinuous (inhomogeneous) yielding. Also known as *Lüder's lines*, Lüder's bands, *Piobert lines*, or *stretcher strains.*

Haynes, Elwood. 1857–1925. An inventor, metallurgist, and automotive pioneer, Elwood Haynes was born in Portland, Indiana. He was little interested in school, preferring as a teenager to spend his time experimenting with chemicals and metals. He built a little furnace in his backyard where he melted metals. He attended Worcester Tech, in Massachusetts, where he majored in chemistry. His senior year thesis was on the effect of tungsten on steel.

When the first automobiles were invented in Europe, Haynes decided to build a "horseless carriage." He assembled a gasoline engine and installed it in a small open carriage. The machine worked, and he asked Elmer Apperson, who owned a machine shop, to go into business with him making automobiles. In 1894, they erected a small factory and the Haynes Apperson Co. was in business. Not satisfied with the life of spark plug points, Haynes spent a good deal of time in his laboratory experimenting with new alloys.

By the turn of the century, the automobile company was prospering, producing high-end 12-cylinder sedans. Haynes at the time had three jobs: officer of the local gas company, researcher in his laboratory, and president of one of the first auto companies in the United States.

Haynes became interested in a cobalt-chromium alloy that was very hard and highly corrosion resistant. He started selling knife blade blanks made out of the alloy to cutlers. He called the alloy Stellite after the Latin *stella,* meaning star, and obtained a patent for the alloy. He changed all of the lathe tools at his auto factory to Stellite, which proved to have at least three times the life of high-speed steels. Haynes had found another business and set up the Haynes Stellite Company.

While working on Stellite in his laboratory, Haynes discovered that iron alloyed with approximately 12% Cr produced an alloy that could be heat treated like steel but that did not rust. He applied for a patent, which was denied because there already existed a virtually identical application, from Harry Brearley of Sheffield, England. Haynes found that the Brearley patent actually had been filed about two weeks *later* than his own. Outraged, Haynes wrote letters to the U.S. Patent Office and to Brearley and finally got a patent. However, Brearley still had his patent, apparently an oversight on the part of the patent office.

Brearley offered to set up a patent-holding company that would hold both patents and others as well. The American Stainless Steel Company was set up in Pittsburgh, with the Firth-Brearley Stainless Steel Syndicate having 40% ownership; Haynes, 30% ownership; and five stainless steel producers equally sharing 30% ownership.

In 1920, Haynes sold the Stellite business to the Union Carbide and Carbon Company of Niagara Falls, New York, with the provision that he would be permitted to use the laboratory. In 1924, the Haynes Apperson business went bankrupt. In 1925, Haynes died at the age of 68. The Haynes Museum is in his house at Kokomo, Indiana, and his horseless carriage is on display at the Smithsonian Institute. Haynes International is a thriving business offering heat-resistant and corrosion-resistant alloys.

H-band steel. A carbon, carbon-boron, or alloy steel produced to specified limits of hardenability; the chemical composition range may be slightly different from that of the corresponding grade of ordinary carbon or alloy steel.

header. A horizontal mechanical press used to make parts from bar stock or tubing by *upset forging,* piercing, bending, or otherwise forming in dies. Also known as an *upsetter.*

heading. *Upsetting* wire, rod, or bar stock in dies to form parts that usually contain portions that are greater in cross-sectional area than the original wire, rod, or bar.

healed-over scratch. A scratch that occurred in an earlier mill operation and was partially masked in subsequent rolling. It may open up during forming.

hearth. The bottom portions of certain furnaces, such as blast furnaces, air furnaces, and other reverberatory furnaces, that support the charge and sometimes collect and hold molten metal.

heat. A generic term denoting a specific lot of steel, based on steelmaking and casting considerations.

heat-affected zone (HAZ). That portion of the base metal that was not melted during brazing, cutting, or welding, but whose microstructure and mechanical properties were altered by the heat.

heat analysis. A chemical analysis determined by the steel producer as being representative of a specific heat of steel.

heat check. A pattern of parallel surface cracks formed by alternate rapid heating and cooling of the extreme surface metal, sometimes found on forging dies and piercing punches. There may be two sets of parallel cracks, one set perpendicular to the other. Also known as *checks,* check marks, checking, and *surface checking.*

heat number. The alpha, numeric, or alpha-numeric designator used to identify a specific heat of steel.

heat-resistant alloy. An alloy developed for very-high-temperature service where relatively high stresses (tensile, thermal, vibratory, or shock) are encountered and where oxidation resistance is frequently required.

heat time. In multiple-impulse or seam welding, the time that the current flows during any one impulse.

heat tinting. The coloration of a metal surface through oxidation by heating it to reveal details of the microstructure.

heat treatable alloy. An alloy that can be hardened by heat treatment.

heat treating film. A thin coating or film, usually an oxide, formed on the surface of a metal during heat treatment.

heat treatment. Heating and cooling a solid metal or alloy in such a way as to obtain desired conditions or properties. Heating for the sole purpose of hot working is excluded from the meaning of this definition.

heavy thickness coils. Hot rolled steel sheet and strip coils with a thickness of 6.0 to 25 mm (0.230 to 1.000 in.). The product types include carbon, commercial, drawing, structural, high-strength low-alloy, high-strength low-alloy with improved formability, and ultrahigh-strength steels (ASTM A1018).

heel. The surface on which a single-point tool rests when held in a tool post. The same as *base* (1). Also see *single-point tool,* Fig. 50.

hemming. The forming of an edge by bending the metal back on itself; a bend of 180° made in two steps.

HERF. A common abbreviation for *high-energy-rate forging* or *high-energy-rate forming.*

herringbone pattern. A fractographic pattern of radial marks (shear ledges) that look like nested Vs. Herringbone patterns typically are found on brittle fracture surfaces in parts whose widths are considerably greater than their thickness. The points of the herringbone patterns can be traced back to the fracture origin. Also known as *chevron pattern.*

Heyn stress. Residual stresses that vary from tension to compression in a distance (presumably approximating the grain size) that is small compared with the gage length in ordinary strain measurements. They are not detectable by dissection methods, but sometimes can be measured from line shift or line broadening in an x-ray diffraction pattern. Also known as *microscopic stresses..*

high brass. A brass that contains 65% Cu and 35% Zn, has a high tensile strength, and is used for screws, springs, and rivets.

high-conductivity copper. Copper that, in the annealed condition, has a minimum electrical conductivity of 100% *IACS* (International Annealed Copper Standard) as determined in accordance with ASTM methods of testing.

high-energy-rate forging (HERF). Producing forgings at extremely high ram velocities resulting from the sudden release of a compressed gas against a free piston. Forging usually is completed in one blow. Also known as HERF processing, high-velocity forging, and high-speed forging.

high-energy-rate forming (HERF). A group of special forming processes in which metal undergoes deformation at high velocity, usually at least ten times the velocity of 0.2 to 6 m/s (0.5 to 20 ft/s) achieved in conventional forming. *Explosive forming,* electrohydraulic forming, and electromagnetic forming are the most common HERF processes.

high-frequency resistance welding. A resistance welding process that produces a coalescence of metals with the heat generated from the resistance of the workpieces to a high-frequency alternating current in the 10 to 500 kHz range and the rapid application of an upsetting force after heating is substantially completed. The path of the current in the workpiece is controlled by the use of the proximity effect (the feed current follows closely the return current conductor).

highlighting. Buffing or polishing selected areas of a complex shape to increase the luster or change the color of those areas.

high residual phosphorus copper. Deoxidized copper with residual phosphorus present in amounts (usually 0.013 to 0.04%) generally sufficient to decrease appreciably the conductivity of the copper.

hindered contraction. A contraction in which the shape will not permit a casting to contract in certain regions in keeping with the coefficient of expansion.

hob. A rotary cutting tool with its teeth arranged along a helical thread, used for generating gear teeth or other evenly spaced forms on the periphery of a cylindrical workpiece. The hob and the workpiece are rotated in a timed relationship to each other while the hob is fed axially or tangentially across or radially into the workpiece. Hobs are not the same as multiple-thread milling cutters, rack cutters, and similar tools, in which the teeth are not arranged along a helical thread.

hogging. Machining a part from bar stock, plate, or a simple forging in which much of the original stock is removed.

holddown. A plate, ring, or fingers used to hold work stationary during forming, blanking, piercing, or shearing.

holding furnace. A small furnace into which molten metal can be transferred to be held at the proper temperature until it can be used to make castings.

hold time. In resistance welding, the time during which pressure is applied to the work after the current ceases.

hole flanging. Forming an integral collar around the periphery of a previously formed hole. See *extruded hole*.

holidays. Discontinuities (such as porosity, cracks, gaps, and similar flaws) in a coating that allow areas of basis metal to be exposed to a corrosive environment that contacts the coated surface.

holmium. A chemical element having atomic number 67, atomic weight 165, and the symbol Ho. *Holmia* is the Latin name for Stockholm, the native city of chemist Per Theodor Cleve, who discovered the metal in approximately 1879. Swiss chemists J. L. Soret and M. Delafontaine previously observed Holmium's absorption spectrum.

homogeneous carburizing. Use of a carburizing process to convert a low-carbon ferrous alloy to one of uniform and higher carbon content throughout the section.

homogenizing. Holding at high temperature to eliminate or decrease chemical segregation by diffusion.

honing. A low-speed finishing process used chiefly to produce uniform high dimensional accuracy and fine finish, most often on inside cylindrical surfaces. In honing, very thin layers of stock are removed by simultaneously rotating and reciprocating a bonded abrasive stone or stick that is pressed against the surface being honed with lighter force than is typical of grinding.

hook. A concavity in a tooth face giving a variation in rack at different points along the tooth face.

Hooker process. The extrusion of a hollow billet or cup through an annulus formed by the die aperture and the mandrel or pilot to form a tube or long cup.

Hooke's law. A principle that states that the stress imposed on a solid is directly proportional to the strain produced. The law applies only below the proportional limit.

Hoopes process. An electrolytic refining process for aluminum, using three liquid layers in the reduction cell.

horizontal-position welding. (1) Making a fillet weld on the upper side of the intersection of a vertical surface and a horizontal surface. (2) Making a horizontal groove weld on a vertical surface.

horizontal-rolled-position welding. The topside welding of a butt joint connecting two horizontal pieces of rotating pipe.

horn. In a resistance welding machine, a cylindrical arm or beam that transmits the electrode pressure and usually conducts the welding current.

horn press. A mechanical press equipped with or arranged for a cantilever block or horn that acts as the die or support for the die, used in forming, piercing, setting down, or riveting hollow cylinders and odd-shaped work.

horn spacing. The distance between adjacent surfaces of the horns of a resistance welding machine.

hot bed. An area adjacent to the *runout table* where hot rolled metal is placed to cool. Sometimes known as a cooling table.

hot chamber machine. A *die casting* machine in which the metal chamber under pressure is immersed in the molten metal in a furnace. The chamber is sometimes called a *gooseneck*, and the machine is sometimes called a gooseneck machine.

hot-cold working. (1) A high-temperature thermomechanical treatment consisting of deforming a metal above its transformation temperature and cooling it quickly enough to preserve some or all of the deformed structure. (2) A general term synonymous with *warm working*.

hot crack. A crack formed in a cast metal due to internal stress that developed on cooling following solidification. A hot crack is less open than a *hot tear* and usually exhibits less oxidation and decarburization along the fracture surface.

hot dip coating. A metallic coating obtained by dipping the basis metal into a molten metal.

hot extrusion. Extrusion at elevated temperature that does not cause strain hardening. See also *extrusion*.

hot forming. Deforming metal plastically at such a temperature and strain rate that recrystallization takes place simultaneously with the deformation, thus avoiding any strain hardening. Also known as *hot working*.

hot isostatic pressing. A process for simultaneously heating and forming a powder metallurgy compact in which metal powder, contained in a sealed flexible mold, is subjected to equal pressure from all directions at a temperature high enough for sintering to take place.

hot isostatic pressure welding. A diffusion welding method that produces a coalescence of materials by heating and applying hot inert gas under pressure.

hot mill. A production line or facility for hot rolling of metals.

hot press forging. Plastically deforming metals between dies in presses at temperatures high enough to avoid strain hardening.

hot pressing. Forming a powder metallurgy compact at a temperature high enough to effect concurrent *sintering*.

hot pressure welding. A solid-state welding process that produces a coalescence of materials with heat and application of pressure sufficient to produce macrodeformation of the base material. Vacuum or other shielding media may be used. See also *forge welding* and *diffusion welding*. Compare with *cold welding*.

hot quenching. An imprecise term used to cover a variety of quenching procedures in which a quenching medium is maintained at a prescribed temperature above 70 °C (160 °F). See *quenching*.

hot rod. Hot rolled coiled stock that is to be cold drawn into wire. Also known as *wire rod*.

hot shortness. A tendency for some alloys to separate along grain boundaries when stressed or deformed at temperatures near the melting point. Hot shortness is caused by a low-melting constituent, often present only in minute amounts, that is segregated at grain boundaries.

hot tear. A fracture formed in a metal during solidification because of *hindered contraction*. Compare with *hot crack*.

hot top. (1) A reservoir, thermally insulated or heated, that holds molten metal on top of a mold for feeding the ingot or casting as it contracts on solidifying, thus preventing formation of pipe or voids (Fig. 33). (2) A refractory-lined steel or iron casting that is inserted into the tip of the mold and is supported at various heights to feed the ingot as it solidifies.

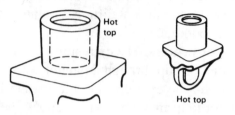

Fig. 33 Hot top

hot trimming. Removing flash or excess metal from a hot part (such as a forging) in a trimming press.

hot working. Deforming metal plastically at such a temperature and strain rate that recrystallization takes place simultaneously with the deformation, thus avoiding any strain hardening.

hubbing. Producing forging die cavities by pressing a male master plug, known as a hub, into a block of metal.

Hull cell. A type of electrodeposition cell used to qualitatively check the condition of an electroplating bath, allowing for the optimization of the current density range and the recognition of impurity effects.

humidity test. A corrosion test involving exposure of specimens at controlled levels of humidity and temperature. Contrast with *salt fog test*.

Huntsman, Benjamin. 1704–1776. An English clockmaker and instrument maker who, in 1740, discovered the long-lost art of making crucible steel, a process used in India around 500 B.C. Huntsman mixed broken pieces of blister steel and slag in a closed pot that was fired for three hours. High quality, but very expensive, steel was made by this process into the 20th century, when it was largely superseded by the electric furnace process.

hydraulic press. A press in which fluid pressure is used to actuate and control the ram.

hydride descaling. Removing the thick layer of oxides formed on some metals at elevated temperatures by the action of a hydride in a fused alkali. See *descaling*.

hydrogen brazing. A term sometimes used to denote brazing in a hydrogen-containing atmosphere, usually in a furnace.

hydrogen damage. A general term for the embrittlement, cracking, blistering, and hydride formation that can occur when hydrogen is present in some metals.

hydrogen embrittlement. A condition of low ductility in metal resulting from the absorption of hydrogen.

hydrogen loss. The loss in weight of metal powder or of a *compact* caused by heating a representative sample for a specified time and temperature in a purified hydrogen atmosphere. Broadly, a measure of the oxygen content of the sample when applied to materials containing only such oxides as are reducible with hydrogen and no hydride-forming element.

hydrogen overvoltage (in electroplating). *Overvoltage* associated with the liberation of hydrogen.

hydrogen-reduced powder. Metal powder produced by hydrogen reduction of a compound.

hydrometallurgy. Industrial *winning* or *refining* of metals using water or an aqueous solution.

hydrostatic tension. Three equal and mutually perpendicular tensile stresses.

hypereutectic alloy. In an alloy system exhibiting a *eutectic*, any alloy whose composition has an excess of alloying element compared with the eutectic composition, and whose equilibrium microstructure contains some eutectic structure.

hypereutectoid alloy. In an alloy system exhibiting a *eutectoid*, any alloy whose composition has an excess of alloying element compared with the eutectoid composition, and whose equilibrium microstructure contains some eutectoid structure.

hypoeutectic alloy. In an alloy system exhibiting a *eutectic*, any alloy whose composition has an excess of base metal compared with the eutectic composition, and whose equilibrium microstructure contains some eutectic structure.

hypoeutectoid alloy. In an alloy system exhibiting a *eutectoid*, any alloy whose composition has an excess of base metal compared with the eutectoid composition, and whose equilibrium microstructure contains some eutectoid structure.

hysteresis, magnetic. The lag of the magnetization of an iron or steel specimen behind any cyclic variation of the applied magnetizing field.

Dictionary of Metals
H.M. Cobb, editor

IACS. International Annealed Copper Standard; a standard reference used in reporting electrical conductivity. The conductivity of a material, in %IACS, is equal to 1724.1 divided by the electrical resistivity of the material in $n\Omega \cdot m$.

idiomorphic crystal. An individual crystal that has grown without restraint so that the habit planes are clearly developed. Compare with *allotriomorphic crystal*.

immersion cleaning. Cleaning in which the work is immersed in a liquid solution.

immersion coating. A coating produced in a solution by chemical or electrochemical action without the use of external current.

immersion plating. Depositing a metallic coating on a metal immersed in a liquid solution, without the aid of an external electric current. Also known as *dip plating*.

impact energy. The amount of energy, typically expressed in joules or foot-pound force, required to fracture a material, usually measured by means of an *Izod test* or a *Charpy test*. The type of specimen and the test conditions affect the values and therefore should be specified.

impact extrusion. The process (or resultant product) in which a punch strikes a slug (usually unheated) in a confining die. The metal flow may be either between punch and die or through another opening. Impact extrusion of unheated slugs often is called cold extrusion. See *extrusion*.

impact line. A blemish on a drawn sheet metal part caused by a slight change in metal thickness, which results from the impact of the punch on the blank. See *recoil line*.

impact strength. A measure of the resiliency or toughness of a solid. The maximum force or energy of a blow (given by a fixed procedure) that can be withstood without fracture, as opposed to fracture strength under a steady applied force. See *impact energy*.

impact test. A test to determine the behavior of materials when subjected to high rates of loading, usually in bending, tension, or torsion. The quantity measured is the energy absorbed in breaking the specimen by a single blow, as in *Charpy* and *Izod tests*.

imperfection. (1) When referring to the physical condition of a part or metal product, any departure of a quality characteristic from its intended level or state. The existence of an imperfection does not imply nonconformance (see *nonconforming*), nor does it have any implication as to the usability of the product. An imperfection must be rated on a scale of severity, in accordance with applicable specifications, to establish whether or not the part or metal product is of acceptable quality. (2) Generally, any departure from an ideal design, state, or condition. (3) In crystallography, any deviation from an ideal space lattice.

impregnation. (1) The treatment of porous castings with a sealing medium to stop pressure leaks. (2) The process of filling the pores of a sintered compact, usually with a liquid such as a lubricant. (3) The process of mixing particles of a nonmetallic substance in a matrix of metal powder, as in diamond-impregnated tools.

impression-die forging. A forging that is formed to the required shape and size by machined impressions in specially prepared dies that exert three-dimensional control on the workpiece.

impurities. Elements or compounds whose presence in a material is undesirable.

inclinable press. A press that can be inclined to facilitate handling of the formed parts. See also *open-back inclinable press*.

inclusions. Particles of foreign material in a metallic matrix. The particles are usually compounds (such as oxides, sulfides, or silicates), but may be of any substance that is foreign to (and essentially insoluble in) the matrix.

inclusion shape control. The addition of alloying elements during secondary steelmaking in order to affect the inclusion morphology.

indentation. In a spot, seam, or projection weld, a depression on the exterior surface of the base metal.

indentation hardness. The resistance of a material to indentation. Indentation hardness testing is the usual type of hardness test performed. The test involves pressing a pointed or rounded indenter into a surface under a substantially static load. Also known as *penetration hardness*.

indication. In inspection, a response to a nondestructive stimulus that implies the presence of an *imperfection*. The indication must be interpreted to determine if (a) it is a true indication or a *false indication* and (b) whether or not a true indication represents an unacceptable deviation.

indicator. A substance that, through some visible change such as color, indicates the condition of a solution or other material as to the presence of free acid, alkali, or other substance.

indirect-arc furnace. An electric-arc furnace in which the metallic charge is not one of the poles of the arc.

indirect extrusion. The conversion of an ingot or billet into lengths of uniform cross section by forcing metal to flow plastically through a die orifice. The die is at the ram end of the stock and the product travels in the direction opposite that of the ram, either around the ram (as in impact extrusion of cylinders such as cases for dry cell batteries) or up through the center of a hollow ram. See *extrusion*. See *extrusion*.

indium. A chemical element having atomic number 49, atomic weight 115, and the symbol In, named for the brilliant indigo line in its spectrum. Indium was discovered in 1863 by German chemists Ferdinand Reich and Hieronymous Theodor Richter while testing for thallium in zinc blende samples from the mines around Freiberg, Saxony. Spectrographic examination of the crude zinc chloride liquor showed a prominent indigo blue line that had not previously been observed. Indium is a member of a group of metals including aluminum, boron, gallium, and thallium. Applications for indium have included bearings, low-melting-point alloys, glass-sealing alloys, dental alloys, magnetic alloys, intermetallic semiconductors, and nuclear applications such as Ag-Cd-In control rods for pressurized water reactors. The development of indium was promoted by the Indium Corporation of America, a company founded in 1934 that supplies products to the electronics, semiconductor, solar, thin film, and thermal management markets.

induction brazing. *Brazing* in which the heat required to join the components is generated by subjecting the workpiece to electromagnetic induction, rather than by using a direct electrical connection.

induction furnace. An alternating current (ac) electric furnace in which the primary conductor is coiled and generates, by electromagnetic induction, a secondary current that develops heat within the metal charge.

induction hardening. A surface-hardening process in which only the surface layer of a suitable ferrous workpiece is heated by electromagnetic induction to a temperature above the upper critical temperature and immediately quenched.

induction heating. Heating by combined electrical resistance and hysteresis losses induced by subjecting a metal to the varying magnetic field surrounding a coil carrying alternating current.

induction melting. Melting in an *induction furnace.*

induction welding. Welding in which the required heat is generated by subjecting the workpiece to electromagnetic induction.

inert anode. An *anode* that is insoluble in the *electrolyte* under the conditions prevailing in the *electrolysis.*

infiltration. The process of filling the pores of a sintered or unsintered powder metallurgy compact with a metal or alloy of lower melting temperature.

ingate. The portion of the runner in a mold through which molten metal enters the mold cavity. Sometimes the generic term is applied to the entire network of connecting channels that conduct metal into the mold cavity. Also known as *gate.*

ingot. A casting of simple shape, suitable for hot working or remelting.

ingot iron. Commercially pure iron.

inhibitor. A substance that retards some specific chemical reaction. Picking inhibitors retard the dissolution of metal without hindering the removal of scale from steel.

inoculation. The addition of a material to molten metal to form nuclei for crystallization.

insert. (1) A part formed from a second material, usually a metal, that is placed in a mold and appears as an integral structural part of the final casting. (2) A removable portion of a die or mold.

insert die. A relatively small die that contains part or all of the impression of a forging, and which is fastened to a master die block.

inserted-blade cutters. Cutters having replaceable blades that are either solid or tipped and are usually adjustable.

intercept method. A quantitative metallographic technique in which the desired quantity, such as grain size or amount of precipitate, is expressed as the number of times per unit length a straight line on a metallographic image crosses particles of the feature being measured.

intercommunicating porosity. In a sintered powder metallurgy compact, a type of porosity in which individual pores are connected in such a way that a fluid may pass from one pore to another throughout the entire compact.

intercrystalline. Between the crystals, or grains, of a metal.

interdendritic corrosion. Corrosive attack that progresses preferentially along interdendritic paths. This type of attack results from local differences in composition, such as coring commonly encountered in alloy castings.

interface. A surface that forms the boundary between phases or systems.

interfacial tension. The contractile force of an interface between two phases.

interference. The difference in lateral dimensions at room temperature between two mating components before assembly by expansion, shrinking, or press fitting. Can be expressed in absolute or in relative terms.

interference fit. Any of various classes of fit between mating parts in which there is nominally a negative or zero allowance between the parts, and where there is either part interference or no gap when the mating parts are made to the respective extremes of individual tolerances that ensure the tightest fit between the parts. Contrast with *clearance fit.*

intergranular corrosion. Corrosion occurring preferentially at grain boundaries, usually with slight or negligible attack on the adjacent grains. See also *interdendritic corrosion.*

intermediate annealing. Annealing wrought metals at one or more stages during manufacture and before final treatment.

intermediate electrode. An *electrode* in an *electrolytic cell* that is not mechanically connected to the power supply, but is placed in the electrolyte, between the *anode* and *cathode*, so that the part nearer the anode becomes cathodic and the part nearer the cathode becomes anodic. Also known as *bipolar electrode.*

intermediate phase. In an alloy or a chemical system, a distinguishable homogeneous phase whose composition range does not extend to any of the pure components of the system.

intermetallic compound. An intermediate phase in an alloy system, having a narrow range of homogeneity and relatively simple stoichiometric proportions; the nature of the atomic binding can be of various types, ranging from metallic to ionic.

intermittent weld. A weld in which the continuity is broken by recurring unwelded spaces.

internal friction. The conversion of energy into heat by a material subjected to fluctuating stress. In free vibration, the internal friction is measured by the *logarithmic decrement.*

internal grinding. Grinding an internal surface such as that inside a cylinder or hole.

internal oxidation. Preferential in situ oxidation of certain components or phases within the bulk of a solid alloy accomplished by diffusion of oxygen into the body; a form of *subsurface corrosion.*

internal stress. See preferred term *residual stress*: Stress present in a body that is free of external forces or thermal gradients.

interpass temperature. In a multipass weld, the lowest temperature of a *pass* before the next one is commenced.

interrupted aging. Aging at two or more temperatures, by steps, and cooling to room temperature after each step. See also *aging,* and compare with *progressive aging* and *step aging.*

interrupted-current plating. Plating in which the flow of current is discontinued for periodic short intervals to decrease anode polarization and elevate the *critical current density.* It is most commonly used in cyanide copper plating.

interrupted quenching. A quenching procedure in which the workpiece is removed from the first quench at a temperature substantially higher than that of the quenchant and is then subjected to a second quenching system having a cooling rate different from that of the first. See *quenching.*

interstitial-free steel. A steel that has essentially all of its carbon and nitrogen chemically combined rather than being present interstitially.

interstitial solid solution. A solid solution in which the solute atoms occupy positions that do not correspond to lattice points of the solvent but instead occupy (interstitial) positions between the atoms in the structure of the solvent. Contrast with *substitutional solid solution.*

intracrystalline. Within or across the crystals or grains of a metal; same as transcrystalline and transgranular.

Invar. An alloy of iron with 36% Ni that has a very low coefficient of thermal expansion. The alloy was invented in 1896 by Swiss scientist Charles Édouard Guillaume, who received the Nobel Prize in Physics in 1920 for his invention.

inverse chill. A condition in an iron casting in which the interior is comprised of chilled or white iron, while the surfaces are either mottled or contain free graphite.

inverse segregation. Segregation in cast metal in which an excess of lower-melting constituents occurs in the earlier freezing portions, apparently the result of liquid metal entering cavities developed in the earlier-solidified metal.

investment casting. (1) Casting metal into a mold produced by surrounding, or investing, an expendable pattern with a refractory slurry coating that sets at room temperature, after which the pattern is removed through the use of heat. Also called *precision casting* or *lost-wax process.* (2) A part made by the investment casting process.

investment compound. A mixture of a graded refractory filler, a binder, and a liquid vehicle, used to make molds for *investment casting.*

ion. An atom, or group of atoms, that has gained or lost one or more outer electrons and thus carries an electric charge. Positive ions, or cations, are deficient in outer electrons. Negative ions, or anions, have an excess of outer electrons.

ion exchange. The reversible interchange of ions between a liquid and a solid, with no substantial structural changes in the solid.

ionic bond. A bond between two or more atoms that is the result of electrostatic attractive forces between positively and negatively charged ions.

ionic crystal. A crystal in which atomic bonds are *ionic bonds.* This type of atomic linkage, also known as (hetero) polar bonding, is characteristic of many compounds, such as sodium chloride.

ionization chamber. An enclosure containing two or more electrodes surrounded by a gas capable of conducting an electric current when it is ionized by x-rays or other ionizing rays. It is commonly used for measuring the intensity of such radiation.

iridium. A chemical element having atomic number 77, atomic weight 192, and the symbol Ir, named for the Latin *iris,* meaning rainbow, because iridium salts are of many different colors. The metal was isolated and identified in 1803 by Smithson Tennant. It is one of the densest metals known, with a specific gravity of 22.65—twice that of lead at 11.43. The metal is seldom used because of its high cost and poor malleability.

iron. A chemical element having atomic number 26, atomic weight 56, and the symbol Fe, for the Latin word *ferrum.* The name iron is from the Anglo-Saxon *yron.* It is said that a wrought iron sickle was found in one of the Egyptian pyramids that was estimated to date from about 5000 B.C. Iron was one of the seven metals of antiquity, along with gold, silver, tin, lead, copper, and mercury. The so-called iron of antiquity was actually a mixture of iron and threads of slag that has been called "wrought iron" since about 1700. Iron, the most important metal, is a heavy, magnetic, whitish metal that is one of the most abundant and most widely distributed of all metals, representing approximately 5% of the earth's crust. Iron has little use in its pure form, which melts at a very high temperature and is softer than any of its alloys.

The following products are forms of iron:

- ARMCO Iron: Commercially-pure iron developed by the former American Rolling Mills Company for the deep drawing of sheet.
- Bar iron: Wrought iron in the form of bars.
- Cast iron: A generic term for a large family of cast ferrous alloys in which the carbon content exceeds the solubility of carbon in austenite at the eutectic temperature. Most cast irons contain at least 2% C, plus silicon and sulfur, and may or may not contain other alloying elements. For the various forms—*gray cast iron, white cast iron, malleable cast iron,* and *ductile cast iron*—the word *cast* is often left out, resulting in gray iron, white iron, malleable iron, and ductile iron, respectively.
- Ingot iron: High-purity iron, containing a small amount of carbon and very small quantities of other elements, produced in an open hearth furnace.
- Pig iron: Impure, high-carbon iron produced by the reduction of iron ore in a blast furnace.
- Puddled iron (wrought iron): Nearly pure iron with up to 5% siliceous slags made, starting in the late 18th century, from pig iron refined with coal.
- Wrought iron (iron): Iron with a very low carbon content, in comparison to steel, and that has fibrous inclusions known as slag. See also *wrought iron.*

For over 5000 years, iron was a major metal used for ornaments, tools, horseshoes, water pipe, steam locomotive parts, bridges, railroad rails, armor plate, ships, and structures. The metal consisted of practically pure iron in which hundreds of threads of slag were embedded, so that the fracture of a bar had a woody appearance. The first record of the use of the term *wrought iron* was in an Act of Parliament in 1703 that read "wares of wrought iron." It is guessed that this usage was to make sure it was not confused with the cast iron that was becoming popular. From that time on, the terms *iron* and *wrought iron* were used. Iron was malleable and strong, having approximately 80% of the strength of steel, which was sold in quantity until Bessemer invented his converter in 1855. Iron was better than steel in some ways, particularly with regard to its corrosion resistance. Iron water pipes lasted for 100 years or more. By 1930, the demand for wrought iron had become so great in the United States that a new plant for producing the metal was set up in Ambridge, Pennsylvania, that would use the Aston Process, a highly mechanized operation that could produce 8000-pound batches of the metal. After World War II, however, wrought iron sales declined, and in 1969 the United States stopped producing the metal.

iron casting. A part made of *cast iron.*

ironing. Increasing the length of a hollow article by decreasing the thickness of the walls and the outside diameter by drawing it between a punch and die.

iron-powder electrode. A welding electrode with a covering containing up to approximately 50% iron powder, some of which becomes part of the deposit.

irradiation. The exposure of a material in a field of radiation; the cumulative exposure.

isostatic pressing. A process for forming a powder metallurgy compact by applying pressure equally from all directions to metal powder contained in a sealed flexible mold. See also *hot isostatic pressing.*

isothermal annealing. Austenitizing a ferrous alloy and then cooling to and holding at a temperature at which austenite transforms to a relatively soft ferrite-carbide aggregate.

isothermal transformation. A change in phase that takes place at a constant temperature. The time required for transformation to be completed, and in some instances the time delay before transformation begins, depends on the amount of supercooling below (or superheating above) the equilibrium temperature for the same transformation.

isotope. One of several different nuclides of an element having the same number of protons in their nuclei and therefore the same atomic number, being the same element chemically but differing in the number of neutrons and therefore in atomic weight.

isotropy. The quality of having identical properties in all directions.

Izod test. A pendulum-type single-blow impact test in which the specimen, usually notched, is fixed at one end and broken off at the other end by a falling pendulum. The energy absorbed by the specimen, as measured by the subsequent rise of the pendulum after impact, is a measure of the impact strength or toughness of the material. Contrast with *Charpy test.*

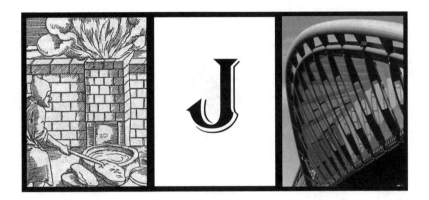

jewelry bronze. An alloy with 86–89% Cu, 0.05% maximum iron, 0.05% maximum lead, and the balance zinc.

jig grinding. Analogous to jig boring, where the holes are ground rather than machined.

joggle. An offset in a flat plane consisting of two parallel bends at the same angle but in opposite directions.

joint. The location where two or more members are to be or have been fastened together mechanically or by brazing or welding.

joint efficiency. The strength of a welded joint expressed as a percentage of the strength of the unwelded base metal.

joint penetration. The minimum depth to which a groove or flange weld extends from its face into the joint, exclusive of reinforcement (Fig. 34). Joint penetration may include *root penetration.*

Jominy test. A laboratory procedure for determining the hardenability of a steel or other ferrous alloy. Hardenability

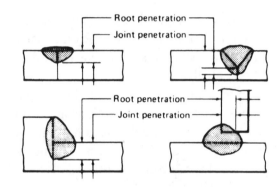

Fig. 34 Examples of weld joint penetration

is determined by heating a standard specimen to above the upper critical temperature, placing the hot specimen in a fixture so that a stream of cold water impinges on one end, and, after cooling to room temperature is completed, measuring the hardness near the surface of the specimen at regularly spaced intervals along its length. The data are normally plotted as hardness versus distance from the quenched end. Also referred to as *end-quench hardability test.*

Dictionary of Metals
H.M. Cobb, editor

keel block. A standard test casting, for steel and other high-shrinkage alloys, consisting of a rectangular bar that resembles the keel of a boat, attached to the bottom of a large riser, or shrinkhead. Keel blocks that have only one bar are often called Y-blocks; keel blocks having two bars, double keel blocks. Test specimens are machined from the rectangular bar, and the shrinkhead is discarded.

Kellering. A shop term. See preferred term, *tracer milling.* Duplicating a three-dimensional form by means of a cutter controlled by a tracer that is directed by a master form.

kerf. The space that was occupied by material removed during cutting; the width of the cut produced during the cutting process.

keyhole specimen. A type of specimen containing a hole-and-slot notch, shaped like a keyhole, usually used in impact bend tests. See *Charpy test* and *Izod test.*

killed steel. Steel treated with a strong deoxidizing agent such as silicon or aluminum in order to reduce the oxygen content to such a level that no reaction occurs between carbon and oxygen during solidification.

kiln. A large furnace used for baking, drying, or burning firebrick or refractories, or for calcining ores or other substances.

kish. Free graphite that forms in molten hypereutectic cast iron as it cools. In castings, the kish may segregate toward the cope surface, where it lodges at or immediately beneath the casting surface.

knockout. (1) A mechanism for freeing formed parts from a die used for stamping, blanking, drawing, forging, or heading operations. See also *stripper punch* and *ejector rod.* (2) A partly pierced hole in a sheet metal part, where the slug remains in the hole and can be forced out by hand if a hole is actually needed. (3) The removal of sand cores from a casting. (4) The jarring of an investment casting mold to remove the casting and investment from the flask.

Knoop hardness test. A microhardness test that determines hardness from the resistance of metal to indentation by a pyramidal diamond indenter that has edge angles of 172°-30′ and 130° and makes a rhombohedral impression with one very long and one very short diagonal.

knurling. Impressing a design into a metallic surface, usually by means of small, hard rollers that carry the corresponding design on their surfaces.

Kroll process. A process for the production of metallic titanium sponge by the reduction of titanium tetrachloride with a more active metal, such as magnesium, yielding titanium in the form of granules or powder.

Dictionary of Metals
H.M. Cobb, editor

ladle. A metal receptacle frequently lined with refractories used for transferring and pouring molten metal.

laminate. (1) A composite metal, usually in the form of sheet or bar, composed of two or more metal layers so bonded that the composite metal forms a structural member. (2) To form a metallic product of two or more bonded layers.

lamination. (1) A type of discontinuity with separation or weakness generally aligned parallel to the worked surface of a metal. May be the result of pipe, blisters, seams, inclusions, or segregation elongated and made directional by working. Laminations also may occur in powder metallurgy compacts. (2) In electrical products such as motors, a blanked piece of electrical sheet that is stacked up with several other identical pieces to make a stator or rotor.

lancing. (1) A press operation in which a single-line cut is made in strip stock without producing a detached slug. Chiefly used to free metal for forming, or to cut partial contours for blanked parts, particularly in progressive dies. (2) A misnomer for *oxyfuel gas cutting*.

land. (1) For profile-sharpened milling cutters, the relieved portion immediately behind the cutting edge. (2) For reamers, drills, and taps, the solid section between the flutes. (3) On punches, the portion adjacent to the nose that is parallel to the axis and of maximum diameter.

lanthanum. A chemical element having atomic number 57, atomic weight 139, and the symbol La, from the Greek *lanthana* and *lanthanein,* meaning being hidden. Lanthanum lies mixed among the rare-earth-bearing minerals. Lanthanum was discovered in 1839 in the mineral cerite by Swedish chemist Carl Gustav Mosander.

lap. A surface imperfection, with the appearance of a seam, caused by hot metal, fins, or sharp corners being folded over and then being rolled or forged into the surface but without being welded.

lap joint. A joint made between two overlapping members (Fig. 35).

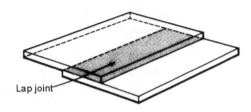

Fig. 35 Example of a lap joint

lapping. Finishing a surface by abrasion with an object, usually made of copper, lead, cast iron, or close-grained wood, having very fine abrasive particles rolled into its surface. Lapping is performed to achieve extreme dimensional accuracy and correction of minor shape imperfections.

laser. A device that emits a concentrated beam of electromagnetic radiation (light). Laser beams are used in metalworking to melt, cut, or weld metals; in less concentrated form they are sometimes used to inspect metal parts.

laser beam cutting. A cutting process that severs materials with the heat obtained by directing a beam from a *laser* against a metal surface. The process can be used with or without an externally supplied shielding gas.

laser beam machining. Removing material by melting and vaporizing the workpiece at the point of impingement of a highly focused beam of coherent monochromatic light (a *laser* beam).

laser beam welding. A welding process that joins metal parts using the heat obtained by directing a beam from a *laser* onto the weld joint.

laten. A metal or alloy, especially brass, made in thin sheets. One of the early names for *brass*. Also spelled "latten," "laton," and "lattyn."

latent heat. Thermal energy absorbed or released when a substance undergoes a phase change.

lateral extrusion. An operation in which the product is extruded sideways through an orifice in the container wall.

lateral runout. For any rotating element, the total variation from a true plane of rotation, taken in a direction parallel to the axis of rotation. The same as axial runout. Compare with *radial runout*.

laton. A metal or alloy, especially brass, made in thin sheets. One of the early names for *brass*. Also spelled "latten," "laton," and "lattyn." .

latten. A metal or alloy, especially brass, made in thin sheets. One of the early names for *brass*. Also spelled "latten," "laton," and "lattyn."

lattice constant. See *lattice parameter*.

lattice parameter. The length of any side of a unit cell of a given crystal structure; if the lengths are unequal, all unequal lengths must be given. The same as *lattice constant*.

lattyn. A metal or alloy, especially brass, made in thin sheets. One of the early names for *brass*. Also spelled "latten," "laton," and "lattyn." .

launder. (1) A channel for conducting molten metal. (2) A box conduit conveying particles suspended in water.

lawrencium. A chemical element having atomic number 103, atomic weight 260, and the symbol Lr, for Ernest O. Lawrence, inventor of the cyclotron particle accelerator. The synthetic element was discovered by a nuclear physics team led by Albert Ghiorso in 1961.

lay. The direction of predominant surface pattern remaining after cutting, grinding, lapping, or other processing.

leaching. Extracting an element or compound from a solid alloy or mixture by preferential dissolution in a suitable liquid.

lead. A chemical element having atomic number 82, atomic weight 207, and the symbol Pb, for the Latin *plumbum,* whose meanings include heavy and liquid silver. The Anglo-Saxon word *lead* is of unknown origin. Lead was one of the seven metals of antiquity, having been discovered by the Egyptians about 3000 B.C. Lead was extensively used by the Romans for plumbing applications and in pewter for drinking vessels and utensils. In the present day, approximately 80% of the lead produced is used for automobile batteries. See *Technical Note 6.*

lead. (1) The axial advance of a helix in one complete turn. (2) The slight bevel at the outer end of a face cutting edge of a face mill.

lead angle. In cutting tools, the helix angle of the flutes.

lead burning. A misnomer for the welding of lead.

leaded brass. An alloy of copper and zinc with a lead addition that gives excellent machinability.

lead-free brass. A brass as defined by California Assembly Bill AB 1953 that contains not more than 0.25% Pb. The

---

TECHNICAL NOTE 6

Lead and Lead Alloys

LEAD was one of the first metals known to man, with the oldest lead artifact dating back to about 3000 B.C. All civilizations, beginning with the ancient Egyptians, Assyrians, and Babylonians, have used lead for many ornamental and structural purposes. Pipe was one of the earliest applications of lead. The Romans produced 15 standard sizes of water pipe in regular 3 m (10 ft) lengths. Presently, battery applications constitute more than 80% of lead alloy use.

Although there are at least 60 known lead-containing minerals, by far the most important is galena (PbS), which is smelted and refined to produce 99.99% pig lead. Recycling of scrap lead (from batteries, lead sheet, and cable sheathing) is also a major source, providing more than half of the lead used in the United States. Considerable tonnages of scrap solder and bearing materials are also recovered and used again.

The properties of lead that make it useful in a wide variety of applications are density, malleability, lubricity, flexibility, electrical conductivity, and coefficient of thermal expansion, all of which are quite high; and elastic modulus, elastic limit, strength, hardness, and melting point, all of which are quite low. Lead also has good resistance to corrosion under a wide variety of conditions. Lead is easily alloyed with many other metals and casts with little difficulty.

TECHNICAL NOTE 6 (*continued*)

Compositions of selected lead alloys for battery grids

UNS designation	Composition, %						
	As max	Ag max	Ca	Pb	Sb	Sn	Other
Calcium-lead alloys							
L50760	0.0005	0.001	0.06–0.08	bal	0.0005 max	0.0005 max	(a)
L50770	0.0005	0.001	0.10 nom	bal	0.0005 max	0.0005 max	(a)
L50775	0.0005	0.001	0.08–0.11	bal	0.0005 max	0.2–0.4	(a)
L50780	0.0005	0.001	0.08–0.11	bal	0.0005 max	0.4–0.6	(a)
L50790	0.0005	0.001	0.08–0.10	bal	0.0005 max	0.9–1.1	(a)
Antimony-lead alloys							
L52760	0.18 nom	...	...	bal	2.75 nom	0.2 nom	...
L52765	0.3 nom	...	...	bal	2.75 nom	0.3 nom	...
L52770	0.15 nom	...	...	bal	2.9 nom	0.3 nom	...
L52840	0.15 nom	...	...	bal	2.9 nom	0.3 nom	...

(a) 0.005% max Bi and 0.0005 % max each for Cu, Zn, Cd, Ni, and Fe

The most significant applications of lead and lead alloys are lead-acid storage batteries (in the grid plates, posts, and connector straps), ammunition, cable sheathing, and building construction materials, such as sheet, pipe, and tin-lead solders. Other important applications include counterweights and cast products, such as bearings, ballast, gaskets, type metal, terneplate, and foil. Lead in various forms and combinations is also used as a material for controlling sound and mechanical vibrations and shielding against x-rays and gamma rays. In addition, lead is used as an alloying element in steel and copper alloys to improve machinability, and it used in fusible alloys. Substantial amounts of lead are also used in the form of lead compounds, including tetraethyl and tetramethyl lead used as antiknock compounds in gasoline engines, litharge (PbO) used in glasses, and various corrosion-inhibiting lead pigments, such as red lead (Pb_3O_4).

Because lead presents a health hazard (toxicity), it should not be used to conduct very soft water for drinking, nor should it come into contact with foods. Inhalation of lead dust and fumes should be avoided, and paints containing lead should be removed from structures.

Selected References

- A.W. Worcester and J.T. O'Reilly, Lead and Lead Alloys, *Metals Handbook*, 10th ed., Vol 2, ASM International, 1990, p 543–556
- J.F. Smith, Corrosion of Lead and Lead Alloys, *Metals Handbook*, 9th ed., Vol 13, ASM International, 1987, p 784–792

intent of the bill was to reduce the amount of lead in specific plumbing fixtures.

lead proof. A casting of the die impression made to confirm the exactness of the impression. Also called *die proof.*

leakage field. The magnetic field that leaves or enters a magnetized part at a magnetic pole.

ledeburite. The eutectic of the iron-carbon system, the constituents of which are *austenite* and *cementite.* The austenite decomposes into *ferrite* and cementite on cooling below Ar_1.

leg of fillet weld. (1) Actual: The distance from the root of the joint to the toe of the fillet weld (Fig. 36). (2) Nominal: The length of a side of the largest right triangle that can be inscribed in the cross section of the weld.

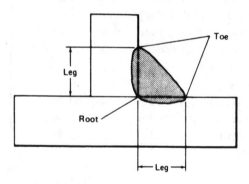

Fig. 36 Leg of a fillet weld

leveler lines. Lines on sheet or strip running transverse to the direction of *roller leveling,* which may be seen on stoning or light sanding after leveling (but before drawing). Usually can be removed by moderate stretching.

leveling. Flattening of rolled sheet, strip, or plate by reducing or eliminating distortions. See also *stretcher leveling* and *roller leveling.*

leveling action. Action exhibited by a plating solution yielding a plated surface smoother than the basis metal.

levigation. (1) Separation of fine powder from coarser material by forming a suspension of the fine material in a liquid. (2) A means of classifying a material as to particle size by the rate of settling from a suspension.

levitation melting. An *induction melting* process in which the metal being melted is suspended by the electromagnetic field and is not in contact with a container.

light drawn. An imprecise term, applied to drawn products such as wire and tubing, that indicates a lesser amount of cold reduction than for *hard drawn* products.

lightly coated electrode. A filler-metal electrode used in arc welding, consisting of a metal wire with a light coating, usually of metal oxides and silicates, applied subsequent to the drawing operation primarily for stabilizing the arc. Contrast with *covered electrode.*

light metal. One of the low-density metals, such as aluminum, magnesium, titanium, beryllium, or their alloys.

limiting current density. The maximum current density that can be used to obtain a desired electrode reaction without undue interference such as from *polarization.*

lineage structure. (1) Deviation from perfect alignment of parallel arms of a columnar dendrite as a result of interdendritic shrinkage during solidification from a liquid. This type of deviation may vary in orientation, from one area to another, from a few minutes to as much as two degrees of arc. (2) A type of substructure consisting of elongated subgrains.

linear elastic fracture mechanics. A method of fracture analysis that can determine the stress (or load) required to induce fracture instability in a structure containing a cracklike flaw of known size and shape. See also *stress-intensity factor.*

linear strain. The change per unit length due to force in an original linear dimension. An increase in length is considered positive. See also *strain.*

liner. (1) The slab of coating metal that is placed on the core alloy and is subsequently rolled down to clad sheet as a composite. (2) In extrusion, a removable alloy steel cylindrical chamber, having an outside longitudinal taper firmly positioned in the container or main body of the press, into which the billet is placed for extrusion.

lip. For a *milling cutter,* the material included between a relieved land and a tooth face.

lip angle. (1) For a *milling cutter,* the included angle between a tooth face and a relieved land. (2) See the figure accompanying the term *single-point tool.*

liquation. The partial melting of an alloy, usually as a result of *coring* or other compositional heterogeneities.

liquation temperature. The lowest temperature at which partial melting can occur in an alloy that exhibits the greatest possible degree of segregation.

liquid honing. Producing a finely polished finish by directing an air-injected chemical emulsion containing fine abrasives against the surface to be finished.

liquid penetrant inspection. A type of nondestructive inspection that locates discontinuities that are open to the surface of a metal by first allowing a penetrating dye or fluorescent liquid to infiltrate the discontinuity, removing the excess penetrant, and then applying a developing agent that causes the penetrant to seep back out of the discontinuity and register as an indication. Suitable for both ferrous and nonferrous materials, inspection is limited to the detection of open surface discontinuities in nonporous solids.

liquid phase sintering. *Sintering* a powder metallurgy compact under conditions that maintain a liquid metallic phase within the compact during all or part of the sintering schedule. The liquid phase may be derived from a component of the green compact or may be infiltrated into the compact from an outside source.

liquid shrinkage. The reduction in volume of liquid metal as it cools to the liquidus. See also *casting shrinkage.*

liquidus. In a *constitution* or *equilibrium diagram,* the locus of points representing the temperatures at which the various compositions in the system begin to freeze on cooling or finish melting on heating. See also *solidus.*

liquor finish. A smooth, bright finish characteristic of wet-drawn wire. Formerly produced by using liquor from fermented grain mash as a drawing lubricant.

live center. A lathe or grinder center that holds, yet rotates with, the work. It is used in either the headstock or tailstock of a machine to prevent wear and to reduce the driving torque.

loading. (1) In cutting, building up of a cutting tool back of the cutting edge by undesired adherence of material removed from the work. (2) In grinding, filling the pores of a grinding wheel with material from the work, usually resulting in a decrease in production and quality of finish. (3) In powder metallurgy, filling of the die cavity with powder.

loam. A molding material consisting of sand, silt, and clay, used over brickwork or other structural backup material for making massive castings, usually of iron or steel.

local action. Corrosion due to the action of "local cells"—that is, *galvanic cells* resulting from inhomogeneities between adjacent areas on a metal surface exposed to an *electrolyte.*

local cell. A *galvanic cell* resulting from inhomogeneities between areas on a metal surface in an *electrolyte.* The inhomogeneities may be of physical or chemical nature in either the metal or its environment.

local current density. Current density at a point or on a small area.

localized precipitation. Precipitation from a supersaturated solid solution similar to *continuous precipitation,* except that the precipitate particles form at preferred locations, such as along slip planes, grain boundaries, or incoherent twin boundaries.

lock. In forging, a condition in which the flash line is not entirely in one plane. When two or more plane changes occur, it is called a compound lock. When a lock is placed in the die to compensate for die shift caused by a steep lock, it is called a counterlock.

logarithmic decrement (log decrement). The natural logarithm of the ratio of successive amplitudes of vibration of a member in free oscillation. It is equal to one-half the specific damping capacity.

longitudinal direction. The direction parallel to the direction of maximum elongation in a worked metal; the principal direction of flow in a worked metal.

longitudinal field. A magnetic field that extends within a magnetized part from one or more poles to one or more other poles, and that is completed through a path external to the part.

long transverse. See *transverse*: Literally, "across," usually signifying a direction or plane perpendicular to the direction of working. In rolled plate or sheet, the direction across the width often is called long transverse, and the direction through the thickness, short transverse.

looping mill. An arrangement of hot rolling stands such that a hot bar, while being discharged from one stand, is fed into a second stand in the opposite direction.

loose metal. An area in a formed panel that is not stiff enough to hold its shape; may be confused with *oil canning.*

lost wax process. An *investment casting* process in which a wax pattern is used.

lot. A finite quantity of a given product manufactured under production conditions that are considered uniform. Often used to describe a finite quantity of product submitted for inspection as a single group. For a bulk product (such as a chemical or powdered metal), the term *batch* is often used synonymously with lot.

low-alloy steels. A category of ferrous alloys that exhibit mechanical properties superior to plain carbon steels as the result of alloying elements such as chromium, nickel, and molybdenum. The total alloy content can range from 2.07% and up.

low brass. A copper-zinc alloy containing 20% Zn and having a light golden color and excellent ductility; used for flexible metal hoses and metal bellows.

lower punch. The lower part of a die, which forms the bottom of the die cavity and which may or may not move in relation to the die body; usually movable in a forging die.

low-hydrogen electrode. A covered arc welding electrode that provides an atmosphere around the arc and molten weld metal that is low in hydrogen.

low-residual-phosphorous copper. Deoxidized copper with residual phosphorous present in amounts (usually 0.004 to 0.012%) generally too small to decrease appreciably the electrical conductivity of the copper.

low shaft furnace. A short shaft-type blast furnace used to produce pig iron and ferroalloys from low-grade ores, using low-grade fuel. The air blast often is enriched with oxygen. Also used for making a variety of other products such as alumina, cement-making slags, and ammonia synthesis gas.

lubricant. Any substance used to reduce friction or wear between two surfaces in relative motion.

Lüders lines. Elongated surface markings or depressions caused by localized plastic deformation that results from discontinuous (inhomogeneous) yielding. Also known as Lüders bands, Hartmann lines, Piobert lines, and *stretcher strains.*

luster finish. A bright as-rolled finish, produced on ground metal rolls; it is suitable for decorative painting or plating, but usually must undergo additional surface preparation after forming.

lute. (1) A mixture of fireclay used to seal cracks between a crucible and its cover, or between container and cover when heat is to be applied. (2) To seal with clay or other plastic material.

machinability. The relative ease of machining metal.

machinability index. A relative measure of the machinability of an engineering material under specified standard conditions.

machine forging. Forging performed in upsetters or horizontal forging machines.

machine welding. Welding with equipment that performs under the continual observation and control of a welding operator. The equipment may or may not load and unload the work. Compare with *automatic welding*.

machining. Removing material from a metal part, usually using a cutting tool, and usually using a power-driven machine.

machining allowance. The amount of excess metal surrounding the intended final configuration of a formed part. Also known as *finish allowance*, *forging envelope*, or *cleanup allowance*.

machining stress. *Residual stress* caused by machining.

macroetching. *Etching* a metal surface to accentuate gross structural details (such as grain flow, segregation, porosity, or cracks) for observation by the unaided eye or at magnifications to approximately 25×.

macrograph. A graphic reproduction of the surface of a prepared specimen at a magnification not exceeding 25×. When photographed, the reproduction is known as a *photomacrograph*.

macroscopic. Visible at magnifications at or below 25×.

macroscopic stress. Residual stress that varies from tension to compression in a distance (presumably many times the grain size) comparable to the gage length in ordinary strain measurements, and hence, detectable by x-ray or dissection methods. Also known as *macrostress*.

macroshrinkage. Isolated, clustered, or interconnected voids in a casting that are detectable macroscopically. Such voids are usually associated with abrupt changes in section size and are caused by feeding that is insufficient to compensate for solidification shrinkage.

macrostress. Residual stress that varies from tension to compression in a distance (presumably many times the grain size) comparable to the gage length in ordinary strain measurements, and hence, detectable by x-ray or dissection methods. Also known as *macroscopic stress*.

macrostructure. The structure of metals as revealed by macroscopic examination of the etched surface of a polished specimen.

magnesium. A chemical element having atomic number 12, atomic weight 24, and the symbol Mg, for Magnesia, a region in Thessaly, Greece. Compounds of magnesium had been known for many years. It was recognized as an element by Joseph Black in 1755 and first isolated by Sir Humphry Davy in England in 1808. With a specific gravity of 1.74, magnesium has approximately two-thirds the density of aluminum and is widely used where weight reduction is important, especially in aircraft, cameras, and cell phones. It is highly pyrophoric in powder and foil form and is used in flares, fireworks, and photographic flash bulbs. It is used in the form of anodes for the corrosion protection of steel. The major use of magnesium is as an alloying element in aluminum, where it increases strength while reducing weight. See *Technical Note 7*.

magnetically hard alloy. A ferromagnetic alloy capable of being magnetized permanently because of its ability to retain induced magnetization and magnetic poles after removal of externally applied fields. An alloy with high coercive force. The name is based on the fact that the quality of the early permanent magnets was related to their hardness.

magnetically soft alloy. A ferromagnetic alloy that becomes magnetized readily upon application of a field and that returns to practically a nonmagnetic condition when the field is removed. An alloy with the properties of high magnetic permeability, low coercive force, and low magnetic hysteresis loss.

magnetic-analysis inspection. A nondestructive method of inspection to determine the existence of variations in magnetic flux in ferromagnetic materials of constant cross section, such as might be caused by discontinuities and variations in hardness. The variations are usually indicated by a change in pattern on an oscilloscope.

magnetic-particle inspection. A nondestructive method of inspection for determining the existence and extent of surface cracks and similar imperfections in ferromagnetic materials. Finely divided magnetic particles, applied to the magnetized part, are attracted to and outline the pattern of any magnetic-leakage fields created by discontinuities.

magnetic pole. The area on a magnetized part at which the magnetic field leaves or enters the part. It is a point of maximum attraction in a magnet.

magnetic separator. A device used to separate magnetic from less magnetic or nonmagnetic materials. The crushed material is conveyed on a belt past a magnet.

magnetic writing. In magnetic-particle inspection, a *false indication* caused by contact between a magnetized part and another piece of magnetic material.

magnetite wheel. A grinding wheel bonded with magnesium oxychloride.

magnetizing force. A force field, resulting from the flow of electric currents or from magnetized bodies, that produces magnetic induction.

Magnesium and Magnesium Alloys

MAGNESIUM is a silvery white metal that is valued chiefly for lightweight components (pure magnesium has a density of approximately 1.7 g/cm^3, versus 2.7 g/cm^3 for aluminum and 7.8 g/cm^3 for steel). Magnesium is produced commercially by the electrolysis of a fused chloride (from brine wells or from seawater) or extracted from mineral ore (most commonly dolomite).

Two major magnesium alloy systems are available. The first includes alloys containing 2–10% Al, combined with minor additions of zinc and manganese. The mechanical properties of these alloys are good to 95–120°C (200–250 °F). Beyond this, the properties deteriorate rapidly with increasing temperature. The second group consists of magnesium alloyed with various elements (rare earths, zinc, thorium, silver, etc.) except aluminum, all containing a small but effective zirconium content (~0.7%) that imparts a fine grain structure and thus improved mechanical properties. These alloys generally also possess much better elevated-temperature properties.

Typical magnesium alloy systems and nominal compositions

Alloy	Element, %(a)							Product form (b)
	Al	Zn	Mn	Ag	Zr	Th	Re	
AM60	6	...	0.2	...	...	...	...	C
AZ31	3	1	0.2	...	...	...	...	W
AZ61	6	1	0.2	...	...	...	...	W
AZ63	6	3	0.2	...	...	...	...	C
AZ80	8	0.5	0.2	...	...	...	...	C, W
AZ91	9	1	0.2	...	...	...	...	C
EZ33	...	2.5	...	...	0.5	...	2.5	C
ZM21	...	2	1	...	...	...	...	W
HK31	...	0.1	...	...	0.5	3	...	C, W
HZ32	...	2	...	...	0.5	3	...	C
QE22	...	...	...	2.5	0.5	...	2	C
QH21	...	...	...	2.5	0.5	1	1	C
ZE41	...	4.5	...	...	0.5	...	1.5	C
ZE63	...	5.5	...	...	0.5	...	2.5	C
ZK40	...	4.0	...	...	0.5	...	...	C, W
ZK60	...	6.0	...	...	0.5	...	...	C, W

(a) For details, see alloying specifications. (b) C, castings; W, wrought products

Magnesium alloys are produced in both cast and wrought forms. Magnesium alloys castings can be produced by nearly all of the conventional casting methods, sand, permanent and semipermanent mold, and shell, investment, and die casting, the latter being the highest in volume. Wrought magnesium alloys are produced as bars, billets, shapes, wire, sheet, plate, and forgings.

TECHNICAL NOTE 7 (*continued*)

Magnesium and magnesium alloys are used in a wide variety of structural and nonstructural applications. Structural applications include automotive, industrial, materials handling, commercial, and aerospace equipment. However, it is with non-structural applications that magnesium finds its greatest use. It is used as an alloying element in alloys of aluminum (the single largest application for magnesium), zinc, lead, and other nonferrous metals. It is used as an oxygen scavenger and desulfur-izer in the manufacture of nickel and copper alloys; as a desulfurizer in the iron and steel industry; and as a reducing agent in the production of beryllium, titanium, zirconium, hafnium, and uranium. Magnesium powders are used to manufacture Grignard reagents, which are organometallic halides used in organic synthesis to produce pharmaceuticals, perfumes, and other chemicals. Magnesium powder also finds some use in pyrotechnics, both as pure magnesium and alloyed with 30% or more aluminum. As a galvanic anode, magnesium provides effective corrosion pro-tection for water heaters, underground pipelines, ship hulls, and ballast tanks. Small, lightweight, high-current-output primary batteries also use magnesium as the anode. Gray iron foundries use magnesium and magnesium-containing alloys as ladle addi-tion agents introduced just before the casting is poured. The magnesium makes the graphite particles nodular and greatly improves the toughness and ductility of the cast iron.

Selected References

- S. Housh, B. Mikucki, and A. Stevenson, Selection and Application of Magnesium and Magnesium Alloys, *Metals Handbook*, 10th ed., Vol 2, ASM International, 1990, p. 455–479
- H. Proffitt, Magnesium and Magnesium Alloys Castings, *Metals Handbook*, 9th ed., Vol 15, ASM International, 1988, p 798–810
- J. Hillis et al., Corrosion of Magnesium and Magnesium Alloys, *Metals Handbook*, 9th ed., Vol 13, ASM International, 1987, p 740–754

---

magnetostriction. The characteristic of a material that is manifest by strain when it is subjected to a magnetic field; or the inverse. Some iron-nickel alloys expand; pure nickel contracts.

malleability. The characteristic of metals that permits deformation in compression without rupture.

malleable cast iron. A cast iron made by prolonged annealing of *white cast iron* in which decarburization or graphitization, or both, take place to eliminate some or all of the cementite. The graphite is in the form of temper carbon. If decarburization is the predominant reaction, the product will ex-hibit a light fracture surface, hence "white-heart malleable"; otherwise, the fracture surface will be dark, hence "blackheart malleable." Ferritic malleable iron has a predominantly ferritic matrix (Fig. 37); pearlitic malleable may contain pearl-ite, spheroidite, or tempered martensite

(Fig. 38), depending on heat treatment and desired hardness. The chemical composition of malleable iron generally conforms to the ranges given in Table 10.

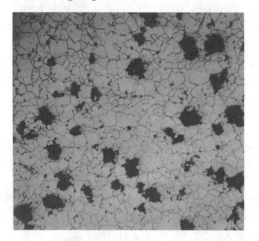

Fig. 37 Structure of annealed ferritic malleable iron showing temper carbon in ferrite. Original magnification: 100×

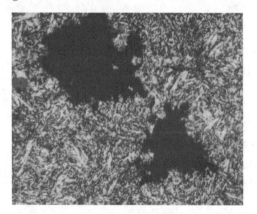

Fig. 38 Pearlitic malleable iron showing graphic nodules (black) in tempered martensite. Original magnification: 500×

Table 10 Typical compositions for malleable iron

Element	Composition, %	
	Ferritic	Pearlitic
Total carbon	2.2–2.9	2.0–2.9
Silicon	0.9–1.9	0.9–1.9
Manganese	0.2–0.6	0.2–1.3
Sulfur	0.02–0.2	0.05–0.2
Phosphorus	0.02–0.2	0.02–0.2

malleable iron. Cast iron that has been toughened by gradual heating or slow cooling. A less common name for wrought iron. Also called *malleable cast iron.* See *wrought iron.*

malleablizing. Annealing *white cast iron* in such a way that some or all of the combined carbon is transformed into graphite or, in some instances, so that part of the carbon is removed completely.

mandrel. (1) A blunt-ended tool or rod used to retain the cavity in a hollow metal product during working. (2) A metal bar around which other metal may be cast, bent, formed, or shaped. (3) A shaft or bar for holding work to be machined. (4) A form, such as a mold or matrix, used as a cathode in electro-forming.

manganese. A chemical element having atomic number 25, atomic weight 55, and the symbol Mn. It was named for *magnes,* the Latin word for magnet, because of the magnetic properties of pyrolusite in which manganese was found. The element was identified in 1774 by Carl Wilhelm Scheele and isolated the same year by Johan Gottlieb Gahn. There are no uses for pure manganese, which is hard and brittle, but it is highly useful as an alloying element. It is a deoxidizer and desulfurizer of steel, where approximately 1% is usually present. In certain austenitic stainless, 3–5% Mg is used as a partial substitute for nickel. It also is used in amounts of approximately 12% to produce austenitic steels commonly known as Hadfield's manganese steels. Manganese is used as a desulfurizer in copper alloys and, when added to brass, imparts a pleasing color while increasing corrosion resistance and strength.

Manganese also forms useful alloys with aluminum, magnesium, and nickel.

manganese brass. A brass used in making golden dollar coins in the United States. It is approximately 70% Cu, 29% Zn, and 1.3% Mg.

Mannesmann mill. Mill used in the *Mannesmann process.*

Mannesmann process. A process used for piercing tube billets in making seamless tubing. The billet is rotated between two heavy rolls mounted at an angle and is forced over a fixed mandrel.

manual welding. Welding wherein the entire welding operation is performed and controlled by hand.

maraging. A precipitation-hardening treatment applied to a special group of iron-base alloys to precipitate one or more intermetallic compounds in a matrix of essentially carbon-free martensite. Note: The first developed series of maraging steels contained, in addition to iron, more than 10% Ni, and one or more supplemental hardening elements. In this series, aging is done at 480 °C (900 °F).

margin. The cylindrical portion of the *land* of a drill that is not cut away to provide clearance.

marquenching. See *martempering.*

martempering. (1) A hardening procedure in which an austenitized ferrous workpiece is quenched in an appropriate medium whose temperature is maintained just above the martensite start temperature (M_s) of the workpiece, held in the medium until its temperature is uniform throughout—but not long enough to permit bainite to form—and then cooled in air. The treatment is frequently followed by tempering. (2) When the process is applied to carburized material, the

controlling M_s is that of the case. This variation of the process is frequently called *marquenching.*

martensite. A generic term for microstructures formed by diffusionless phase transformation in which the parent and product phases have a specific crystallographic relationship. Martensite is characterized by an acicular pattern in the microstructure in both ferrous and nonferrous alloys. In alloys where the solute atoms occupy interstitial positions in the martensitic lattice (such as carbon in iron), the structure is hard and highly strained; but where the solute atoms occupy substitutional positions (such as nickel in iron), the martensite is soft and ductile. The amount of high-temperature phase that transforms to martensite on cooling depends to a large extent on the lowest temperature attained, there being a rather distinct beginning temperature (M_s) and a temperature at which the transformation is essentially complete (M_f).

martensite range. The temperature interval between the martensite start (M_s) and the martensite finish (M_f) temperatures.

martensitic transformation. A reaction that takes place in some metals on cooling, with the formation of an acicular structure called *martensite.*

mash resistance seam welding. Resistance *seam welding* in which the weld is made in a lap joint, the thickness at the lap being reduced plastically to approximately the thickness of one of the lapped parts.

masking tape. A tape used as a *resist* for stopping-off purposes.

master alloy. An alloy, rich in one or more desired addition elements, that is added to a melt to raise the percentage of a desired constituent.

match. A condition in which a point in one forging die half is aligned properly with the corresponding point in the opposite die half within specified tolerance.

matched edges. Two edges of a forging die face that are machined at exactly 90° to each other, and from which all dimensions are taken in laying out the die impression and aligning the dies in the forging equipment. Same as *match lines*.

match lines. Two edges of a forging die face that are machined at exactly 90° to each other, and from which all dimensions are taken in laying out the die impression and aligning the dies in the forging equipment. Also known as *matched edges*.

match plate. A plate of metal or other material on which patterns for metal casting are mounted (or formed as an integral part) so as to facilitate molding (Fig. 39). The pattern is divided along its *parting plane* by the plate.

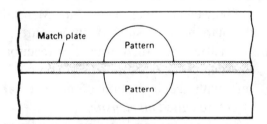

Fig. 39 Metal match plate

matrix. (1) The principal phase or aggregate in which another constituent is embedded. (2) In electroforming, a form used as a cathode.

matte. An intermediate product of *smelting;* an impure metallic sulfide mixture made by melting a roasted sulfide ore, such as an ore of copper, lead, or nickel.

matte dip. An etching solution used to produce a dull finish on metal.

matte finish. (1) A dull texture produced by rolling sheet or strip between rolls that have been roughened by blasting. (2) A dull finish characteristic of some electrodeposits, such as cadmium or tin. Also spelled mat finish.

McQuaid-Ehn test. A test for revealing grain size after heating into the austenitic temperature range. Eight standard McQuaid-Ehn grain sizes are used for rating structures, No. 8 being the finest, No. 1, the coarsest.

mean stress. (1) In fatigue loading, the algebraic mean of the maximum and minimum stresses in one cy cle. Also called the steady stress component. (2) In any multiaxial stress system, the algebraic mean of three *principal stresses;* more correctly called mean normal stress.

mechanical equation of state. Any equation relating stress, strain, strain rate, and temperature based on the concept that the instantaneous value of any one of these quantities is a single-valued function of the others, regardless of the prior history of the deformation.

mechanical hysteresis. Energy absorbed in a complete cycle of loading and unloading within the elastic limit and represented by the closed loop of the stress-strain curves for loading and unloading. Sometimes referred to as elastic, but more properly, mechanical.

mechanical metallurgy. The science and technology dealing with the behavior of metals when subjected to applied forces; often considered to be restricted to plastic working or shaping of metals.

mechanical plating. Plating wherein fine metal powders are peened onto the work by *tumbling* or other means.

mechanical press. A press whose slide is operated by a crank, eccentric, cam, toggle links, or other mechanical device.

mechanical properties. The properties of a metal or alloy that reveal its elastic and inelastic behavior when force is applied, thereby indicating its suitability for mechanical applications; for example, modulus of elasticity, tensile strength, elongation, hardness, and fatigue limit. Compare with *physical properties.*

mechanical testing. The determination of *mechanical properties.*

mechanical twin. A *twin* formed in a crystal by simple shear under external loading.

mechanical working. Subjecting metal to pressure, exerted by rolls, hammers, or presses, in order to change the shape or physical properties of the metal.

melting point. The temperature at which a pure metal, compound, or eutectic changes from solid to liquid; the temperature at which the liquid and the solid are in equilibrium.

melting range. The range of temperature over which an alloy other than a compound or eutectic changes from solid to liquid; the range of temperature from *solidus* to *liquidus* at any given composition on a *constitution diagram.*

melting rate. In electric arc welding, the weight or length of an electrode melted in a unit of time. Sometimes called *melt-off rate* or *burnoff* rate.

melt-off rate. See *melting rate.*

Mendeleev, Dimitri. 1834–1907. A Russian chemist, professor, and editor who formulated the Periodic Law according to which all known chemical elements were arranged in order of increasing atomic weights. Mendeleev was born in Tobolsk, Siberia, the youngest of at least 14 children. The Periodic Table of the Elements, published by Mendeleev in 1869 to limited acceptance, had gaps, but Mendeleev predicted elements would be found to fill the gaps. Mendeleev also wrote *The Principles of Chemistry,* which became a classic textbook.

mendelevium. A chemical element having atomic number 101, atomic weight 258, and the symbol Md, for Dimitri Mendeleev, the creator of the Periodic Table of the Elements. The element was first synthesized by Albert Ghiorso, Glenn Seaborg, and colleagues in 1955.

merchant mill. (obsolete) A mill, consisting of a group of stands of three rolls each arranged in a straight line and driven by one power unit, used to roll rounds, squares, or flats of smaller dimensions than would be rolled on a bar mill.

merchant quality steel bars. Merchant quality is the lowest quality of carbon steel bars. Merchant quality bars are produced to specified sizes, with appropriate control of the chemistry limits or mechanical properties for noncritical uses. Bars of this quality are usually rolled from uninspected and unconditioned billets. The size ranges are limited, and the type of steel applied is at the option of the producer, that is, rimmed, capped, semikilled, or killed steel.

The bars are designated as M1008, M1010, M1012, M1015, M1017, M1020, M1023, M1025, M1031, and M1044 in ASTM A575.

Merchant quality steel bars are produced for a wide range of uses such as structural and similar miscellaneous applications involving mild cold bending, mild hot forming, and welding as used in the production of noncritical parts for bridges, buildings, ships, agricultural implements, road building equipment, railway equipment, and general machinery. This quality is not suitable for applications that involve forging, heat treating, and cold drawing, where internal soundness is required. Internal porosity, surface seams, and other surface irregularities may be present in this quality bar.

mesh. The screen number of the finest screen of a specified standard screen scale through which almost all of the particles of a powder sample will pass. Also called mesh size.

metal. (1) An opaque lustrous elemental chemical substance that is a good conductor of heat and electricity and, when polished, a good reflector of light. Most elemental metals are malleable and ductile and are, in general, denser than the other elemental substances. (2) As to structure, metals may be distinguished from nonmetals by their atomic binding and electron availability. Metallic atoms tend to lose electrons from the outer shells, the positive ions thus formed being held together by the electron gas produced by the separation. The ability of these "free electrons" to carry an electric current, and the fact that this ability decreases as temperature increases, establish the prime distinctions of a metallic solid. (3) From the chemical viewpoint, an elemental substance whose hydroxide is alkaline. (4) An *alloy.*

metal-arc cutting. Any of a group of arc cutting processes in which metals are severed by being melted with the heat of an arc between a metal electrode and the work.

metal-arc welding. Any of a group of arc welding processes in which metals are fused together using the heat of an arc between a metal electrode and the work. Use of the specific process name is preferred.

metal inert-gas welding. *Gas metal arc welding* using an inert gas such as argon as the shielding gas.

metal leaf. Thin metal sheet, usually thinner than foil, and traditionally produced by beating rather than by rolling.

metallic bond. The principal bond between metal atoms, which arises from the increased spatial extension of valence-electron wave functions when an aggregate of metal atoms is brought close together. See *covalent bond* and *ionic bond.*

metallic glass. A noncrystalline metal or alloy, commonly produced by drastic supercooling of a molten alloy by electrodeposition or by vapor deposition. Also known as amorphous alloy.

metallizing. (1) Forming a metallic coating by atomized spraying with molten metal or by *vacuum deposition.* Also called *spray metallizing.* (2) Applying an electrically conductive metallic layer to the surface of a nonconductor.

metallograph. An optical instrument designed for both visual observation and photomicrography of prepared surfaces of opaque materials at magnifications ranging in diameter from 25 to approximately 2000×. The instrument consists of a high-intensity illuminating source, a microscope, and a camera bellows. On some instruments, provisions are made for examination of specimen surfaces with polarized light, phase contrast, oblique illumination, dark-field illumination, and customary bright-field illumination.

metallography. The science dealing with the constitution and structure of metals and alloys as revealed by the unaided eye or by such tools as low-powered magnification, optical microscopy, electron microscopy, and diffraction or x-ray techniques.

metalloid. A chemical element with properties that are in between or a mixture of those of metals and nonmetals. The metalloids, also known as semimetals, fall into a group in the periodic table of the elements separating the metals from the nonmetals. The seven elements commonly recognized as metalloids are boron, silicon, germanium, arsenic, antimony, tellurium, and polonium. Sometimes selenium and bismuth are included as metalloids.

metallurgical coke. A coke, usually low in sulfur, having a very high compressive strength at elevated temperatures; used in metallurgical furnaces not only as fuel, but also to support the weight of the charge.

metallurgy. The science and technology of metals and alloys. Process metallurgy is concerned with the extraction of metals from their ores and the refining of metals; physical metallurgy, with physical and mechanical properties of metals as affected by composition, processing, and environmental conditions; and mechanical metallurgy, with the response of metals to applied forces.

metal penetration. A surface condition in castings in which metal or metal oxides have filled voids between sand grains without displacing them.

metal spraying. Coating metal objects by spraying molten metal against their surfaces. See also *thermal spraying* and *flame spraying.*

metastable. Possessing a state of pseudo-equilibrium that has a higher free energy than that of the true equilibrium state.

M_f temperature. For any alloy system, the temperature at which martensite formation on cooling is essentially finished. See also *transformation temperature* for the definition applicable to ferrous alloys.

microalloyed steel. A low-alloy steel that conforms to a specification requiring the presence of one or more carbide, nitride, or carbonitride elements, generally in concentrations less than 0.15 mass percent, to enhance strength (ASTM A941).

microfissure. A crack of microscopic proportions.

micrograph. A graphic reproduction of the surface of a prepared specimen, usually etched, at a magnification greater than $25 \times$ the diameter. If produced by photographic means it is called a *photomicrograph* (not a microphotograph).

microhardness. The hardness of a material as determined by forcing an indenter such as a Vickers or Knoop indenter into the surface of the material under very light load; usually, the indentations are so small that they must be measured with a microscope. Capable of determining the hardnesses of different microconstituents within a structure, or of measuring steep hardness gradients such as those encountered in *case hardening.*

microprobe. An instrument for the selective analysis of a microscopic component or feature in which an electron beam bombards the point of interest in a vacuum at a given energy level. Scanning of a larger area permits determination of the distribution of selected elements. The analysis is made by measuring the wavelengths and intensities of secondary electromagnetic radiation resulting from the bombardment. Preferred term is *electron beam microprobe analyzer.*

microradiography. The technique of passing x-rays through a thin section of an alloy in contact with a fine-grained photographic film and then viewing the radiograph at 50 to 100× to observe the distribution of alloying constituents and voids.

microscopic. Visible only at a magnification greater than 25× the diameter.

microscopic stress. Residual stress that varies from tension to compression in a distance (presumably approximating the grain size) that is small compared to the gage length in ordinary strain measurements. Microscopic stress is not detectable by dissection methods but can sometimes be measured from line shift or line broadening in an x-ray diffraction pattern.

microsegregation. *Segregation* within a grain, crystal, or small particle. See also *coring.*

microshrinkage. A casting imperfection, not detectable microscopically, consisting of interdendritic voids. Microshrinkage results from contraction during solidification where the opportunity to supply filler material is inadequate to compensate for shrinkage. Alloys with wide ranges in solidification temperature are particularly susceptible.

microstress. Residual stress that varies from tension to compression in a distance (presumably approximating the grain size) that is small compared to the gage length in ordinary strain measurements.

Microstress is not detectable by dissection methods, but sometimes can be measured from line shift or line broadening in an x-ray diffraction pattern. Also known as *microscopic stress.*

microstructure. The structure of a metal as revealed by microscopic examination of the etched surface of a polished specimen.

middling. A product intermediate between concentrate and tailing and containing enough of a valuable mineral to make retreatment profitable.

migration. Movement of entities (such as electrons, ions, atoms, molecules, vacancies, and grain boundaries) from one place to another under the influence of a driving force (such as an electrical potential or a concentration gradient).

MIG welding. *Gas metal arc welding* using an inert gas such as argon as the shielding gas. Same as *metal inert-gas welding.*

mil. An English measurement equal to one thousandth of an inch (0.001 in.).

mild steel. *Carbon steel* with a maximum of approximately 0.25% C.

mill. (1) A factory in which metals are hot worked, cold worked, or melted and cast into standard shapes suitable for secondary fabrication into commercial products. (2) A production line, usually of four or more stands, for hot or cold rolling metal into standard shapes such as bar, rod, plate, sheet, or strip. (3) A single machine for hot rolling, cold rolling, or extruding metal; examples include *blooming mill, cluster mill, four-high mill,* and *Sendzimir mill.* (4) A shop term for a *milling cutter.* (5) A machine or group of machines for grinding or crushing ores and other minerals; see *ball mill, milling* (2).

mill edge. The normal edge produced in hot rolling. This edge is customarily removed when hot rolled sheets are further processed into cold rolled sheets.

Miller indices. A system for identifying planes and directions in any crystal system by means of sets of integers. The indices of a plane are related to the intercepts of that plane with the axes of a unit cell; the indices of a direction, to the multiples of lattice parameter that represent the coordinates of a point on a line parallel to the direction and passing through the arbitrarily chosen origin of a unit cell.

mill finish. A nonstandard (and typically nonuniform) surface finish on mill products that are delivered without being subjected to a special surface treatment (other than a corrosion-preventive treatment) after the final working or heat treating step.

milling. (1) Removing metal with a *milling cutter.* (2) The mechanical treatment of material, as in a *ball mill,* to produce particles or alter their size or shape, or to coat one component of a powder mixture with another.

milling cutter. A rotary cutting tool having one or more cutting elements, called teeth, which intermittently engage the workpiece and remove material by relative movement of the workpiece and cutter.

mill product. A commercial product of a *mill.*

mill scale. The heavy oxide layer formed during the hot fabrication or heat treatment of metals.

mineral dressing. The physical and chemical concentration of raw ore into a product from which a metal can be recovered at a profit.

minimized spangle. A hot dip galvanized coating of very small grain size, which makes the *spangle* less visible when the part is subsequently painted.

minimum bend radius. The minimum radius over which a metal product can be bent to a given angle without fracture.

minus sieve. The portion of a sample of a granular substance (such as metal powder) that passes through a standard sieve of a specified number. Contrast with *plus sieve.*

mischmetal. A natural mixture of rare-earth elements (having atomic numbers 57 through 71) in metallic form. It contains approximately 50% Ce, the remainder being principally lanthanum and neodymium.

mismatch. The error in register between forged surfaces formed by opposing dies (Fig. 40).

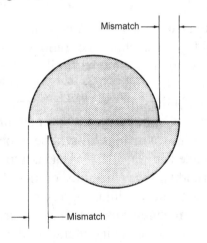

Fig. 40 Schematic of mismatch

mispickel. A mineral that is the most common ore of arsenic. It is a sulfide of iron and copper. Also known as arsenical pyrite and arsenopyrite.

misrun. A casting defect characterized by the casting not being fully formed, resulting from the metal solidifying before the mold is filled.

mixed dislocation. Any combination of an edge dislocation and a screw dislocation, the two basic types of a *dislocation,* which is a linear imperfection in a crystalline array of atoms.

mixing. In powder metallurgy, the thorough intermingling of powders of two or more different materials (not *blending*).

mixing chamber. The part of a torch or furnace burner in which gases are mixed.

modification. The treatment of molten hypoeutectic (8–13% Si) or hypereutectic (13–19% Si) aluminum-silicon alloys to improve the mechanical properties of the solid alloy by refinement of the size and distribution of the silicon phase. It involves additions of small percentages of sodium, strontium, or calcium (hypoeutectic alloys) or of phosphorus (hypereutectic alloys).

modulus of elasticity. A measure of the rigidity of metal. The ratio of stress, below the proportional limit, to the corresponding strain. Specifically, the modulus obtained in tension or compression is Young's modulus, stretch modulus, or modulus of extensibility; the modulus obtained in torsion or shear is modulus of rigidity, shear modulus, or modulus of torsion; the modulus covering the ratio of the mean normal stress to the change in volume per unit volume is the bulk modulus. The tangent modulus and secant modulus are not restricted within the proportional limit; the former is the slope of the stress-strain curve at a specified point; the latter is the slope of a line from the origin to a specified point on the stress-strain curve. Also called *elastic modulus* and *coefficient of elasticity.*

modulus of rigidity. The modulus obtained in torsion or shear. Also called

shear modulus and modulus of torsion. See *modulus of elasticity.*

modulus of rupture. The nominal stress at a fracture in a bend test or torsion test. In bending, the modulus of rupture is the bending moment at fracture divided by the section modulus. In torsion, the modulus of rupture is the torque at fracture divided by the polar section modulus.

modulus of strain hardening. The rate of change of true *stress* with respect to true *strain* in the plastic range. See preferred term *rate of strain hardening.*

Mohs scale. A scratch hardness rating, developed by German geologist/mineralogist Frederich Mohs, for determining comparative hardness using 10 standard minerals from diamond (the hardest, number 10) to talc (the softest, number 1).

mold. (1) A form made of sand, metal, or other material that contains a cavity into which molten metal is poured to produce a casting of definite shape and outline. (2) A *die.*

molding machine. A machine for making sand molds by mechanically compacting sand around a pattern.

molding press. A press used to form powder metallurgy *compacts.*

mold jacket. Wood or metal form that is slipped over a sand mold for support during the pouring of a casting (Fig. 41).

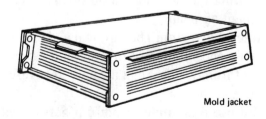

Mold jacket

Fig. 41 A mold jacket

mold wash. An aqueous or alcoholic emulsion or suspension of various materials used to coat the surface of a casting mold cavity.

molybdenum. A chemical element having atomic number 42, atomic weight 96, and the symbol Mo. It was named from the Greek *molybdos*, meaning lead, because any dark, heavy metal that marked paper was called lead in ancient times. Carl Wilhelm Scheele identified molybdenum in 1778, and Peter Jakob Hjelm first isolated it about 1782. The largest and best known deposit of molybdenum sulfide ore is in Colorado, where a mountain is gradually being removed to recover the ore that contains approximately 8.62 kg (19 lb) of metallic material per 907.18 kg (1 ton) of ore. The largest use of molybdenum is as an alloying element for steel. The hardenability of carbon and low-alloy steel is increased by adding as little as 0.2% of the metal. Molybdenum on the order of 5% is used in most high-speed steels where hot hardness is required. When approximately 3% Mo is added to austenitic stainless steels, the resistance of the metal to corrosive acids and marine atmospheres is enhanced considerably. Molybdenum is also added to cast irons to improve strength and corrosion resistance.

Mond process. A process for extracting and purifying nickel. The main features consist of forming nickel carbonyl by reaction of finely divided reduced metal with carbon monoxide, then decomposing the nickel carbonyl to deposit purified nickel as small nickel pellets.

monotectic. An isothermal reversible reaction in a binary system, in which a liquid on cooling decomposes into a second liquid of a different composition and a solid. It differs from a *eutectic* in that only one of the two products of the reaction is below its freezing range.

monotron hardness test. (obsolete) A method of determining the *indentation hardness* of a metal by measuring the load required to force a spherical penetrator into the metal to a specified depth.

monotropism. The ability of a solid to exist in two or more forms (crystal structures), but in which one form is the stable modification at all temperatures and pressures. *Ferrite* and *martensite* are a monotropic pair below the temperature at which *austenite* begins to form: Ac_1 in steels, for example. Also spelled "monotrophism."

mosaic structure. In crystals, a substructure in which neighboring regions have only slightly differing orientations.

M_s temperature. For any alloy system, the temperature at which martensite starts to form on cooling. See *transformation temperature* for the definition applicable to ferrous alloys.

mulling. Mixing sand and clay particles with water by kneading, rolling, rubbing, or stirring to develop suitable properties for molding.

multiaxial stresses. Any stress state in which two or three principal stresses are not zero.

multiple. A piece of stock cut from a longer *mill product* to provide the exact amount of material needed for a single workpiece.

multiple heat. Two or more primary *heats,* in whole or in part, combined in a common ladle or in a common non-oscillating mold.

multiple-impulse welding. Spot, projection, or upset welding with more than one impulse of current during a single machine cycle. Sometimes called *pulsation welding*.

multiple-pass weld. A weld made by depositing filler metal with two or more successive passes.

multiple-slide press. A press with individual slides, built into the main slide or connected to individual eccentrics on the main shaft, that can be adjusted so as to give variations in length of stroke and in timing.

multiple spot welding. Spot welding in which several spots are made during one complete cycle of the welding machine.

Muntz metal. A commonly used brass alloy with approximately 60% Cu and 40% Zn. Also called yellow metal and 60/40.

Dictionary of Metals
H.M. Cobb, editor

native metal. (1) A deposit in the earth's crust consisting of uncombined metal. (2) The metal in such a deposit.

natural aging. The spontaneous aging of a supersaturated solid solution at room temperature. See also *aging*, and compare with *artificial aging*.

natural strain. While *strain* is a measure of the relative change in the size or shape of a body, natural strain (or true strain) is the natural logarithm of the ratio of the length at the moment of observation to the original gage length.

naval brass. A copper alloy, similar to admiralty brass, with 40% Zn and 1% Sn.

necking. (1) Reducing the cross-sectional area of a metal in a localized area by stretching. (2) Reducing the diameter of a portion of the length of a cylindrical shell or tube.

necking down. The localized reduction in the area of a specimen during tensile deformation.

necking strain. The strain occurring prior to the beginning of localization of strain (necking); the strain to maximum load in the tension test. Also known as *uniform strain.*

negative rake. A tooth face in rotation whose cutting edge lags the surface of the tooth face. See *face mill* (Fig. 26).

neodymium. A chemical element having atomic number 60, atomic weight 144, and the symbol Nd. Its name is from the Greek *neo,* meaning new, and *didymos,* meaning twin, because didymium (Greek for twin element) is a mixture of the elements neodymium and praseodymium. The discovery of neodymium was made in 1885 by Austrian chemist Carl Auer von Welsbach. One of the rare earth metals, the principal use of neodymium is in permanent magnets such as in hybrid electric car motors.

network structure. A metallic structure in which one constituent occurs primarily at the grain boundaries, thus partially or completely enveloping the grains of the other constituents.

Neumann band. A *mechanical twin*—a *twin* formed in a crystal by simple shear under external loading—in ferrite. Also called Neumann line. Neumann lines are named after Johann G. Neumann, who discovered them in 1848 in the iron meteorite Braunau.

neutral flame. In gas welding, a flame in which there is no excess of either fuel or oxygen in the inner flame. Oxygen from ambient air is used to complete the combustion of CO_2 and H_2 produced in the inner flame. The flame is neither reducing nor oxidizing in its effect on the workpiece.

neutron. An elementary nuclear particle that has a mass approximately the same as that of a hydrogen atom and that is electrically neutral; its mass is 1.67495×10^{-27} kg.

neutron embrittlement. *Embrittlement* resulting from bombardment with neutrons, usually encountered in metals that have been exposed to a neutron flux in the core of a reactor. In steels, neutron embrittlement is evidenced by a rise in the ductile-to-brittle transition temperature.

nibbling. The contour cutting of sheet metal by use of a rapidly reciprocating punch that makes numerous small cuts.

nickel. A chemical element having atomic number 28, atomic weight 59, and the symbol Ni, for the German *kupfer-nickel,* meaning copper nickel, the ore in which nickel was discovered. The medieval German miners who found the red mineral, however, had expected to find copper and, when they were unable to extract any copper from it, believed they had been tricked by a devil of German mythology, Nickel. (The term *Old Nick* is associated with this name.) In 1751, Baron Axel Friedrik Cronstedt, a young Swedish mineralogist, extracted nickel from kopparnickel, which means copper demon. For more information on nickel see *Technical Note 8.*

niobium. Also known as *columbium.* A chemical element having atomic number 41, atomic weight 93, and the symbol Nb. Around 1734, John Winthrop the Younger, the first governor of Connecticut, discovered a new mineral that he called columbite (Columbia is a synonym for America), a sample of which was sent to the British Museum. The sample finally was examined in 1801 by the British chemist and manufacturer, Charles Hatchett, who discovered in it a new element that he called columbium. Columbium was rediscovered and renamed niobium around 1844 by Heinrich Rose. The principal ore is columbite, which is often associated with tantalite. The metal is exceptionally corrosion resistant, being attacked only by aqua regia. Small additions of columbium to the 18%Cr–8%Ni alloy are made to create the stable type 347 alloy that is not subject to carbide precipitation and intergranular corrosion. The addition of up to 5% Nb improves the creep strength of nickel-base high-temperature alloys. The metal is used as a cladding (canning) metal for fuel elements in nuclear reactors and is a principal element in *superconductor* alloys.

nitriding. Introducing nitrogen into the surface layer of a solid ferrous alloy by holding at a suitable temperature (below Ac_1 for ferritic steels) in contact with a nitrogenous material, usually ammonia or molten cyanide of appropriate composition. Quenching is not required to produce a hard case.

TECHNICAL NOTE 8

Nickel and Nickel Alloys

NICKEL AND NICKEL-BASE ALLOYS are vitally important to modern industry because of their ability to withstand a wide variety of severe operating conditions involving corrosive environments, high temperatures, high stresses, and combinations of these factors. There are several reasons for these capabilities. Pure nickel is ductile and tough, because it possesses a face-centered cubic (fcc) structure up to its melting point. Therefore, nickel and its alloys are readily fabricated by conventional methods (wrought, cast, and powder metallurgy products are available), and they offer freedom from the ductile-to-brittle transition behavior of most body-centered cubic (bcc) and noncubic metals. Nickel has good resistance to corrosion in the normal atmosphere, in natural freshwaters, and in deaerated nonoxidizing acids, and it has excellent resistance to corrosion by caustic alkalis. Therefore, nickel offers very useful corrosion resistance itself, and it is an excellent base on which to develop specialized alloys. Its atomic size and nearly complete 3d electron shell enable it to receive large amounts of alloying additions before encountering phase instabilities. This allows a wide variety of alloys to be fashioned in a manner that can adequately capitalize on the unique properties of specific alloying elements. Finally, unique intermetallic phases can form between nickel and some of its alloying elements; this enables the formulation of alloys with very high strengths for both low- and high-temperature services. See also *Superalloys* (Technical Note 13).

Nickel is extracted from sulfide ores, mined principally in Canada, or oxide ores. These ores, which contain approximately 1–3% total nickel, are smelted and refined electrolytically. The single largest use for nickel is as an alloying element in stainless steels. Commercial nickel-base alloys, which account for approximately 13% of all nickel consumed, are divided into groups or families by their major elemental constituents—for example, nickel-copper, nickel-chromium, nickel-chromium-iron, etc. Other uses for nickel are listed in the table at right.

Use	Amount consumed, %
Stainless steel	57
Alloy steel	9.5
Nickel-base alloys	13
Copper-base alloys	2.3
Plating	10.4
Foundry	4.4
Other	3.3

Source: Nickel Development Institute

Nickel and nickel alloys are used for a wide variety of applications, the majority of which involve corrosion resistance and/or heat resistance. Some of these include aircraft gas turbines, steam turbine power plants, turbochargers and valves in reciprocating engines, medical applications (prosthetic devices), heat treating equipment, components used in the chemical and petrochemical industries, pollution control equipment, coal gasification and liquefaction systems, and parts used in pulp and paper mills. A number of other applications for nickel alloys involve the unique physical properties of special-purpose alloys, such as low-expansion alloys, electrical resistance alloys, soft magnetic alloys, and shape memory alloys.

TECHNICAL NOTE 8 (continued)

Nominal chemical compositions of some typical nickel-base alloys

Common alloy designation	UNS designation	C(a)	Nb	Cr	Cu	Fe	Mo	Ni	Si(a)	Ti	W	Other
Nickel												
200	N02200	0.1	...	...	0.25 max	0.4 max	...	99.2 min	0.15	0.1 max	...	...
201	N02201	0.02	...	...	0.25 max	0.4 max	...	99.0 min	0.15	0.1 max	...	...
Nickel-copper												
400	N04400	0.15	...	...	31.5	1.25	...	bal	0.5	...	...	...
R-405	N04405	0.15	...	...	31.5	1.25	...	bal	0.5	...	...	0.0435
Nickel-molybdenum												
B-2	N10665	0.01	...	1.0 max	...	2.0 max	28	bal	0.1	...	...	...
B	N10001	0.05	...	1.0 max	...	5.0	28	bal	1.0	...	...	...
Nickel-chromium-iron												
600	N06600	0.08	...	16.0	0.5 max	8.0	...	bal	0.5	0.3 max	...	...
601	N06601	...	...	23.0	...	14.1	...	bal	...	...	...	1.35Al
800	N08800	0.1	...	21.0	0.75 max	44.0	...	32.5	1.0	0.38	...	...
800H	N08810	0.08	...	21.0	0.75 max	44.0	...	32.5	1.0	0.38	...	...
Nickel-chromium-iron-molybdenum												
825	N08825	0.05	...	21.5	2.0	29.0	3.0	42	0.5	1.0	...	...
G	N06007	0.05	2.0	22.0	2.0	19.5	6.5	43	1.0	...	1.0 max	...
G-2/2550	N06975	0.03	...	24.5	1.0	20.0	6.0	48	1.0	1.0	...	...
G-3	N06985	0.015	0.8	22.0	2.0	19.5	7.0	44	1.0	...	1.5 max	...
H	...	0.03	...	22.0	...	19.0	9.0	42	1.0	...	2.0	...
G-30	N06030	0.03	0.8	29.5	2.0	15.0	5.5	43	1.0	...	2.5	...
Nickel-chromium-molybdenum-tungsten												
N	N10003	0.06	...	7.0	0.35 max	5.0 max	16.5	71	1.0	0.5 max	0.5 max	...
W	N10004	0.12	...	5.0	...	6.0	24.0	63	1.0	...	...	...
625	N06625	0.1	4.0	21.5	...	5.0 max	9.0	62	0.5	...	...	...
690	N06690	0.02	...	29.0	...	10.0	...	61	...	...	...	...
C-276	N10276	0.01	...	15.5	...	5.5	16.0	57	0.08	0.3	4.0	...
C-4	N06455	0.01	...	16.0	...	3.0 max	15.5	65	0.08	...	...	...
C-22	N06022	0.015	...	22.0	...	3.0 max	13.0	56	0.08	...	3.0	...

(a) Maximum

Selected References

- W.L Mankins and S. Lamb, Nickel and Nickel Alloys, *Metals Handbook*, 10th ed., Vol 1, ASM International, 1990, p 429–445

- A.I. Asphahani et al., Corrosion of Nickel-Base Alloys, *Metals Handbook*, 9th ed., Vol 13, ASM International, 1987, p 641–657

nitrocarburizing. Any of several processes in which both nitrogen and carbon are absorbed into the surface layers of a ferrous material at temperatures below the lower critical temperature and, by diffusion, create a concentration gradient. Nitrocarburizing is conducted mainly to provide an antiscuffing surface layer and to improve fatigue resistance. Compare with *carbonitriding.*

noble metal. (1) A metal whose potential is highly positive relative to the hydrogen electrode. (2) A metal with marked resistance to chemical reaction, particularly to oxidation and to solution by inorganic acids. The term as often used is synonymous with *precious metal.* Contrast with *base metal* (4).

noble potential. A potential more cathodic (positive) than the standard hydrogen potential.

no-draft forging. A forging with extremely close tolerances and little or no *draft,* requiring a minimum of machining to produce the final part. Mechanical properties can be enhanced by close control of grain flow and retention of surface material in the final component.

nodular cast iron. A *cast iron* that has been treated while molten with an element such as magnesium or cerium to induce the formation of free graphite as nodules of spherulites, which imparts a measurable degree of ductility to the cast metal. Preferred term is *ductile cast iron.* Also known as *spherulitic graphite cast iron* and SG iron.

nodular powder. Irregular particles of a metal powder that have knotted, rounded, or other similar shapes.

nominal stress. The stress computed by simple elasticity formulas, ignoring stress raisers and disregarding plastic flow. In a notch bend test, for example, it is bending moment divided by minimum section modulus.

nonconforming. A quality control term describing a unit of product or service that does not meet normal acceptance criteria for the specific product or service. A nonconforming unit is not necessarily *defective.*

nondestructive inspection (NDI). Inspection by a method that neither destroys the part nor impairs its serviceability.

nondestructive testing (NDT). Same as *nondestructive inspection,* but implying the use of a method in which the part is stimulated and its response is measured quantitatively or semiquantitatively.

nonmetallic inclusions. Chemical compounds and nonmetals, in the form of oxides and sulfides, that are present in steel and alloys. They are the product of chemical reactions, physical effects, and contamination that occurs during the melting and pouring process. See *inclusions.*

Nordic gold. An alloy used for 10-, 20-, and 50-cent Euro coins containing 89% Cu, 5% Al, 5% Zn, and 1% Sn.

normalizing. Heating a ferrous alloy to a suitable temperature above the transformation range and then cooling in air to a temperature substantially below the transformation range.

normal segregation. The concentration of alloying constituents that have low melting points in those portions of a casting that solidify last. Compare with *inverse segregation.*

normal stress. The stress component that is perpendicular to the plane on which the forces act. Normal stress can be either tensile or compressive.

nose radius. The radius of the rounded portion of the cutting edge of a tool. See the figure accompanying the term *single-point tool*.

nosing. Closing in the end of a tubular shape to a desired curve contour.

notch acuity. Relates to the severity of the stress concentration produced by a given notch in a particular structure. If the depth of the notch is very small compared with the width (or diameter) of the narrowest cross section, the acuity may be expressed as the ratio of the notch depth to the notch root radius. Otherwise, the acuity is defined as the ratio of one-half the width (or diameter) of the narrowest cross section to the notch root radius.

notch brittleness. The susceptibility of a material to brittle fracture at points of stress concentration. For example, in a notch tensile test, the material is said to be notch brittle if the *notch strength* is less than the tensile strength of an un-notched specimen. Otherwise, it is said to be notch ductile.

notch depth. The distance from the surface of a test specimen to the bottom of the notch. In a cylindrical test specimen, the percentage of the original cross-sectional area removed by machining an annular groove.

notch ductile. The characteristic of a material, determined in a notch tensile test, whereby the *notch strength* is greater than the tensile strength of an un-notched specimen. See *notch brittleness*.

notch ductility. The percentage reduction in area after complete separation of the metal in a tensile test of a notched specimen.

notching. Cutting out various shapes from the edge of a strip, blank, or part.

notching press. A mechanical press used for notching internal and external circumferences and for notching along a straight line. The press is equipped with automatic feeds because only one notch is made per stroke.

notch rupture strength. The ratio of applied load to original area of the minimum cross section in a *stress-rupture test* of a notched specimen.

notch sensitivity. A measure of the reduction in strength of a metal caused by the presence of a stress concentration. Values can be obtained from static, impact, or fatigue tests.

notch sharpness. See *notch acuity*.

notch strength. The maximum load on a notched tensile-test specimen divided by the minimum cross-sectional area (the area at the root of the notch). Also called notch tensile strength.

nucleation. The initiation of a phase transformation at discrete sites, the new phase growing on nuclei. See *nucleus* (1).

nucleus. (1) The first structurally stable particle capable of initiating recrystallization of a phase or the growth of a new phase and possessing an interface with the parent matrix. The term also is applied to a foreign particle that initiates such action. (2) The heavy central core of an atom, in which most of the mass and the total positive electric charge are concentrated.

nugget. (1) A small mass of metal, such as gold or silver, found free in nature. (2) The weld metal in a spot, seam, or projection weld.

octahedral plane. In cubic crystals, a plane with equal intercepts on all three axes.

offal. The material trimmed from blanks or formed panels.

offhand grinding. Grinding in which the operator manually forces the wheel against the work, or vice versa. It often implies casual manipulation of either grinder or work to achieve the desired result. Dimensions and tolerances frequently are not specified, or are only loosely specified; the operator mainly relies on visual inspection to determine how much grinding should be done. Contrast with *precision grinding*.

offset. The distance along the strain coordinate between the initial portion of a stress-strain curve and a parallel line that intersects the stress-strain curve at a value of stress that is used as a measure of the *yield strength*. It is used for materials that have no obvious *yield point*. A value of 0.2% is commonly used.

off time. In resistance welding, the time that the electrodes are off the work. This term is generally applied where the welding cycle is repetitive.

oil canning. (1) A dished distortion in a flat or nearly flat surface. (2) Enclosing a highly reactive metal within a relatively inert one for the purpose of hot working without undue oxidation of the active metal. Also known as *canning*.

oilstone. A natural or manufactured abrasive stone, generally impregnated with oil, used for sharpening keen-edged tools.

Olsen ductility test. A cupping test in which a piece of sheet metal, restrained except at the center, is deformed by a standard steel ball until fracture occurs. The height of the cup (in thousandths of an inch) at the time of fracture is a measure of the ductility.

open-back inclinable press. A vertical crank press that can be inclined so that the bed will have an inclination generally varying from 0 to 30°. The formed parts slide off through an opening in the back. It is often referred to as an OBI press.

open-die forging. The hot mechanical forming of metals between flat or shaped dies in which metal flow is not completely restricted. Also known as *flat-die forging*, *hand forging*, and *smith forging*.

open dies. Forging or forming impression dies in which there is little or no restriction to lateral flow of metal to the cavity within the die set. Contrast with *closed dies*.

open-gap upset welding. A form of *forge welding* in which the weld interfaces are heated with a fuel gas flame, then forced into intimate contact by the application of force. Not the same as *upset welding*, which is a resistance welding process.

open hearth furnace. A reverberatory melting furnace with a shallow hearth and a low roof. The flame passes over the charge on the hearth, causing the charge to be heated both by direct flame and by radiation from the roof and sidewalls of the furnace. In the ferrous industry, the furnace is regenerative.

open rod press. A hydraulic press in which the slide is guided by vertical, cylindrical rods (usually four) that also serve to hold the crown and bed in position.

operating stress. The stress to which a structural unit is subjected in service.

optical pyrometer. An instrument for measuring the temperature of heated material by comparing the intensity of light emitted with a known intensity of an incandescent lamp filament.

oralloy. A code name given to enriched uranium during the Manhattan Project at the Oak Ridge National Laboratory. A shortened version of "Oak Ridge alloy," the term still is used occasionally to refer to enriched uranium.

orange peel. A surface roughening in a pebble-grained pattern that occurs when a metal of unusually coarse grain size is stressed beyond its elastic limit. Also known as *pebbles* and alligator skin.

ordering. Forming a *superlattice*. Contrast with *disordering*.

ore. A natural mineral that may be mined and treated for the extraction of any of its components, metallic or otherwise, at a profit.

ore dressing. Also known as *mineral dressing*.

orientation. Arrangement in space of the axes of the lattice of a crystal with respect to a chosen reference or coordinate system. See also *preferred orientation*.

oscillating die press. A small high-speed press in which the die and punch move horizontally with the strip during the working stroke. Through a reciprocating motion, the die and punch return to their original positions to begin the next stroke.

osmium. Atomic number 76, atomic weight 190, symbol, Os. For the Latin word osme, meaning smell. Osmium was discovered in 1803 by Smithson Tennant in England. Osmium is a member of the platinum group of metals which includes Platinum, Rhodium, Iridium, Ruthenium, Palladium, and Osmium. Osmium is a brittle metal and has the least attractive properties of the group. It has the lowest potential for usefulness.

ounce metal. A cast leaded red brass also known as 85-5-5-5 and *red brass*. The alloy, which is used for valves, flanges, and fittings, has a nominal chemical composition of 85% Cu, 5% Sn, 5% Pb, and 5% Zn. The UNS number is C83600. (See ASTM B62.)

overaging. *Aging* under conditions of time and temperature greater than those required to obtain maximum change in a certain property, so that the property is altered in the direction of the initial value.

overbending. Bending metal through a greater arc than that required in the finished part, to compensate for springback.

overdraft. A condition wherein a metal curves upward on leaving the rolls because of the higher speed of the lower roll.

overhauling. Cutting surface layers from castings or slabs to remove scale and surface imperfections. Sometimes called *scalping* or *slab milling*.

overhead-drive press. A mechanical press with the driving mechanism mounted in or on the crown or upper parts of the uprights.

overhead-position welding. Welding that is performed from the underside of the joint.

overheating. Heating a metal or alloy to such a high temperature that its properties are impaired. When the original properties cannot be restored by further heat treating, by mechanical working, or by a combination of working and heat treating, the overheating is known as *burning*.

overlap. (1) Protrusion of weld metal beyond the toe, face, or root of a weld. (2) In resistance seam welding, the area in a given weld remelted by the succeeding weld.

oversize powder. Particles of a powdered metal coarser than the maximum permitted by a given specification for particle size.

overstressing. In fatigue testing, cycling at a stress level higher than that used at the end of the test.

overvoltage. The difference between the actual electrode potential when appreciable electrolysis begins and the reversible electrode potential.

oxidation. (1) A reaction in which there is an increase in valence resulting from a loss of electrons. Contrast with *reduction*.

(2) A corrosion reaction in which the corroded metal forms an oxide; usually applied to reaction with a gas containing elemental oxygen, such as air.

oxidized steel surface. Surface having a thin, tightly adhering, oxidized skin (from straw to blue in color), extending in from the edge of a coil or sheet. Sometimes called annealing border.

oxidizing agent. A compound that causes *oxidation*, thereby itself being reduced.

oxidizing flame. A gas flame produced with excess oxygen in the inner flame that has an oxidizing effect.

oxyacetylene cutting. An *oxyfuel gas cutting* process in which the fuel gas is acetylene.

oxyacetylene welding. An *oxyfuel gas welding* process in which the fuel gas is acetylene.

oxyfuel gas cutting (OFC). Any of a group of processes used to sever metals by means of chemical reaction between hot base metal and a fine stream of oxygen. The necessary metal temperature is maintained by gas flames resulting from combustion of a specific fuel gas such as acetylene, hydrogen, natural gas, or propane. See also *oxygen cutting*.

oxyfuel gas welding (OFW). Any of a group of processes used to fuse metals together by heating them with gas flames resulting from combustion of a specific fuel gas such as acetylene, hydrogen, natural gas, or propane. The process may be used with or without the application of pressure to the joint, and with or without the addition of any filler metal.

oxygen cutting. Metal cutting by directing a fine stream of oxygen against a hot metal. The chemical reaction between oxygen and the base metal furnishes heat for localized melting, hence, cutting.

oxygen deficiency. A form of *crevice corrosion* in which galvanic corrosion proceeds because oxygen is prevented from diffusing into the crevice.

oxygen-free copper. Electrolytic copper free from cuprous oxide, produced without the use of residual metallic or metalloidal deoxidizers.

oxygen gouging. Oxygen cutting in which a chamfer or groove is formed.

oxygen lance. A length of pipe used to convey oxygen, either to the point of cutting in oxygen lance cutting, or beneath the surface of the melt in a steelmaking furnace.

oxyhydrogen cutting. An *oxyfuel gas cutting* process in which the fuel gas is hydrogen.

oxyhydrogen welding. An *oxyfuel gas welding* process in which the fuel gas is hydrogen.

oxynatural gas cutting. An *oxyfuel gas cutting* process in which the fuel gas is natural gas.

oxynatural gas welding. An *oxyfuel gas welding* process in which the fuel gas is natural gas.

oxypropane cutting. An *oxyfuel gas cutting* process in which the fuel gas is propane.

oxypropane welding. An *oxyfuel gas welding* process in which the fuel gas is propane.

packing material. Any material in which powder metallurgy compacts are embedded during the presintering or sintering operation.

pack rolling. Hot rolling a pack of two or more sheets of metal; scale prevents their being welded together.

palladium. A chemical element having atomic number 46, atomic weight 106, and the symbol Pd. It was named for the asteroid Pallas, discovered in 1802, and which itself was named for Pallas Athena, the Greek goddess of art and wisdom. The metal was identified by William Hyde Wollaston in 1803. Palladium is the least expensive of the platinum group of metals, making it useful for jewelry, one of its most important applications. It is used as an alloying element with silver, molybdenum, and other metals of the platinum group. The largest uses of palladium are as a catalyst in powder form and as fine wire gauze.

pancake forging. A rough forged shape, usually flat, that may be obtained quickly with a minimum of tooling. It usually requires considerable machining to attain finish size.

paramagnetic material. A material whose specific permeability is greater than unity and is practically independent of the magnetizing force. Compare with *diamagnetic material* and *ferromagnetic material*.

Parkes process. A process used to recover precious metals from lead and based on the principle that if 1–2% Zn is stirred into the molten lead, a compound of zinc with gold and silver separates out and can be skimmed off.

partial annealing. An imprecise term used to denote a treatment given to cold worked material to reduce its strength to a controlled level or to effect stress relief. To be meaningful, the type of material, the degree of cold work, and the time-temperature schedule must be stated.

particle size. The controlling lineal dimension of an individual particle, such as of a powdered metal, as determined by analysis with screens or other suitable instruments.

particle size distribution. The percentage, by weight or by number, of each fraction into which a powder sample has been classified with respect to sieve number or *particle size*. Preferred usage distinguishes particle size distribution by weight or particle size distribution by frequency.

parting. (1) In the recovery of precious metals, the separation of silver from gold. (2) The zone of separation between *cope* and *drag* portions of the mold or flask in sand casting. (3) A composition sometimes used in sand molding to facilitate the removal of the pattern. (4) Cutting simultaneously along two parallel lines or along two lines that balance each other in side thrust. (5) A shearing operation used to produce two or more parts from a stamping.

parting line. (1) The intersection of the parting plane of a casting mold, or the parting plane between forging dies, with the mold or die cavity. (2) A raised line or projection on the surface of a casting or forging that corresponds to said intersection.

parting plane. (1) In forging, the dividing plane between dies. Contrast with *forging plane*. (2) In casting, the dividing plane between mold halves (Fig. 42).

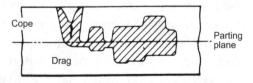

Fig. 42 The parting plane in a casting mold

parting sand. Fine sand for dusting on sand mold surfaces that are to be separated.

parts former. A type of *upsetter* designed to work on short billets instead of bars and tubes, usually for cold forging.

pass. (1) A single transfer of metal through a *stand* of rolls. (2) The open space between two grooved rolls through which metal is processed. (3) The weld metal deposited in one trip along the axis of a weld.

passivation. The changing of a chemically active surface of a metal to a much less reactive state. Contrast with *activation*.

passivation treatment. The chemical treatment of a stainless steel with a mild oxidant for the removal of free iron or other foreign matter (ASTM A967). When oxide scale and other exogenous matter are removed, a protective passive film forms on the stainless steel, creating a shell against corrosion.

passivity. A condition in which one piece of metal, because of an impervious covering of oxide or other compound, has a potential much more positive than that of the metal in the active state.

patenting. In wiremaking, a heat treatment applied to medium-carbon or high-carbon steel before drawing of wire or between drafts. This process consists of heating to a temperature above the transformation range and then cooling to a temperature below Ae_1 in air or in a bath of molten lead or salt.

patent leveling. Leveling a piece of metal (that is, removing warp and distortion) by gripping it at both ends and subjecting it to a stress higher than its yield strength. Also known as *stretcher leveling*.

pattern. (1) A form of wood, metal, or other material around which molding material is placed to make a mold for casting metals. (2) A full-scale reproduction of a part used as a guide in cutting.

Pattinson process. A process used to separate silver from lead, in which the molten lead is slowly cooled so that crystals poorer in silver solidify out and are removed, leaving the melt richer in silver.

pearlite. A metastable lamellar aggregate of *ferrite* and *cementite* resulting from the transformation of *austenite* at temperatures above the *bainite* range.

pearlitic malleable. A malleable cast iron that may contain *pearlite, spheroidite*, or tempered *martensite* depending on heat treatment and desired hardness. See *malleable cast iron*.

pebbles. A surface roughening in a pebble-grained pattern that occurs when a metal of unusually coarse grain size is stressed beyond its elastic limit. Also known as *orange peel* and alligator skin.

peeling. The detaching of one layer of a coating from another, or from the basis metal, because of poor adherence.

peening. Mechanical working of metal by hammer blows or shot impingement.

penetrant. A liquid with low surface tension used in *liquid penetrant inspection* to flow into surface openings of parts being inspected.

penetrant inspection. A type of nondestructive inspection that locates discontinuities that are open to the surface of a metal by first allowing a penetrating dye or fluorescent liquid to infiltrate the discontinuity, removing the excess penetrant, and then applying a developing agent that causes the penetrant to seep back out of the discontinuity and register as an indication. Suitable for both ferrous and nonferrous materials, but is limited to the detection of open surface discontinuities in nonporous solids. See preferred term *liquid penetrant inspection*.

penetration. (1) In founding, an *imperfection* on a casting surface caused by metal running into voids between sand grains; usually referred to as *metal penetration*. (2) In welding, the distance from the original surface of the base metal to that point at which fusion ceased. See also *joint penetration*.

penetration hardness. The resistance of a material to incur an indentation. Indentation hardness testing is the usual type of hardness testing performed. The test involves pressing a pointed or rounded indenter into a surface under a substantially static load. Also known as *indentation hardness*.

percussion welding. Resistance welding in which abutting surfaces are heated by an intense spark between them, welding being consummated by applying a hammerlike blow during or immediately after the electrical discharge.

perforating. Piercing holes of desired shapes arranged in a definite pattern in sheets, blanks, or formed parts.

periodic reverse. The process that involves periodic changes in the direction of the flow of the current in electrolysis. Also, the machine that controls the time for both directions.

peripheral clearance angle. The angle between a plane containing the flank of the tool and a plane passing through the cutting edge in the direction of relative motion between the cutting edge and the work. Also known as *clearance angle*. See also *face mill* (Fig. 26).

peripheral milling. Milling a surface parallel to the axis of the cutter.

peripheral speed. The linear speed of relative motion between the tool and the workpiece in the principal direction of cutting. See preferred term *cutting speed*.

peritectic. An isothermal reversible reaction in which a liquid phase reacts with a solid phase to produce a single (and different) solid phase on cooling.

peritectoid. An isothermal reversible reaction in which a solid phase reacts with a second solid phase to produce a single (and different) solid phase on cooling.

permanent mold. A metal, graphite, or ceramic mold (other than an ingot mold) of two or more parts that is used repeatedly for the production of many *castings* of the same form. Liquid metal is poured in by gravity.

permanent set. Plastic deformation that remains after the release of the applied load that produces the deformation.

permeability. (1) In founding, the characteristics of molding materials that permit gases to pass through them. "Permeability number" is determined by a standard test. (2) In powder metallurgy, a property measured as the rate of passage, under specified conditions, of a liquid or gas through a compact. (3) A general term used to express various relationships between magnetic induction and magnetizing force. These relationships are either "absolute permeability," which is a change in magnetic

induction divided by the corresponding change in magnetizing force, or "specific (relative) permeability," the ratio of the absolute permeability to the permeability of free space.

pewter. Any of various alloys in which tin is the chief constituent; especially an alloy of tin and lead formerly used for domestic utensils. Table 11 lists chemical compositions of modern pewter alloys.

pH. The negative logarithm of the hydrogen-ion activity; it denotes the degree of acidity or basicity of a solution. At 25 °C (77 °F), 7.0 is the neutral value. Decreasing values below 7.0 indicate increasing acidity; increasing values above 7.0, increasing basicity.

phase. A physically homogeneous and distinct portion of a material system.

phase diagram. A graphical representation of the temperature and composition limits of phase fields in an alloy system as they actually exist under the specific conditions of heating or cooling. A phase diagram may be an equilibrium diagram, an approximation to an equilibrium diagram, or a representation of metastable conditions or phases. Also known as *constitution diagram*. See also *equilibrium diagram*.

Table 11   Chemical composition limits for modern pewter

Specification	Composition, %							
	Sn	Sb	Cu	Pb max	As max	Fe max	Zn max	Cd max
ASTM B 560								
Type 1(a)	90–93	6–8	0.25–2.0	0.05	0.05	0.015	0.005	...
Type 2(b)	90–93	5–7.5	1.5–3.0	0.05	0.05	0.015	0.005	...
Type 3(c)	95–98	1.0–3.0	1.0–2.0	0.05	0.05	0.015	0.005	...
BS 5140	bal	5–7	1.0–2.5	0.5	...	...	...	0.05
		3–5	1.0–2.5	0.5	...	...	...	0.05
DIN 17810	bal	1–3	1–2	0.5	...	...	...	...
		3.1–7.0	1–2	0.5	...	...	...	...

(a) Casting alloy, nominal composition 92Sn-7.5Sb-0.5Cu. (b) Sheet alloy, nominal composition 91Sn-7Sb-2Cu. (c) Special-purpose alloy

phosphating. Forming an adherent phosphate coating on a metal by immersion in a suitable aqueous phosphate solution. Also called phosphatizing.

phosphorized copper. General term applied to copper deoxidized with phosphorous. The most commonly used deoxidized copper.

photoelasticity. An optical method for evaluating the magnitude and distribution of stresses, using a transparent model of a part, or a thick film of photoelastic material bonded to a real part.

photomacrograph. A *macrograph*, which is a graphic reproduction of the surface of a prepared specimen at a magnification not exceeding 25×, produced by photographic means.

photomicrograph. A *micrograph*, which is a graphic reproduction of the surface of a prepared specimen, usually etched, diameter magnification greater than 25×, produced by photographic means.

photon. The smallest possible quantity of electromagnetic radiation that can be characterized by a definite frequency.

physical metallurgy. The science and technology dealing with the properties of metals and alloys, and of the effects of composition, processing, and environment on those properties.

physical properties. Properties of a metal or alloy that are relatively insensitive to structure and can be measured without the application of force; for example, density, electrical conductivity, coefficient of thermal expansion, magnetic permeability, and lattice parameter. Does not include chemical reactivity. Compare with *mechanical properties*.

physical testing. The determination of *physical properties*.

pickle liquor. A spent acid-pickling bath.

pickle patch. A tightly adhering oxide or scale coating not properly removed during *pickling*.

pickle stain. Discoloration of metal due to chemical cleaning without adequate washing and drying.

pickling. Removing surface oxides from metals by chemical or electrochemical reaction.

pickoff. An automatic device for removing a finished part from the press die after it has been stripped.

pickup. Transfer of metal from tools to part or from part to tools during a forming operation. See *galling*.

Pidgeon process. A process for production of magnesium by reduction of magnesium oxide with ferrosilicon.

piezoelectric effect. The reversible interaction, exhibited by some crystalline materials, between an elastic strain and an electric field. The direction of the strain depends on the polarity of the field or vice versa. Compare with *electrostrictive effect*.

pig. A metal casting used in remelting.

pig iron. (1) High-carbon iron made by the reduction of iron ore in the blast furnace. (2) Cast iron in the form of a *pig*.

Pilger tube-reducing process. A process used to reduce the diameter and wall thickness of tubing with a mandrel and a pair of rolls with tapered grooves. A uniform rod (broach) reciprocates with the tubing, and the fixed rolls rotate continuously. During the gap in each revolution, the tubing is advanced and rotated and then, on roll contact, reduced and partially returned.

pinchers. Surface disturbances that result from rolling processes and that ordinarily

appear as fernlike ripples running diagonally to the direction of rolling.

pinch pass. A pass of sheet material
through rolls to effect a very small reduction in thickness.

pinch trimming. Trimming the edge of a
tubular part or shell by pushing or pinch-
ing the flange or lip over the cutting edge
of a stationary punch or over the cutting
edge of a draw punch.

pine-tree crystal. A type of *dendrite*.

pin expansion test. A test for determining
the ability of a tube to be expanded, or
for revealing the presence of cracks or
other longitudinal weaknesses, made by
forcing a tapered pin into the open end
of a tube.

pinhead blister. A very small *blister*, which
is a raised area, often dome-shaped, resulting from (a) loss of adhesion between
a coating or deposit and the basis metal
or (b) delamination under the pressure of
expanding gas trapped in a metal near a
subsurface zone. Also known as a pepper
blister.

pinhole porosity. Porosity consisting of
numerous small gas holes distributed
throughout a metal; found in weld metal,
castings, and electrodeposited metal.

pinion. The smaller of two mating gears.

Piobert lines. Elongated surface markings
or depressions caused by localized plastic deformation that results from discontinuous (inhomogeneous) yielding. Also
known as *Lüders lines*, Lüders bands,
Hartmann lines, and *stretcher strains*.

pipe. (1) The central cavity formed by a
contraction in metal, especially ingots,
during solidification (Fig. 43). (2) An
imperfection in wrought or cast products
resulting from such a cavity. (3) A tubular metal product, cast or wrought.

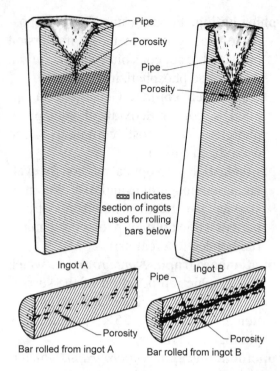

Fig. 43 Longitudinal sections of two types of ingots showing typical pipe and porosity

pipe tap. A *tap* for making internal *pipe
threads* within pipe fittings or holes.

pipe threads. Internal or external machine
threads, usually tapered, of a design intended for making pressure-tight mechanical joints in piping systems.

pitch. The shape of the solidifying surface of a metal, especially copper, with
respect to concavity or convexity. Also
may be called *set*.

pitting. Forming small sharp cavities in a
metal surface by nonuniform *electrodeposition* or by corrosion.

plain carbon steels. Steels containing approximately 0.05–0.2% C with no intentionally added alloying elements,
with the possible exception of elements
added to improve machinability, such
as sulfur, lead, and tellurium, which act
as chip breakers. These steels have too

little carbon for any increase in hardness to be produced by thermal treatment. Examples are the AISI steels 1005, 1010, and 1020. Depending on the method of manufacture and the quality, these steels exhibit good ductility with little work hardening in the lower carbon range.

planchet. A metal disk with milled edges, ready for coining.

plane strain. The stress condition in *linear elastic fracture mechanics* in which there is zero strain in a direction normal to both the axis of applied tensile stress and the direction of crack growth (i.e., parallel to the crack front); most nearly achieved in loading thick plates along a direction parallel to the plate surface. Under plane-strain conditions, the plane of fracture instability is normal to the axis of the principal tensile stress.

plane stress. The stress condition in *linear elastic fracture mechanics* in which the stress in the thickness direction is zero; most nearly achieved in loading very thin sheet along a direction parallel to the surface of the sheet. Under plane-stress conditions, the plane of fracture instability is inclined 45° to the axis of the principal tensile stress.

planimetric method. A method of measuring grain size in which the grains within a definite area are counted.

planing. Producing flat surfaces by linear reciprocal motion of work and the table to which it is attached, relative to a stationary single-point cutting tool.

planishing. Producing a smooth surface finish on metal by a rapid succession of blows delivered by highly polished dies or by a hammer designed for the purpose, or by rolling in a planishing mill.

plasma-arc cutting. An arc cutting process that severs metals by melting a localized area with heat from a constricted arc and removing the molten metal with a high-velocity jet of hot, ionized gas issuing from the plasma torch.

plasma arc welding (PAW). An arc welding process that produces coalescence of metals by heating them with a constricted arc between an electrode and the work piece (transferred arc) or the electrode and the constricting nozzle (nontransferred arc). Shielding is obtained from hot, ionized gas issuing from an orifice surrounding the electrode and may be supplemented by an auxiliary source of shielding gas, which may be an inert gas or a mixture of gases. Pressure may or may not be used, and filler metal may or may not be supplied.

plasma spraying. A *thermal spraying* process in which the coating material is melted with heat from a plasma torch that generates a nontransferred arc (defined in *plasma arc welding*); molten coating material is propelled against the basis metal by the hot, ionized gas issuing from the torch.

plaster molding. Molding wherein a gypsum-bonded aggregate flour in the form of a water slurry is poured over a pattern, permitted to harden, and, after removal of the pattern, thoroughly dried. This technique is used to make smooth nonferrous castings of accurate size.

plastic deformation. Deformation that remains permanent after removal of the load that caused it.

plastic flow. The phenomenon that takes place when metals are stretched or compressed permanently without rupture.

plasticity. The ability of a metal to deform nonelastically without rupture.

plate. A flat-rolled metal product of some minimum thickness and width arbitrarily dependent on the type of metal.

plate-as-rolled. The quantity of plate product rolled at one time, either from an individual slab or directly from an ingot.

platelets. Flat particles of metal powder having considerable thickness. Contrast with *flake powder*.

platen. (1) The face of a bolster, slide, or ram to which a tool assembly is attached. (2) A part of a resistance welding, mechanical testing, or other machine with a flat surface to which dies, fixtures, backups, or electrode holders are attached, and that transmits pressure or force.

plating. Forming an adherent layer of metal on an object; often used as a shop term for *electroplating*.

plating rack. A fixture used to hold work and conduct current to it during electroplating.

plating range. The current-density range over which a satisfactory electroplate can be deposited.

platinum. A chemical element having atomic number 78, atomic weight 195, and the symbol Pt. It was discovered in 1748 by Spanish mathematician Antonio de Ulloa, who named it *platina*, meaning little silver. Platinum is one of the precious metals.

platinum black. A finely divided form of platinum of a dull black color, usually but not necessarily produced by the reduction of salts in an aqueous solution.

plug. (1) A rod or mandrel over which a pierced tube is forced. (2) A rod or mandrel that fills a tube as it is drawn through a die. (3) A punch or mandrel over which

a cup is drawn. (4) A protruding portion of a die impression for forming a corresponding recess in the forging. (5) A false bottom in a die. Also called a peg.

plug tap. A *tap* with *chamfer* extending from three to five threads.

plug weld. A circular weld made by either arc or gas welding through one member of a lap or tee joint. If a hole is used, it may be only partly filled. Neither a fillet-welded hole nor a spot weld is to be construed as a plug weld.

plumbago. A special quality of powdered graphite used to coat molds and, in a mixture with clay, to make crucibles.

plunge grinding. *Grinding* wherein the only relative motion of the wheel is radially toward the work.

plus sieve. The portion of a sample of a granular substance (such as metal powder) retained on a standard sieve of specified number. Contrast with *minus sieve*.

plutonium. A chemical element having atomic number 94, atomic weight 244, and the symbol Pu, named for the planet Pluto (Pluto was the Roman god of the underworld). Plutonium is a man-made element, and the second element after uranium. It was first synthesized by Glenn T. Seaborg, Edwin McMillan, Joseph Kennedy, and Arthur Wahl in 1940. Plutonium is used for producing nuclear energy.

plymetal. Sheet consisting of bonded layers of dissimilar metals.

P/M. The acronym for *powder metallurgy* the art of producing metal powders and of using metal powders for the production of massive materials and shaped objects.

point angle. In general, the angle at the point of a cutting tool. Most commonly, the included angle at the point of a twist drill, the general-purpose angle being 118°.

pointing. (1) Reducing the diameter of wire, rod, or tubing over a short length at the end by swaging or hammer forging, turning, or squeezing to facilitate entry into a drawing die and gripping in the drawhead. (2) The operation in automatic machines of chamfering or rounding the threaded end or the head of a bolt.

Poisson's ratio. The absolute value of the ratio of the transverse strain to the corresponding axial strain in a body subjected to uniaxial stress; usually applied to elastic conditions.

poke welding. Spot or projection welding in which the force is applied manually to one electrode, and the work or a backing bar takes the place of the other electrode. Also known as *push welding*.

polar bond. A bond between two or more atoms in which electrons are shared unequally. See *ionic bond*.

polarization. A change in the potential of an electrode during electrolysis, such that the potential of an *anode* becomes more noble, and that of a *cathode* more active, than their respective reversible potentials. Often accomplished by formation of a film on the electrode surface.

pole. (1) A means of designating the orientation of a crystal plane by stereographically plotting its normal. For example, the north pole defines the equatorial plane. (2) Either of the two regions of a permanent magnet or electromagnet where most of the lines of induction enter or leave.

pole figure. A stereographic projection representing the statistical average distribution of poles of a specific crystalline plane in a polycrystalline metal, with reference to an external system of axes. In an isotropic metal, that is, in one having a completely random distribution of orientations, the pole density is stereographically uniform; preferred orientation is shown by an increased density of poles in certain areas.

poling. A step in the fire refining of copper to reduce the oxygen content to tolerable limits by covering the bath with coal or coke and thrusting green wood poles below the surface. There is a vigorous evolution of reducing gases, which combine with the oxygen contained in the metal.

polishing. Smoothing metal surfaces, often to a high luster, by rubbing the surface with a fine abrasive, usually contained in a cloth or other soft lap. Results in microscopic flow of some surface metal together with actual removal of a small amount of surface metal. Term may be extended to include *electropolishing*. Contrast with *buffing* and *burnishing*.

polycrystalline. Pertaining to a solid composed of many crystals.

polymorphism. The ability of a solid to exist in more than one form. In metals, alloys, and similar substances, this usually means the ability to exist in two or more crystal structures, or in an amorphous state and at least one crystal structure. See also *enantiotropy* and *monotropism*.

pop-off. The loss of small portions of a porcelain enamel coating. The usual cause is outgassing of hydrogen or other gases from the basis metal during firing, but pop-off may also occur because of oxide particles or other debris on the surface of the basis metal. Usually, the pits are minute and cone shaped, but when pop-off is the result of severe *fishscale*, the pits may be much larger and irregular.

porcelain enameling steels. Steel sheets used for vitreous (porcelain) enameling with very low carbon, phosphorus, sulfur, and silicon contents. When the steel is to be used for deep drawing prior to enameling, the carbon content is limited to 0.008 to 0.05% maximum, depending upon the application. (ASTM A424). Also known as *enameling steels*.

pores. (1) Small voids in the body of a metal. (2) Minute cavities in a powder metallurgy compact, sometimes intentional. (3) Minute perforations in an electroplated coating.

porosity. Fine holes or pores within a metal.

porthole die. A multiple-section extrusion die capable of producing tubing or intricate hollow shapes without the use of a separate mandrel. Metal is extruded in separate streams through holes in each section and is rewelded by extrusion pressure before it leaves the die. The same as *spider die*. Compare with *bridge die*.

positioned weld. A weld made in a joint that has been oriented to facilitate making the weld.

positive rake. A tooth face in rotation whose cutting edge leads the surface of the tooth face. See the figure accompanying the term *face mill*.

postheating. Heating weldments immediately after welding, to provide tempering, stress relieving, or a controlled rate of cooling to prevent formation of a hard or brittle microstructure. The same as *post-weld heat treatment*.

post-weld heat treatment. Heating weldments immediately after welding, to provide tempering, stress relieving, or a controlled rate of cooling to prevent formation of a hard or brittle microstructure. Also known as *postheating*.

pot. (1) A vessel for holding molten metal. (2) The electrolytic reduction cell used to make such metals as aluminum from a fused electrolyte.

pot annealing. Annealing a metal or alloy in a sealed container under conditions that minimize oxidation. In pot annealing a ferrous alloy, the charge usually is heated slowly to a temperature below the transformation range, but sometimes above or within it, and then is cooled slowly. Also known as *close annealing* and *box annealing*. See also *black annealing*.

potassium. A chemical element having atomic number 19, atomic weight 39, and the symbol K. from the neo-Latin *kalium*. Kalium is taken from the word *alkali*, which in turn was derived from the Arab *algoli*, meaning plant ashes. The English word *potassium* was named for pot ash or the Dutch *pot-aschen*: ashes from vegetation that is burned in a pot produce sodium and potassium carbonate, the two alkaline compounds used to make soap. The word *potassium* is used in the English, Celtic, and Roman languages, whereas *kalium* is used in most other languages.

Potassium was identified as an element and isolated by Sir Humphry Davy in 1807. It is a very light, soft, silvery-white metal with a density of 0.86 g/cc and a melting point of only 64 °C (147 °F). It is highly reactive and must be stored out of contact with oxygen. The element is widely distributed in nature, primarily as potassium aluminum silicates, sylvite (KCl) and salt peter (KNO_3). There are no commercial uses in the pure form, although a sodium-potassium alloy has been used in thermometry and as a heat transfer medium. Potassium has been used in small amounts as a hardener in lead. The principal applications are as compounds, such as the use of potassium cyanide in the case hardening of steels. It is used in electroplating and the carbonate is used to make hard glass.

pot die forming. Forming products from sheet or plate through the use of a hollow die and internal pressure that causes the preformed workpiece to assume the contour of the die.

poultice corrosion. A term used in the automotive industry to describe the corrosion of vehicle body parts due to the collection of road salts and debris on ledges and in pockets that are kept moist by weather and washing.

pouring. Transferring molten metal from a furnace to a ladle or a ladle to a mold.

pouring basin. A basin on top of a mold that receives the molten metal before it enters the sprue or downgate.

powder. Particles of a solid characterized by small size, nominally within the range of 0.1 to 1000 μm.

powder lubricant. An agent mixed with or incorporated in a powder to facilitate pressing and ejection of a powder metallurgy compact.

powder metallurgy (P/M). The technology and art of producing and using metal powders for the production of massive materials and shaped objects.

powder metallurgy forging. Plastically deforming a powder metallurgy *compact* or preform into a fully dense finished shape using compressive force; usually done hot and within closed dies.

power reel. A reel that is driven by an electric motor or some other source of power, used to wind or coil strip or wire as it is drawn through a continuous normalizing furnace, through a die, or through rolls.

praseodymium. A chemical element having atomic number 59, atomic weight 141, and the symbol Pr, from the Greek *prasios*, meaning green, and *didymos*, meaning twin. The earth, *didymia*, was divided into two salts: praseodymium, having green salts, and neodymium, having rose-colored salts. In 1885, Carl Auer von Welsbach separated the two salts. A rare earth metal, praseodymium is a common coloring pigment. Another use is as an alloying agent to create high-strength metals for aircraft engines.

precharge. In forming, the pressure introduced into the cavity prior to the forming of the part.

precious metal. One of the relatively scarce and valuable metals: gold, silver, and the platinum-group metals. See *Technical Note 9*.

precipitation hardening. Hardening caused by the precipitation of a constituent from a supersaturated solid solution. See also *age hardening* and *aging*.

precipitation heat treatment. *Artificial aging* in which a constituent precipitates from a supersaturated solid solution.

precision. The closeness of approach of each of a number of similar measurements to the arithmetic mean, the sources of error not necessarily being considered critically. Accuracy demands precision, but precision does not ensure accuracy.

precision casting. A metal casting of reproducible accurate dimensions, regardless of how it is made.

precision grinding. Machine grinding to specified dimensions and low *tolerances*. Contrast with *offhand grinding*.

precoat. (1) In investment casting, a special refractory slurry applied to a wax or plastic expendable pattern to form a thin

TECHNICAL NOTE 9

Precious Metals

THE EIGHT PRECIOUS METALS, listed in order of their atomic number as found in periods 5 and 6 (groups VIII and Ib) of the periodic table, are ruthenium, rhodium, palladium, silver, osmium, iridium, platinum, and gold. Precious metals, also referred to as noble metals, are of inestimable value to modern civilization. Their functions in coins, jewelry, and bullion, and as catalysts in devices to control auto exhaust emissions, are widely understood. But in certain applications, their functions are not as spectacular and, although vital to the application, are largely unknown except to the users. For example, precious metals are used in dental restorations and dental fillings; thin precious metal films are used to form electronic circuits; and certain organometallic compounds containing platinum are significant drugs for cancer chemotherapy.

Silver is a bright, white metal that, next to gold, is the most easily fabricated metal in the periodic table. It is very soft and ductile in the annealed condition. Silver does not oxidize at room temperature, but it is attacked by sulfur. Nitric, hydrochloric, and sulfuric acids attack silver, but the metal is resistant to many organic acids and to sodium and potassium hydroxide. The primary application for silver (approximately 50% of the silver demand) is its use for photographic emulsions. The use of silver in photography is based on the ability of exposed silver halide salts to undergo a secondary image amplification process called development. The second largest use is in the electrical and electronic industries for electrical contacts, conductors, and in primary batteries. Other applications include brazing alloys, dental alloys, electroplated ware, sterling ware (see *sterling silver*), and jewelry and coins.

Gold is a bright, yellow, soft, and very ductile metal. Its special properties include corrosion resistance, good reflectance, resistance to sulfidation and oxidation, and high electrical and thermal conductivity. Because gold is easy to fashion, has a bright pleasing color, is nonallergenic, and remains tarnish-free indefinitely, it is used extensively in jewelry (approximately 55% of the gold market). For much the same reasons, it is used in dental alloys and appliances. Gold is also used to a considerable extent in electronic devices, particularly in printed circuit boards, connectors, keyboard contactors, and miniaturized circuitry. Other applications include gold films used as a reflector of infrared radiation in thermal barrier windows for large buildings and spaced vehicles, fired-on gold organometallic compounds used for decorating glass and china, sliding electrical contacts, and brazing alloys.

The six remaining precious metals are referred to as the platinum-group metals because they are closely related and commonly occur together in nature. Ruthenium, rhodium, and palladium each have a density of approximately 12 g/cm^3; osmium, iridium, and platinum each have a density of about 22 g/cm^3. The most distinctive trait of the platinum-group metals is their exceptional resistance to

TECHNICAL NOTE 9 (continued)

Nominal composition and solidification temperatures for silver-base brazing filler metals

AWS designation (a)	UNS No.	Composition, wt% Ag	Cu	Zn	Cd	Ni	Sn	Li	Mn	Other elements, total (b)	Solidus °C	Solidus °F	Liquidus °C	Liquidus °F	Brazing temperature range °C	Brazing temperature range °F
BAg-1	P07450	44.0–46.0	14.0–16.0	14.0–18.0	23.0–25.0	…	…	…	…	0.15	607	1125	618	1145	618–760	1145–1400
BAg-1a	P07500	49.0–51.0	14.5–16.5	14.5–18.5	17.0–19.0	…	…	…	…	0.15	627	1160	635	1175	635–760	1175–1400
BAg-2	P07350	34.0–36.0	25.0–27.0	19.0–23.0	17.0–19.0	…	…	…	…	0.15	607	1125	702	1295	702–843	1295–1550
BAg-2a	P07300	29.0–31.0	26.0–28.0	21.0–25.0	19.0–21.0	…	…	…	…	0.15	607	1125	710	1310	710–843	1310–1550
BAg-3	P07501	49.0–51.0	14.5–16.5	13.5–17.5	15.0–17.0	2.5–3.5	…	…	…	0.15	632	1170	688	1270	688–816	1270–1500
BAg-4	P07400	39.0–41.0	29.0–31.0	26.0–30.0	…	1.5–2.5	…	…	…	0.15	671	1240	779	1435	779–899	1435–1650
BAg-5	P07453	44.0–46.0	29.0–31.0	23.0–27.0	…	…	…	…	…	0.15	663	1225	743	1370	743–843	1370–1550
BAg-6	P07503	49.0–51.0	33.0–35.0	14.0–18.0	…	…	…	…	…	0.15	688	1270	774	1425	774–871	1425–1600
BAg-7	P07563	55.0–57.0	21.0–23.0	15.0–19.0	…	…	4.5–5.5	…	…	0.15	618	1145	652	1205	652–760	1205–1400
BAg-8	P07720	71.0–73.0	bal	…	…	…	…	…	…	0.15	779	1435	779	1435	779–899	1435–1650
BAg-8a	P07723	71.0–73.0	bal	…	…	…	…	0.25–0.50	…	0.15	766	1410	766	1410	766–871	1410–1600
BAg-9	P07650	64.0–66.0	19.0–21.0	13.0–17.0	…	…	…	…	…	0.15	671	1240	718	1325	718–843	1325–1550
BAg-10	P07700	69.0–71.0	19.0–21.0	8.0–12.0	…	…	…	…	…	0.15	691	1275	738	1360	738–843	1360–1550
BAg-13	P07540	53.0–55.0	bal	4.0–6.0	…	0.5–1.5	…	…	…	0.15	718	1325	857	1575	857–968	1575–1775
BAg-13a	P07560	55.0–57.0	bal	…	…	1.5–2.5	…	…	…	0.15	771	1420	893	1640	871–982	1600–1800
BAg-18	P07600	59.0–61.0	bal	…	…	…	9.5–10.5	…	…	0.15	602	1115	718	1325	718–843	1325–1550
BAg-19	P07925	92.0–93.0	bal	…	…	…	…	0.15–0.30	…	0.15	760	1400	891	1635	877–982	1610–1800
BAg-20	P07301	29.0–31.0	37.0–39.0	30.0–34.0	…	…	…	…	…	0.15	677	1250	766	1410	766–871	1410–1600
BAg-21	P07630	62.0–64.0	27.5–29.5	…	…	2.0–3.0	5.0–7.0	…	…	0.15	691	1275	802	1475	802–899	1475–1650
BAg-22	P07490	48.0–50.0	15.0–17.0	21.0–25.0	…	4.0–5.0	…	…	7.0–8.0	0.15	680	1260	699	1290	699–830	1290–1525
BAg-23	P07850	84.0–86.0	…	…	…	…	…	…	bal	0.15	960	1760	970	1780	970–1038	1780–1900
BAg-24	P07505	49.0–51.0	19.0–21.0	26.0–30.0	…	1.5–2.5	…	…	…	0.15	660	1220	705	1305	705–843	1305–1550
BAg-26	P07250	24.0–26.0	37.0–39.0	31.0–35.0	…	1.5–2.5	…	…	1.5–2.5	0.15	705	1305	800	1475	800–870	1475–1600
BAg-27	P07251	24.0–26.0	34.0–36.0	24.5–28.5	12.5–14.5	…	…	…	…	0.15	605	1125	745	1375	745–860	1375–1575
BAg-28	P07401	39.0–41.0	29.0–31.0	26.0–30.0	…	…	1.5–2.5	…	…	0.15	650	1200	710	1310	710–843	1310–1550
BAg-33	P07252	24.0–26.0	29.0–31.0	26.5–28.5	16.5–18.5	…	…	…	…	0.15	607	1125	682	1260	682–760	1260–1400
BAg-34	P07380	37.0–39.0	31.0–33.0	26.0–30.0	…	…	1.5–2.5	…	…	0.15	650	1200	721	1330	721–843	1330–1550

(a) AWS, American Welding Society. (b) The brazing alloy shall be analyzed for the specific elements for which values are shown in this table. If the presence of other elements is indicated in the course of this work, the amount of those elements shall be determined to ensure that their total does not exceed the limit specified for other elements.

---

coating that serves as a desirable base for application of the main slurry. (2) To make the thin coating. (3) The thin coating itself.

precoated metal products. Mill products that have a metallic, organic, or conversion coating applied to their surfaces before they are fabricated into parts.

preferred orientation. A condition of a polycrystalline aggregate in which the crystal orientations are not random, but rather exhibit a tendency for alignment with a specific direction in the bulk material, commonly related to the direction of working; also called *texture*.

preforming. (1) The initial pressing of a metal powder to form a compact that is to be subjected to a subsequent pressing operation other than coining or sizing. Also the preliminary shaping of a refractory metal compact after presintering and before final sintering. (2) Preliminary forming operations, especially for impression-die forging.

preheating. (1) Heating before some further thermal or mechanical treatment. For tool steel, heating to an intermediate temperature immediately before final austenitizing. For some nonferrous alloys, heating to a high temperature for a long time, in order to homogenize the structure before working. (2) In welding and related processes, heating to an intermediate temperature for a short time immediately before welding, brazing, soldering, cutting, or thermal spraying.

presintering. Heating a powder metallurgy compact to a temperature lower than the normal temperature for final sintering, usually to increase ease of handling or forming, or to remove a lubricant or binder before sintering.

press. A machine tool having a stationary bed and a slide or ram that has reciprocating motion at right angles to the bed surface, the slide being guided in the frame of the machine.

---

press brake. An open-frame single-action press used to bend, blank, corrugate, curl, notch, perforate, pierce, or punch sheet metal or plate.

pressed density. The density of an unsintered powder metallurgy compact. Sometimes called green density.

press fit. An interference or *force fit* made through the use of a *press*.

press forging. *Forging* metal, usually hot, between dies in a press.

pressing. (1) In metalworking, the product or process of shallow drawing of sheet or plate. (2) Forming a powder metal part with compressive force.

pressing area. The clear distance (left to right) between housings, stops, gibs, gibways, or shoulders of strain rods, multiplied by the total distance from front to back on the bed of a *press*. Sometimes called working area.

pressing crack. A rupture in a green powder metallurgy compact that develops during ejection of the compact from the die; see also *capping* and *lamination*. Sometimes referred to as a slip crack.

pressure casting. (1) Making castings with pressure on the molten or plastic metal, as in injection molding, *die casting, centrifugal casting*, and cold-chamber pressure casting. (2) A casting made with pressure applied to the molten or plastic metal.

pressure gas welding. An oxyfuel gas welding process that produces coalescence simultaneously over the entire area of abutting surfaces by heating them with gas flames obtained from combustion of a fuel gas with oxygen and by application of pressure, without the use of filler metal.

primary creep. The first, or initial, stage of *creep*, or time-dependent deformation.

primary crystal. The first type of crystal that separates from a melt on cooling.

primary current distribution. The current distribution in an *electrolytic cell* that is free of *polarization*.

primary heat. The product of a single cycle of a batch melting process (ASTM A941).

primary metal. Metal extracted from minerals and free of reclaimed metal scrap. Compare with *secondary metal* and *native metal*.

primary mill. A mill for rolling ingots or the rolled products of ingots to blooms, billets, or slabs. This type of mill often is called a *blooming mill* and sometimes a cogging mill.

primes. Metal products, principally sheet and plate, of the highest quality and free from blemishes or other visible imperfections.

principal stresses. The normal stresses on three mutually perpendicular planes on which there are no shear stresses.

prismatic plane. In noncubic crystals, any plane that is parallel to the principal axis (*c* axis).

process annealing. An imprecise term denoting various treatments used to improve workability. For the term to be meaningful, the condition of the material and the time-temperature cycle used must be stated.

process metallurgy. The science and technology of winning metals from their ores and purifying metals; sometimes referred to as chemical metallurgy. Its two chief branches are *extractive metallurgy* and *refining*.

process tolerance. The dimensional variations of a part characteristic of a specific process, once the setup is made.

product analysis. A chemical analysis of a specimen taken from the semifinished product or from the finished product (ASTM A941).

profiling. Any operation that produces an irregular contour on a workpiece, for which a tracer or template-controlled duplicating equipment usually is employed.

progressive aging. Aging by increasing the temperature in steps or continuously during the aging cycle. See also *aging*, and compare with *interrupted aging* and *step aging*.

progressive die. A *die* in which two or more sequential operations are performed at two or more positions, the work being moved from station to station.

progressive forming. Sequential forming at consecutive stations either with a single die or with separate dies.

projection welding. A resistance welding process similar to spot welding, but in which the welds are localized at projections, embossments, or intersections.

promethium. A chemical element having atomic number 61, atomic weight 147, and the symbol Pm. The element was produced synthetically in 1945 by Charles Coryell, Jacob A. Marinsky, Lawrence E. Glendenin, and Harold G. Richter. They named the element for the mythological Prometheus who stole fire from the gods and was punished.

proof. Any reproduction of a die impression in any material; frequently a lead or plaster cast. See *die proof.*

proof load. A predetermined load, generally some multiple of the service load, to which a specimen or structure is submitted before acceptance for use.

proof stress. (1) The stress that will cause a specified small permanent set in a material. (2) A specified stress to be applied to a member or structure to indicate its ability to withstand service loads.

proportional limit. The maximum stress at which strain remains directly proportional to stress.

protactinium. A chemical element having atomic number 91, atomic weight 231, and the symbol Pa. Originally named protoactinium from the Greek *proto*, meaning first, and the element actinium (the element decays to actinium), which together mean "the parent of actinium." It was identified by Kasimir Fajans and Otto H. Gohring around 1915.

pseudobinary system. (1) A three-component or ternary alloy system in which an intermediate phase acts as a component. (2) A vertical section through a ternary diagram.

pseudocarburizing. Simulating the carburizing operation without introducing carbon. This usually is accomplished by using an inert material in place of the carburizing agent, or by applying a suitable protective coating to the ferrous alloy. Also known as *blank carburizing.*

pseudonitriding. Simulating the nitriding operation without introducing nitrogen. This usually is accomplished by using an inert material in place of the nitriding agent, or by applying a suitable protective coating to the ferrous alloy. Also known as *blank nitriding.*

puckering. Wrinkling or buckling in a drawn shell in an area originally inside the draw ring.

puddled iron. A type of wrought iron, made using a now obsolete technique that was employed in the Industrial Revolution, that had a very low carbon content and less slag and sulfur. By the late 18th century there was a demand for pig iron, an impure form of cast iron, to be refined with coal as fuel— which resulted in puddled iron. The iron was kept separate from the fire in a re-verberatory furnace to prevent phosphorus and sulfur from contaminating the finished iron. Puddled iron, although variable in its properties, was generally more consistent than the earlier irons, and the method lent itself to the production of larger quantities. By 1876, the annual production of puddled iron in Great Britain was over 3.6 million metric tons (4 million short tons).

pull cracks. In a casting, cracks that are caused by residual stresses produced during cooling, and that result from the shape of the object.

pulsation welding. Sometimes used as a synonym for *multiple-impulse welding.*

pulsed power welding. Any arc welding process in which the power is cyclically varied to give short-duration pulses of either voltage or current that are signifi-cantly different from the average value.

pulverization. Reducing metal to pow-der by mechanical means. The same as *comminution.*

punch. (1) The movable tool that forces material into the die in powder mold-ing and most metal forming operations. (2) The movable die in a trimming press or a forging machine. (3) The tool that forces the stock through the die in rod and tube extrusion, and forms the inter-nal surface in can or cup extrusion.

punching. Producing a hole by die shear-ing, in which the shape of the hole is controlled by the shape of the punch and its mating die; piercing. Multiple punch-ing of small holes is called *perforating.*

punch press. (1) In general, any mechani-cal press. (2) In particular, an endwheel gap-frame press with a fixed bed, used in piercing.

punch radius. The radius on the end of the punch that first contacts the work. Also known as *nose radius.*

punch-to-die clearance. The clearance be-tween a mated punch and die; commonly expressed as clearance per side Also known as *clearance* and *die clearance.*

push angle. The angle between a welding electrode and a line normal to the face of the weld when the electrode is pointing forward along the weld joint. See *fore-hand welding* (Fig. 30).

push bench. Equipment used for drawing moderately heavy-gage tubes by cupping sheet metal and forcing it through a die by pressure exerted against the inside bottom of the cup.

pusher furnace. A type of continuous furnace in which parts to be heated are periodically charged into the furnace in containers, which are pushed along the hearth against a line of previously charged containers, thus advancing the containers toward the discharge end of the furnace where they are removed.

push fit. A loosely defined fit similar to a *snug fit.*

push welding. Spot or projection welding in which the force is applied manually to one electrode, and the work or a backing bar takes the place of the other electrode. Also known as *poke welding.*

pyramidal plane. In noncubic crystals, any plane that intersects all three axes.

pyrometallurgy. High-temperature *win-ning* or *refining* of metals.

pyrometer. A device for measuring tem-peratures above the range of liquid thermometers.

quality. (1) The totality of features and characteristics of a product or service that bear on its ability to satisfy a given need (fitness-for-use concept of quality). (2) The degree of excellence of a product or service (comparative concept). Often determined subjectively by comparison against an ideal standard or against similar products or services available from other sources. (3) A quantitative evaluation of the features and characteristics of a product or service (quantitative concept).

quality characteristic. Any dimension, mechanical property, physical property, functional characteristic, or appearance characteristic that can be used as a basis for measuring the quality of a unit of product or service.

quantitative metallography. The determination of specific characteristics of a microstructure by conducting quantitative measurements on micrographs or metallographic images. Quantities so measured include volume concentration of phases, grain size, particle size, mean free path between like particles or secondary phases, and surface-area-to-volume ratio of microconstituents, particles, or grains.

quarter hard. A *temper* of nonferrous alloys and some ferrous alloys characterized by tensile strength about midway between those of *dead soft* and *half hard* tempers.

quasi-binary system. In a ternary or higher-order system, a linear composition series between two substances each of which exhibits congruent melting, wherein all equilibria, at all temperatures or pressures, involve only phases having compositions occurring in the linear series, so that the series may be represented as binary on a *constitution diagram*.

quench-age embrittlement. The *embrittlement* of low-carbon steel evidenced by a loss of ductility on aging at room temperature following rapid cooling from a temperature below the lower critical temperature.

quench aging. Aging induced by rapid cooling after *solution heat treatment.*

quench annealing. Annealing an austenitic ferrous alloy by *solution heat treatment* followed by rapid quenching.

quench cracking. The fracturing of a metal during quenching from elevated temperature. It is most frequently observed in hardened carbon steel, alloy steel, or tool steel parts of high hardness and low toughness. Cracks often emanate from fillets, holes, corners, or other stress raisers and result from high stresses due to the volume changes accompanying transformation to *martensite.*

quench hardening. (1) Hardening suitable α-β alloys (most often certain copper or titanium alloys) by solution treating and quenching to develop a martensite-like structure. (2) In ferrous alloys, hardening by *austenitizing* and then cooling at a rate such that a substantial amount of *austenite* transforms to *martensite.*

quenching. The rapid cooling of a metal (often steel) from a suitable elevated temperature. Quenching is typically done by immersing the metal in water, oil, a polymer solution, or salt. When applicable, the following more specific terms should be used: *direct quenching, fog quenching, hot quenching, interrupted quenching, selective quenching, spray quenching, and time quenching.*

quench time. In resistance welding, the time from the finish of the welding operation to the beginning of tempering. Also called chill time.

quill. (1) A hollow or tubular shaft, designed to slide or revolve, carrying a rotating member within itself. (2) A removable spindle projection for supporting a cutting tool or grinding wheel.

rabbit ear. A recess in the corner of a die to allow for wrinkling or folding of the blank.

rabble. A hoe-like bladed tool or similar device used for stirring molten metal.

radial draw forming. The forming of metals by the simultaneous application of tangential stretch and radial compression forces, which is performed gradually by tangential contact with the die member. This type of forming is characterized by very close dimensional control.

radial forging. A process using two or more moving anvils or dies for producing shafts with constant or varying diameters along their length, or tubes with internal or external variations in diameter; also known as *draw forging* or rotary swaging.

radial marks. Lines on a fracture surface that radiate from the fracture origin and are visible to the unaided eye or at low magnification. Radial marks result from the intersection and connection of brittle fractures propagating at different levels. Also called *shear ledges*. See also *chevron pattern.*

radial rake. The angle between the tooth face and a radial line passing through the cutting edge in a plane perpendicular to the cutter axis. See the figure accompanying the term *face mill.*

radial runout. For any rotating element, the total variation from true radial position, taken in a plane perpendicular to the axis of rotation. Compare with *lateral runout.*

radiation damage. A general term for the alteration of properties of a material arising from exposure to ionizing radiation (penetrating radiation) such as x-rays, gamma rays, neutrons, heavy-particle radiation, or fission fragments in nuclear fuel material. See also *neutron embrittlement.*

radiation dose. The accumulated exposure to ionizing radiation during a specified period of time.

radiation energy. The energy of a given photon or particle in a beam of radiation, often expressed in electron volts.

radiation gage. An instrument for measuring the intensity and quantity of ionizing radiation.

radiation intensity. In general, the quantity of radiant energy at a specified location passing perpendicularly through unit area in unit time. It may be expressed as number of particles or photons per square centimeter per second, or in energy units as $J/m^2 \cdot s$ or Rhm.

radiation monitoring. The continuous or periodic measurement of the intensity of radiation received by personnel or present in any particular area.

radiation quality. The spectrum of radiation produced by a radiation source, with respect to its penetrating power or its suitability for a given application.

radioactive element. An element that has at least one isotope that undergoes spontaneous nuclear disintegration to emit positive α particles, negative β particles, or γ rays.

radioactivity. The spontaneous nuclear disintegration with emission of corpuscular or electromagnetic radiation.

radiograph. A photographic shadow image resulting from uneven absorption of penetrating radiation in a test object.

radiography. A method of nondestructive inspection in which a test object is exposed to a beam of x-rays or gamma rays, and the resulting shadow image of the object is recorded on photographic film placed behind the object. Internal discontinuities are detected by observing and interpreting variations in the image caused by differences in thickness, density, or absorption within the test object. Variations of radiography include *fluoroscopy,* electron radiography, and neutron radiography.

radioisotope. An isotope that emits ionizing radiation during its spontaneous decay.

radionuclide tracer element. A radioactive isotope of an element used to study the movement and behavior of atoms by observing the distribution and intensity of radioactivity.

radium. A chemical element having atomic number 88, atomic weight 226, and the symbol Ra, named for the Latin *radius,* meaning ray. Radium was isolated and identified by Marie and Pierre Curie in 1898. They named the element radium because of its unusual radioactive properties. Radium has no important commercial uses, and its use as a source of x-rays has been replaced by the use of isotopes of cobalt, tantalum, and iridium.

rake. The angular relationship between the tooth face, or a tangent to the tooth face at a given point, and a given reference plane or line. See *face mill* (Fig. 26) and *single-point tool* (Fig. 50) .

ram. The moving member of a hammer, machine, or press to which a tool is fastened.

ramming. Packing sand, refractory, or other material into a compact mass.

ramoff. A casting imperfection resulting from the movement of sand away from the pattern because of improper ramming.

random sequence. A longitudinal welding sequence wherein the weld-bead increments are deposited at random to minimize distortion. The same as *wandering sequence.*

range. In inspection, the difference between the highest and lowest values of a given *quality characteristic* within a single *sample.*

rare earth metal. One of a group of chemically similar metals. The rare earth metals are the 15 lanthanides having atomic numbers 57 through 71—lanthanum, cerium, praseodymium, neodymium, promethium, samarium, europium, gadolinium, terbium, dysprosium, holmium, erbium, thulium, ytterbium, and lutetium—as well as the Group IIIA elements scandium and yttrium.

The rare earth metals have become extremely valuable in recent years for use in compact fluorescent light bulbs, computer hard drives, magnets in the motors of electric vehicles, and wind turbines. See *Technical Note 10*.

ratcheting. Progressive cyclic inelastic deformation (growth, for example) that occurs when a component or structure is subjected to a cyclic secondary stress superimposed on a sustained primary stress. The process is called thermal ratcheting when cyclic strain is induced by cyclic changes in temperature, and isothermal ratcheting when cyclic strain is mechanical in origin (even though accompanied by cyclic changes in temperature).

ratchet marks. Lines or markings on a fatigue fracture surface that result from the intersection and connection of fatigue fractures propagating from multiple origins. Ratchet marks are parallel to the overall direction of crack propagation and are visible to the unaided eye or at low magnification.

rate of strain hardening. The rate of change of *true stress* with respect to *true strain* in the plastic range.

rattail. A surface imperfection on a casting, occurring as one or more irregular lines, caused by expansion of sand in the mold. Compare with *buckle* (2).

RE. Abbreviation for rare earth, as in *rare earth metal* or elements.

reactive metal. A metal that readily combines with oxygen at elevated temperatures to form very stable and undesirable oxides. Examples include titanium, hafnium, beryllium, and zirconium. Reactive metals also may become embrittled by the interstitial absorption of oxygen, hydrogen, and nitrogen.

reamed extrusion ingot. A cast hollow extrusion ingot that has been machined to remove the original inside surface.

reamer. A rotary cutting tool with one or more cutting elements called teeth, used for enlarging a hole to desired size and contour. It is principally supported by the metal around the hole it cuts.

recalescence. A phenomenon, associated with the transformation of γ iron to α iron on cooling (supercooling) of iron or steel, revealed by the brightening (reglowing) of the metal surface owing to the sudden increase in temperature caused by the fast liberation of the latent heat of transformation. Contrast with *decalescence*.

recarburize. (1) To increase the carbon content of molten cast iron or steel by adding carbonaceous material, high-carbon pig iron, or a high-carbon alloy. (2) To carburize a metal part to return surface carbon lost in processing; also known as *carbon restoration*.

recess. A groove or depression in a surface.

reclaim rinse. A nonflowing rinse used to recover *drag-out*.

recoil line. A blemish on a drawn sheet metal part caused by a slight change in metal thickness, which results from the transfer of the blank from the die to the punch during forming, or from a reaction to the blank being pulled sharply through the draw ring. See *impact line*.

recovery. (1) The reduction or removal of work-hardening effects in metals, without motion of large-angle grain boundaries. (2) The proportion of the desired component obtained by processing an ore, usually expressed as a percentage.

TECHNICAL NOTE 10

Rare Earth Metals

THE RARE EARTH METALS include the Group IIIA elements scandium, yttrium, and the lanthanide elements (lanthanum, cerium, praseodymium, neodymium, promethium, samarium, europium, gadolinium, terbium, dysprosium, holmium, erbium, thulium, ytterbium, and lutetium) in the periodic table of the elements. The term *rare* implies that these elements are scarce; in fact, the rare earths are quite abundant. Of the 83 naturally occurring elements, the 16 naturally occurring rare earths as a group lie in the 50th percentile of elemental abundances. Cerium, the most abundant, ranks 28th; thulium, the least abundant, ranks 63rd.

Despite the view by some researchers that the rare earth elements are so chemically similar to one another that collectively they can be considered as one element, a closer examination reveals vast differences in their behaviors and properties. For example, the melting points of the lanthanide elements vary by a factor of almost two between lanthanum (918 °C, or 1684 °F) and lutetium (1663 °C, or 3025 °F). In addition, the modulus of elasticity for these elements varies from as low as ~18 GPa (2.6×10^6 psi) for europium to more than 74 GPa (10.7×10^6 psi) for thulium.

Rare earth elements are found in nature intimately mixed in varying proportions depending on the ore. Separation into pure component rare earths is done on a large scale by liquid-liquid extraction and by ion exchange on a smaller scale. Because impurities significantly influence the properties of rare earth metals, impurity levels must be kept as low as possible. Research-grade metals are usually ≥99.8 at.% pure, although ≥99.95 at.% metals can be prepared. Commercial-grade rare earth metals are approximately 98 at.% pure, but occasionally can be as low as 95 at.% pure. Hydrogen and oxygen are the major impurities in both grades.

The primary application for rare earth metals is as an alloying additive. In many of these applications, the rare earths are added in the form of *mischmetal*, which has the approximate rare earth distribution of 50% Ce, 30% La, 15% Nd, and 5% Pr. Additions of pure metals or finely dispersed rare earth-based oxides (primarily Y_2O_3) are also used. Materials that rare earths are added to include ductile iron (modify carbon morphology), superalloys (increase operating temperatures) magnesium alloys (improve creep resistance), aluminum alloys (improve tensile strength and corrosion resistance), oxygen-free high-conductivity copper (improve oxidation resistance), and dispersion-strengthened materials (improve high-temperature properties). Other key applications include their use as lighter flints, permanent magnet materials (samarium-cobalt and neodymium-iron-boron magnets), magnetooptical materials, and hydrogen storage batteries.

Selected References

- K.A. Gschneidner, B.J. Beaudry, and J. Cappellen, Rare Earth Metals, *Metals Handbook*, 10th ed., Vol 2, ASM International, 1990, p 720–732
- K.A. Gschneidner and B.J. Beaudry, Properties of the Rare Earth Metals, *Metals Handbook*, 10th ed., Vol 2, ASM International, 1990, p 1178–1189

Commercial alloys containing rare earth metals

Designation	Alloy type	Rare earth	Composition, wt%	Remarks
AiResist 13	Co superalloy	Y	0.1	High-temperature parts
AiResist 213	Co superalloy	Y	0.1	Hot corrosion resistance
AiResist 215	Co superalloy	Y	0.17	Hot corrosion resistance
FSX 418	Co superalloy	Y	0.15	Oxidation resistance
FSX 430	Co superalloy	Y	0.03–0.1	Oxidation and hot corrosion resistance
Haynes 188	Co superalloy	La	0.05	Oxidation resistance, strength
Haynes 1002	Co superalloy	La	0.05	…
Melco 2	Co superalloy	Y	0.15	…
Melco 9	Co superalloy	Y	0.13	…
Melco 10	Co superalloy	Y	0.10	…
Melco 14	Co superalloy	Y	0.18	…
C-207	Cr	Y	0.15	…
Cl-41	Cr	Y + La	0.1 (total)	…
253	Fe superalloy	Ce	0.055	…
GE 1541	Fe superalloy	Y	1.0	…
GE 2541	Fe superalloy	Y	1.0	…
Haynes 556	Fe superalloy	La	0.02	High temperature, up to 1095 °C
ICF 42	High-strength steel	R	…	(a)
ICF 45	High-strength steel	R	…	(a)
ICF 50	High-strength steel	R	…	(a)
VAN 50	High-strength steel	Ce	…	(a)
VAN 60	High-strength steel	Ce	…	(a)
VAN 70	High-strength steel	Ce	…	(a)
VAN 80	High-strength steel	Ce	…	(a)
EK 30A	Mg (Zr, Zn)	R	3.0	Creep resistance
EK 41A	Mg (Zr, Zn)	R	4.0	Creep resistance
EZ 33A	Mg (Zr, Zn)	R	3.0	Creep resistance
QE 22A	Mg (Zr, Ag)	R	1.2–3.0	Creep resistance
QE 222A	Mg (Zr, Ag)	Dm(b)	2	Creep resistance
WE 54	Mg (Zr)	Y + R	5.25 + 3.5	High strength, weldability
ZE 10A	Mg (Zn)	R	0.17	Creep resistance
ZE 41A	Mg (Zr, Zn)	R	1.2	…
ZE 63A	Mg (Zr, Zn)	R	2–3	Creep resistance
ZE 63B	Mg (Zr, Zn, Ag)	R	2–3	Creep resistance
C129Y	Nb	Y	0.1	…
Hastelloy N	Ni superalloy	Y	0.26	…
Hastelloy S	Ni superalloy	La	0.05	High stability
Hastelloy T	Ni superalloy	La	0.02	Low thermal expansion
Haynes 214	Ni superalloy	Y	0.02	Oxidation resistance
Haynes 230	Ni superalloy	La	0.5	High-temperature strength
Melni 19	Ni superalloy	La	0.17	…
Melni 22	Ni superalloy	La	0.16	…
René Y	Ni superalloy	La	0.05–0.3	…
Udimet 500 + Ce	Ni superalloy	Ce	…	…
Unimet 700 + Ce	Ni superalloy	Ce	0.2–0.5	…

(a) Rare earth (R) or cerium added for inclusion shape control (b) Dm, Didymium, alloy of 80Nd-20Pr.

recrystallization. (1) The formation of a new, strain-free grain structure from that existing in cold worked metal, usually accomplished by heating. (2) The change from one crystal structure to another, as occurs on heating or cooling through a critical temperature.

recrystallization annealing. Annealing cold worked metal to produce a new grain structure without phase change.

recrystallization temperature. The approximate minimum temperature at which complete recrystallization of a cold worked metal occurs within a specified time.

recuperator. Equipment for transferring heat from gaseous products of combustion to incoming air or fuel. The incoming material passes through pipes surrounded by a chamber through which the outgoing gases pass.

red brass. An American term sometimes used for the Cu-Zn-Sn alloy known as gunmetal, which technically is not a brass. It is also referred to as *ounce metal.*

red mud. A residue, containing a high percentage of iron oxide, obtained in purifying bauxite in the production of alumina in the *Bayer process.*

redrawing. Drawing metal after a previous cupping or drawing operation.

reducing agent. A substance that causes reduction. See *reduction* (3).

reducing flame. A gas flame produced with excess fuel in the inner flame.

reduction. (1) In cupping and deep drawing, a measure of the percentage decrease from blank diameter to cup diameter, or of diameter reduction in redrawing. (2) In forging, rolling, and drawing, either the ratio of the original to final cross-sectional area or the percentage decrease in cross-sectional area. (3) A reaction in which there is a decrease in valence resulting from a gain in electrons. Contrast with *oxidation.*

reduction cell. A pot or tank in which either a water solution of a salt or a fused salt is reduced electrolytically to form free metals or other substances.

reduction of area. (1) Commonly, the difference, expressed as a percentage of original area, between the original cross-sectional area of a tensile test specimen and the minimum cross-sectional area measured after complete separation. (2) The difference, expressed as a percentage of original area, between the original cross-sectional area and that after straining of the specimen.

reeding. The operation of forming serrations and corrugations in metals by coining or embossing.

reel. (1) A spool or hub for coiling or feeding wire or strip. (2) To straighten and planish a round bar by passing it between contoured rolls.

reel breaks. Transverse breaks or ridges on successive inner laps of a coil that result from the crimping of the lead end of the coil into a gripping segmented mandrel. Also called reel kinks.

reference plane. (1) The plane that contains the cutter axis and the point of the cutting edge. See *face mill* (Fig. 26). (2) A plane from which measurements are made.

refining. The branch of *process metallurgy* that deals with the purification of crude or impure metals. Compare with *extractive metallurgy.*

reflector sheet. A clad product consisting of a facing layer of high-purity aluminum capable of taking a high polish, for reflecting heat or light, and a base of commercially pure aluminum or an aluminum-manganese alloy, for strength and formability.

reflowing. The melting of an electrodeposit followed by solidification. The surface has the appearance and physical characteristics of a hot dipped surface (especially tin or tin alloy plates). Also called *flow brightening.*

refractory. (1) A material of very high melting point with properties that make it suitable for such uses as furnace linings and kiln construction. (2) The quality of resisting heat.

refractory alloy. (1) A heat-resistant alloy. (2) An alloy having an extremely high melting point. See also *refractory metal.* (3) An alloy difficult to work at elevated temperatures.

refractory metal. A metal having an extremely high melting point; for example, tungsten, molybdenum, tantalum, niobium (columbium), chromium, vanadium, and rhenium. Broadly, metals having melting points above the range for iron, cobalt, and nickel.

regenerator. Same as *recuperator* except that the gaseous products of combustion heat brick checkerwork in a chamber connected to the exhaust side of the furnace while the incoming air and fuel are being heated by the brick checkerwork in a second chamber, connected to the entrance side. At intervals, the gas flow is reversed so that incoming air and fuel contact hot checkerwork while that in the second chamber is being reheated by exhaust gases.

regulus. The impure button, globule, or mass of metal formed beneath the slag in the smelting and reduction of ores. The name was first applied by alchemists to metallic antimony because it readily alloyed with gold.

rejectable. A characteristic of a product or service that does not meet normal quality criteria for that product or service. See preferred term *nonconforming.*

reliability. A quantitative measure of the ability of a product or service to fulfill its intended function for a specified period of time.

relief. The result of the removal of tool material behind or adjacent to the cutting edge, to provide clearance and to prevent rubbing (heel drag). See *single-point tool* (Fig. 50) .

relief angle. The angle between a relieved surface and a given plane tangent to a cutting edge or to a point on a cutting edge. See *single-point tool* (Fig. 50).

relieving. Buffing or other abrasive treatment of the high points of an embossed metal surface to produce highlights that contrast with the finish in the recesses.

remanence. The magnetic induction remaining in a magnetic circuit after removal of the applied magnetizing force. It sometimes is called remanent induction.

remelted heat. The product of the remelting of a primary heat, in whole or in part.

re-pressing. The application of pressure to a previously pressed and sintered powder metallurgy compact, usually for the purpose of improving some physical or mechanical property, or for dimensional accuracy.

residual elements. Elements present in an alloy in small quantities, but not added intentionally.

residual field. The magnetic field that remains in a part after the magnetizing force has been removed. Also known as *residual magnetic field.*

residual magnetic field. The magnetic field that remains in a part after the magnetizing force has been removed. Also known as *residual field.*

residual method. The method of *magnetic-particle inspection* in which the particles are applied after the magnetizing force has been removed.

residual stress. The stress present in a body that is free of external forces or thermal gradients.

resilience. (1) The amount of energy per unit volume released on unloading. (2) The capacity of a metal, by virtue of high yield strength and low elastic modulus, to exhibit considerable elastic recovery on release of load.

resinoid wheel. A grinding wheel bonded with a synthetic resin.

resist. (1) A material applied to a part of a cathode or plating rack to render the surface nonconductive. (2) A material applied to a part of the surface of an article to prevent reaction of metal from that area during chemical or electrochemical processes. (3) A material applied to prevent flow of brazing filler metal into unwanted areas.

resistance alloys. See *Technical Note 11.*

resistance brazing. Brazing by resistance heating, the joint being part of the electrical circuit.

resistance soldering. Soldering in which the joint is heated by electrical resistance. Filler metal is either face fed into the joint or preplaced in the joint.

resistance welding. Welding with resistance heating and pressure, the work being part of the electrical circuit; for example, resistance *spot welding,* resistance *seam welding, projection welding,* and *flash butt welding.*

resistance welding die. The part of a resistance welding machine, usually shaped to the work contour, in which the parts being welded are held and that conducts the welding current.

resolution. The ability to separate closely spaced forms or entities using a given test method; also a quantitative measure of the degree to which they can be discriminated.

restraint. Any external mechanical force that prevents a part from moving to accommodate changes in dimensions due to thermal expansion or contraction. Often applied to weldments made while clamped in a fixture. Compare with *constraint.*

restriking. (1) Striking a trimmed but slightly misaligned or otherwise faulty forging with one or more blows to improve alignment, improve surface condition, maintain close tolerances, increase hardness, or effect other improvements. (2) A *sizing* operation in which coining or stretching is used to correct or alter profiles and to counteract distortion.

resultant field. The magnetic field that is the result of two or more magnetizing forces impressed on the same area of a magnetizable object. Also known as *vector field.*

resultant rake. The angle between the tooth face and an axial plane through the tooth point measured in a plane perpendicular to the cutting edge. The resultant rake of a cutter is a function of three other angles: radial rake, axial rake, and corner angle. See *face mill* (Fig. 26).

retentivity. The capacity of a material to retain a portion of the magnetic field set up in it after the magnetizing force has been removed.

TECHNICAL NOTE 11

Resistance Alloys

ELECTRICAL RESISTANCE ALLOYS include both the types used in instruments and control equipment to measure and regulate electrical characteristics and those used in furnaces and appliances to generate heat. In the former applications, properties near ambient temperature are of primary interest; in the latter, elevated-temperature characteristics are of prime importance. In common commercial terminology, electrical resistance alloys used for control or regulation of electrical properties are called resistance alloys, and those used for generation of heat are referred to as resistance heating alloys.

The primary requirements for resistance alloys are uniform resistivity, stable resistance (no time-dependent aging effects), reproducible temperature coefficient of resistance, and low thermoelectric potential versus copper. Properties of secondary importance are coefficient of expansion, mechanical strength, ductility, corrosion resistance, and ability to be joined to other metals by soldering, brazing, or welding. Alloys must be strong enough to withstand fabrication operations, and it must be easy to procure an alloy that has consistently reproducible properties in order to ensure resistor accuracy.

Resistors for electrical and electronic devices may be divided into two arbitrary classifications: those employed in precision instruments in which overall error is considerably less than 1%, and those employed where less precision is needed. The choice of alloy for a specific resistor application depends on the variation in properties that can be tolerated. Materials for resistors include: copper-nickel (2-22% Ni) alloys, generally referred to as radio alloys; copper-manganese-nickel alloys (10-13% Mn and 4% Ni), generally referred to as manganins; constantin alloys, whose compositions vary from 50Cu-50Ni to 65Cu-35Ni; nickel-chromium-aluminum alloys that nominally contain 20% Cr, 3% Al, and 2-5% of copper, iron, and/or manganese; 80Ni-20Cr alloys; and iron-chromium-aluminum alloys (nominally 73Fe-22Cr-5Al).

Resistance heating elements are used in many varied applications—from small household appliances to large industrial process heating systems and furnaces that may operate continuously at temperatures of 1300 °C (2350 °F) or higher. The primary requirements of materials used for heating elements are high melting point, high electrical resistivity, reproducible temperature coefficient of resistance, good oxidation resistance, absence of volatile components, and resistance to contamination. Other desirable properties are good creep strength, high emissivity, low thermal expansion and low modulus (both of which help to minimize thermal fatigue), good resistance to thermal shock, and good strength and ductility at fabricating temperature.

The most commonly used resistance heating alloys are nickel-chromium and nickel-chromium-iron alloys (see table above). Other materials include iron-chromium-aluminum alloys similar in composition to resistor alloys,

TECHNICAL NOTE 11 (continued)

Typical properties of resistance heating materials

Basic composition	Resistivity(a), Ω·mm²/m(b)	Average change in resistance(c), %, from 20 °C to:				Thermal expansion, µm/m·°C, from 20 °C to:			Tensile strength		Density	
		260 °C	540 °C	815 °C	1095 °C	100 °C	540 °C	815 °C	MPa	ksi	g/cm³	lb/in.³
Nickel-chromium and nickel-chromium-iron alloys												
78.5Ni-20Cr-1.5Si (80-20)	1.080	4.5	7.0	6.3	7.6	13.5	15.1	17.6	655–1380	95–200	8.41	0.30
77.5Ni-20Cr-1.5Si:1Nb	1.080	4.6	7.0	6.4	7.8	13.5	15.1	17.6	655–1380	95–200	8.41	0.30
68.5Ni-30Cr-1.5Si (70-30)	1.180	2.1	4.8	7.6	9.8	12.2	...	...	825–1380	120–200	8.12	0.29
68Ni-20Cr-8.5Fe-2Si	1.165	3.9	6.7	6.0	7.1	...	12.6	...	895–1240	130–180	8.33	0.30
60Ni-16Cr-22Fe-1.5Si	1.120	3.6	6.5	7.6	10.2	13.5	15.1	17.6	655–1205	95–175	8.25	0.30
37Ni-21Cr-40Fe-2Si	1.08	7.0	15.0	20.0	23.0	14.4	16.5	18.6	585–1135	85–165	7.96	0.288
35Ni-20Cr-43Fe-1.5Si	1.00	8.0	15.4	20.6	23.5	15.7	15.7	...	550–1205	80–175	7.95	0.287
35Ni-20Cr-42.5Fe-1.5Si:1Nb	1.00	8.0	15.4	20.6	23.5	15.7	15.7	...	550–1205	80–175	7.95	0.287
Iron-chromium-aluminum alloys												
83.5Fe-13Cr-3.25Al	1.120	7.0	15.5	...	...	10.6	...	...	620–1035	90–150	7.30	0.26
81Fe-14.5Cr-4.25Al	1.25	3.0	9.7	16.5	...	10.8	11.5	12.2	620–1170	90–170	7.28	0.26
73.5Fe-22Cr-4.5Al	1.35	0.3	2.9	4.3	4.9	10.8	12.6	13.1	620–1035	90–150	7.15	0.26
72.5Fe-22Cr-5.5Al	1.45	0.2	1.0	2.8	4.0	11.3	12.8	14.0	620–1035	90–150	7.10	0.26
Pure metals												
Molybdenum	0.052	110	238	366	508	4.8	5.8	...	690–2160	100–313	10.2	0.369
Platinum	0.105	85	175	257	305	9.0	9.7	10.1	345	50	21.5	0.775
Tantalum	0.125	82	169	243	317	6.5	6.6	...	345–1240	50–180	16.6	0.600
Tungsten	0.055	91	244	396	550	4.3	4.6	4.6	3380–6480	490–940	19.3	0.697
Nonmetallic heating-element materials												
Silicon carbide	0.995–1.995	–33	–33	–28	–13	4.7	...	...	28	4	3.2	0.114
Molybdenum disilicide	0.370	105	222	375	523	9.2	...	...	185	27	6.24	0.225
MoSi₂ + 10% ceramic additives	0.270	167	370	597	853	13.1	14.2	14.8	...	...	5.6	0.202
Graphite	9.100	–16	–18	–13	–8	1.3	...	...	1.8	0.26	1.6	0.057

(a)At 20 °C (68 °F), (b) To convert to Ω-circ mil/ft, multiply by 601.53. (c) Changes in resistances may vary somewhat, depending on cooling rate.

TECHNICAL NOTE 11 (*continued*)

high-melting-temperature pure metals, and nonmetallic materials, which can be used effectively at temperatures as high as 1900 °C (3450 °F).

Selected Reference

- R.A. Watson et al., Electrical Resistance Alloys, *Metals Handbook*, 10th ed., Vol 2, ASM International, 1990, p 822–839

---

retort. A vessel used for the distillation of volatile materials, as in the separation of some metals and in the destructive distillation of coal.

reverberatory furnace. A furnace with a shallow hearth, usually nonregenerative, having a roof that deflects the flame and radiates heat toward the hearth or the surface of the charge.

reverse-current cleaning. *Electrolytic cleaning* in which the work is the anode. Also known as *anodic cleaning.*

reverse drawing. *Redrawing* of a sheet metal part in a direction opposite to that of the original drawing.

reverse flange. A sheet metal flange made by shrinking, as opposed to one formed by stretching.

reverse polarity. Direct-current arc welding circuit arrangement in which the electrode is connected to the positive terminal. Contrast with *straight polarity.*

reverse redrawing. A second drawing operation in a direction opposite to that of the original drawing.

rhenium. A chemical element having atomic number 75, atomic weight 186, and the symbol Re. Rhenium was discovered in Germany in 1925 by Walter Noddack, Ida Tacke-Noddack, and Otto Berg, who named the metal in honor of the River Rhine (*Rhein* in German). The high hardness of the metal achieved during cold working allows it to be used in pen nibs and as pivot bearing points in scientific instruments. It is readily applied by electroplating and used to coat the inside of tanks for transporting and storing acids.

rheology. The science of deformation and the flow of matter.

rheotropic brittleness. That portion of the brittleness characteristic of non-face-centered cubic metals, when tested in the presence of a stress concentration or at low temperatures or high strain rates, that may be eliminated by prestraining under milder conditions.

rhodium. A chemical element having atomic number 45, atomic weight 103, and the symbol Rh, named for the Greek *rhodon,* meaning rose, because of its rose-colored salts. The element was identified and named by William Hyde Wollaston in 1803. Rhodium is one of the six metals of the platinum group and is always found in association with metals of that group, which also includes platinum, osmium, iridium, palladium, and ruthenium.

Rhodium plating is used for jewelry and has high hardness and excellent light reflectivity, making it an attractive, economical alternative to platinum. Rhodium coatings are used in searchlight reflectors and to concentrate heat in infrared ovens. Chemical equipment and apparatuses now are commonly rhodium plated because its chemical resistance is comparable to that of platinum and it is much harder. Electrical contacts that are rhodium plated on both surfaces have a zero electrochemical potential and will withstand severe hammering and high corrosive conditions up to red heat. Rhodium-platinum alloys are used for high temperature furnace windings, potentiometers, and in gauze and powder used as catalysts in chemical engineering plants.

rich low brass. A copper alloy with 15% Zn, known as "Tombac" and having the designation UNS 23000. It is often used in jewelry applications.

riddle. A sieve used to separate foundry sand or other granular materials into various particle-size grades or to free such a material of undesirable foreign matter.

rigging. The engineering design, layout, and fabrication of pattern equipment for producing castings; including a study of the casting solidification program, feeding and gating, risering, skimmers, and fitting flasks.

right-hand cutting tool. A cutter all of whose flutes twist away in a clockwise direction when viewed from either end.

rimmed steel. A low-carbon steel containing sufficient iron oxide to give a continuous evolution of carbon monoxide while the ingot is solidifying, resulting in a case or rim of metal virtually free of voids. Sheet and strip products made from rimmed steel ingots have very good surface quality.

ring and circle shear. A cutting or shearing machine with two rotary-disk cutters driven in unison and equipped with a circle attachment for cutting inside circles or rings from sheet metal, where it is impossible to start the cut at the edge of the sheet. One cutter shaft is inclined to the other to provide cutting clearance so that the outside section remains flat and usable. See also *circle shear.*

ringing. The audible or ultrasonic tone produced in a mechanical part by shock and having the natural frequency or frequencies of the part. The quality, amplitude, or decay rate of the tone may sometimes be used to indicate quality or soundness. See also *sonic testing* and *ultrasonic testing.*

ring riser. A *riser block* with openings matching those in the press bed.

ring rolling. The process of shaping weldless rings from pierced disks, or thick-walled, ring-shaped blanks between rolls that control wall thickness, ring diameter, height, and contour.

rinsability. The relative ease with which a substance can be removed from a metal surface with a liquid such as water.

riser. A reservoir of molten metal connected to a casting to provide additional metal to the casting, required as the result of shrinkage before and during solidification.

riser blocks. (1) Plates or pieces inserted between the top of a press bed or bolster and the die to decrease the height of the die space. (2) Spacers placed between bed and housings to increase *shut height* on a four-piece tie-rod straight-side press.

river pattern. A term used in *fractography* to describe a characteristic pattern of cleavage steps running parallel to the local direction of crack propagation on the fracture surfaces of grains that have separated by *cleavage.*

riveting. The joining of two or more members of a structure by means of metal rivets, the unheaded end being upset after the rivet is in place.

roasting. Heating an ore to effect some chemical change that will facilitate smelting.

robber. An extra cathode or cathode extension that reduces the current density on what would otherwise be a high-current-density area on work being electroplated.

Roberts-Austen, William Chandler. 1843–1902. A chemist and metallurgist, Roberts-Austen became a chemist and assayer at the Royal Mint, a position he retained throughout his career. He also became a metallurgist and was especially interested in the new field of metallography that Sorby had introduced. Until late in the 19th century, photographs of crystalline structures at high magnifications were very rare. Roberts-Austen devoted a great deal of effort to developing the art at a time when men were giving their names to the various crystal structures found in steel, including sorbite, Sorby's structure; martensite for the German, Martens; and ledeburite for Ledebur. It is said that a colleague of Austen honored him by giving the elevated-temperature steel structure the name *austenite*.

Rochelle copper. (1) A copper electrodeposit obtained from copper cyanide plating solution to which Rochelle salt (sodium potassium tartrate) has been added for grain refinement, better anode corrosion, and cathode efficiency. (2) The solution from which a Rochelle copper electrodeposit is obtained.

rock candy fracture. A fracture that exhibits separated-grain facets; most often used to describe intergranular fractures in large-grained metals.

rocking shear. A type of guillotine shear that uses a curved blade to shear sheet metal progressively from side to side by a rocker motion.

Rockrite tube-reducing process. Reducing both the diameter and wall thickness of tubing with a mandrel and a pair of rolls with tapered grooves. The Rockrite process uses a fixed tapered mandrel, and the rolls reciprocate along the tubing with corresponding reversal in rotation. Roll reliefs at the initial and final diameters permit, respectively, advance and rotation of the tubing. See also *tube reducing*.

Rockwell hardness test. An indentation hardness test based on the depth of penetration of a specified penetrator into a specimen under certain arbitrarily fixed conditions. The Rockwell hardness tester was co-invented and patented by Connecticut natives Hugh Rockwell and Stanley Rockwell, not direct relations, who, at the time of invention, worked for a Connecticut ball bearing manufacturer.

rod mill. (1) A *hot mill* for rolling rod. (2) A mill for fine grinding, somewhat similar to a *ball mill*, but employing long steel rods instead of balls to effect grinding.

roll bending. Curving sheets, bars, and sections by means of rolls. See *bending rolls*.

roll compacting. The progressive compacting of metal powders by the use of a rolling mill.

rolled gold. The same as *gold filled* except that the proportion of gold alloy to the weight of the entire article may be less than $\frac{1}{20}$th. Fineness of the gold alloy may not be less than 10 k.

roller leveler breaks. Obvious transverse *breaks* typically approximately 3 to 6 mm

(⅛ to ¼ in.) apart caused by the sheet metal fluting during *roller leveling*. These will not be removed by stretching.

roller leveler lines. Lines on sheet or strip running transverse to the direction of *roller leveling*; they may be seen on stoning or light sanding after leveling (but before drawing). They usually can be removed by moderate stretching. Also known as *leveler lines.*

roller leveling. *Leveling* by passing flat sheet metal stock through a machine having a series of small-diameter staggered rolls that are adjusted to produce repeated reverse bending.

roller stamping die. An engraved roller used for impressing designs and markings on sheet metal.

roll flattening. The flattening of metal sheets that have been rolled in packs by passing them separately through a two-high cold mill, there being virtually no deformation. Not the same as *roller leveling.*

roll forging. Forging with rotating dies that are not full round, the desired shape— either straight or tapered—being produced by grooves in the dies.

roll forming. The forming of flat rolled metal by the use of power-driven rolls whose contours determine the shape of the product. Roll forming is used extensively to make metal window frames, drapery rods, and similar products from metal strip. The term is sometimes used to describe power *spinning.*

rolling. Reducing the cross-sectional area of metal stock, or otherwise shaping metal products, through the use of rotating rolls.

rolling mills. Machines used to decrease the cross-sectional area of metal stock and produce certain desired shapes as the metal passes between rotating rolls mounted in a framework comprising a basic unit called a *stand.* Cylindrical rolls produce flat shapes, and grooved rolls produce rounds, squares, and structural shapes. Types of rolling mills include the billet mill, blooming mill, breakdown mill, plate mill, sheet mill, slabbing mill, strip mill, and temper mill.

roll resistance spot welding. The process for making separated resistance spot welds with one or more rotating circular electrodes. The rotation of the electrodes may or may not be stopped during the making of a weld.

roll straightening. The straightening of metal stock of various shapes by (a) passing it through a series of staggered rolls, the rolls usually being in horizontal and vertical planes; or (b) reeling in two-roll straightening machines.

roll table. A conveyor table on which rolls furnish the contact surface.

roll threading. Making threads by rolling the piece between two grooved die plates, one of which is in motion, or between rotating grooved circular rolls.

roll welding. Solid-state welding in which metals are heated, then welded together by applying pressure, with rolls, sufficient to cause deformation at the faying surfaces. See also *forge welding.*

root crack. A crack in either the weld or heat-affected zone at the root of a weld.

root face. The portion of a weld groove face adjacent to the root of the joint.

root of joint. The portion of a weld joint where the members are closest to each other before welding (Fig. 44). In cross section, this may be a point, a line, or an area.

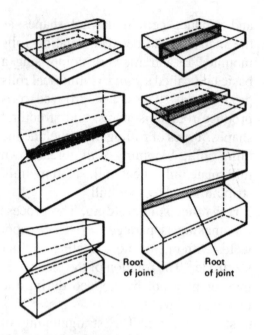

Fig. 44 Examples of weld joint root geometries

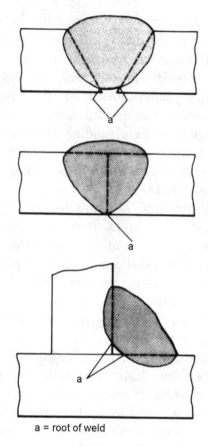

a = root of weld

Fig. 45 Examples of weld root geometries

root of weld. The points, as shown in cross section, at which the weld bead intersects the base-metal surfaces either nearest to or coincident with the *root of joint* (Fig. 45).

root opening. In a weldment, the separation between the members at the *root of joint* prior to the welding (Fig. 46).

root pass. The first bead of a *multiple-pass weld,* laid in the *root of joint.*

root penetration. The depth that a weld extends into the *root of joint,* measured on the centerline of the root cross section. See *joint penetration* (Fig. 34).

rosebuds. Concentric rings of distorted coating, giving the effect of an opened rosebud. Noted only on *minimized spangle.*

rosette. (1) Rounded configuration of microconstituents in metals arranged in whorls or radiating from a center. (2) Strain gages arranged to indicate at a single position strains in three different directions.

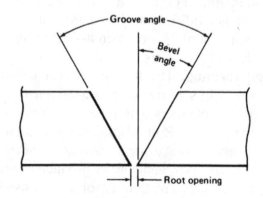

Fig. 46 Root opening in a weld

rotary forging. A process in which the workpiece is pressed between a flat anvil and a swiveling die with a conical working face; the platens move toward each other during forging.

rotary furnace. A circular furnace constructed so that the hearth and workpieces rotate around the axis of the furnace during heating.

rotary shear. A sheet metal cutting machine with two rotating-disk cutters mounted on parallel shafts driven in unison.

rotary swager. A swaging machine consisting of a power-driven ring that revolves at high speed, causing rollers to engage cam surfaces and force the dies to deliver hammerlike blows on the work at high frequency. Both straight and tapered sections can be produced.

rouge finish. A highly reflective finish produced with rouge or other very fine abrasive, similar in appearance to the bright polish or mirror finish on sterling silver utensils.

rough grinding. Grinding without regard to finish, usually to be followed by a subsequent operation.

roughing stand. The first stand of rolls through which a reheated billet passes, or the last stand in front of the finishing rolls.

rough machining. Machining without regard to finish, usually to be followed by a subsequent operation.

roughness. Relatively finely spaced surface irregularities, the heights, widths, and directions of which establish the predominant surface pattern.

roughness-width cutoff. The maximum width (in inches) of surface irregularities to be included in the measurement of roughness height.

rubber blanket. A sheet of rubber or other resilient material used as an auxiliary tool in forming.

rubber forming. Forming sheet metal wherein rubber or another resilient material is used as a functional die part.

Processes in which rubber is employed only to contain the hydraulic fluid are not classified as rubber forming.

rubber-pad forming. A sheet metal forming operation for shallow parts wherein a pad of rubber or other resilient material is attached to the press slide and becomes the mating die for a punch, or group of punches, that has been placed on the press bed or baseplate. Also known as *Guerin forming.*

rubber wheel. A grinding wheel made with a rubber bond.

rubidium. A chemical element having atomic number 37, atomic weight 86, and the symbol Rb, named from the Latin *rubidus,* meaning dark red, because of the color of its spectral lines. The lines were used by German chemists Gustav Kirchhoff and Robert Bunsen to identify it in 1861 employing the newly developed method of flame spectroscopy. Rubidium is one of the alkali metals along with lithium, sodium, potassium, caesium, and francium.

runner. (1) A channel through which molten metal flows from one receptacle to another. (2) The portion of the gate assembly of a casting that connects the sprue with the gate(s). (3) Parts of patterns and finished castings corresponding to the portion of the gate assembly described in (2).

runner box. A distribution box that divides molten metal into several streams before it enters the mold cavity.

running fit. Any *clearance fit* in the range used for parts that rotate relative to each other. Actual values of clearance resulting from stated shaft and hole tolerances are given in ANSI standards.

runout. (1) The unintentional escape of molten metal from a mold, crucible, or furnace. (2) An imperfection in a casting caused by the escape of metal from the mold. (3) See *axial runout* and *radial runout*.

runout table. A *roll table* used to receive a rolled or extruded section.

rust. A corrosion product consisting of hydrated oxides of iron. Applied only to ferrous alloys.

ruthenium. A chemical element having atomic number 44, atomic weight 101, and the symbol Ru, named for the Latin *Ruthenia,* meaning Russian. The metal was discovered by Polish chemist Jedrzej Andrei Śniadecki in 1808 and isolated by Russian chemist Karl Klaus in 1844. Ruthenium is a member of the platinum group. The principal use of ruthenium is as a hardening element for platinum and palladium. Ruthenium is second to osmium in having the least practical use in the platinum group.

sacrificial protection. The reduction of the extent of corrosion of a metal in an electrolyte by coupling it to another metal that is electrochemically more active in the environment.

saddling. Forming a seamless ring by forging a pierced disk over a mandrel (or saddle).

sag. An increase or decrease in the section thickness of a casting caused by insufficient strength of the mold sand of the cope or of the core.

salt fog test. An *accelerated corrosion test* in which specimens are exposed to a fine mist of a solution usually containing sodium chloride but sometimes modified with other chemicals. For testing details see ASTM B117. Also known as a *salt spray test*.

salting out. Precipitating a substance in a solution by adding a second substance, usually a salt, without any chemical reaction such as a double decomposition taking place.

salt spray test. See preferred term *salt fog test*.

samarium. A chemical element having atomic number 62, atomic weight 150, and the symbol Sm, named for the mineral samarskite in which it was found. Samarskite was named for Colonel Vasili Samarsky-Bykhovets, a Russian engineer who discovered the metal. Samarium was identified as a rare earth metal in 1879 by French chemist Paul Émile Le Coq de Boisbaudran.

sample. One or more units of a product (or a relatively small quantity of a bulk material) withdrawn from a *lot* or process stream and then tested or inspected to provide information about the properties, dimensions, or other quality characteristics of the lot or process stream. Not the same as *specimen*.

sand. A granular material naturally or artificially produced by the disintegration or crushing of rocks or mineral deposits. In casting, the term denotes an aggregate, with an individual particle (grain) size of 0.06 to 2 mm (0.002 to 0.08 in.) in diameter, that is largely free of finer constituents such as silt and clay, which often are present in natural sand deposits. The most commonly used foundry sand is silica; however, zircon, olivine, chromite, alumina, and other crushed ceramics are used for special applications.

sandblasting. Abrasive blasting with sand. See also *blasting*, and compare with *shot blasting*.

sand control. The testing and regulation of the chemical, physical, and mechanical properties of foundry sand mixtures and their components.

sand hole. A pit in the surface of a sand casting resulting from a deposit of loose sand in the mold cavity.

Sandvik 2RE10. The brand name of an austenitic low-carbon 25-20 alloy with an extremely low impurity content developed for tubing. It has excellent resistance to corrosion in nitric acid, excellent resistance to intergranular corrosion, good resistance to pitting corrosion, and good weldability. 2RE10 is also known by the designations UNS S31002, EN 1.4335, and X1CrNi 25-20, and is described in ASTM A213.

sandwich rolling. Rolling two or more strips of metal in a pack, sometimes to form a roll-welded composite.

saponification. The alkaline hydrolysis of fats whereby a soap is formed; more generally, hydrolysis of an ester by an alkali with the formation of an alcohol and a salt of the acid portion.

satin finish. A diffusely reflecting surface finish on metals, lustrous but not mirror-like. One type is a *butler finish*.

Sauveur, Albert. 1863–1939. A Belgian-born U.S. metallurgist whose microscopic and metallographic studies of metal structures made him one of the founders of physical metallurgy. His work on the heat treatment of metals is regarded as a scientific landmark. Sauveur wrote *Metallography and Heat Treatment of Iron and Steel* in 1912, and he became the Gordon McMay professor of mining and metallurgy at Harvard In 1924.

saw gumming. In saw manufacture, grinding away of punch marks or milling marks in the gullets (spaces between the teeth) and, in some cases, simultaneous sharpening of the teeth; in the reconditioning of worn saws, the restoration of the original gullet size and shape.

sawing. Cutting a workpiece with a band, blade, or circular disk having teeth.

scab. An imperfection consisting of a thin, flat piece of metal attached to the surface of a sand casting. A sand scab is usually separated from the casting proper by a thin layer of sand and is joined to the casting along one edge. An erosion scab is similar in appearance to a *cut* or wash.

scale pit. (1) A surface depression formed on a forging due to scale remaining in the dies during the forging operation. (2) A pit in the ground in which scale (such as that carried off by cooling water from rolling mills) is allowed to settle out as one step in the treatment of effluent waste water.

scaling. (1) Forming a thick layer of oxidation products on metals at high temperature. (2) Depositing water-insoluble constituents on a metal surface, as in cooling tubes and water boilers.

scalped extrusion ingot. A cast, solid, or hollow extrusion ingot that has been machined on the outside surface.

scalping. Removing surface layers from ingots, billets, or slabs. See also *die scalping*.

scandium. A chemical element having atomic number 21, atomic weight 45, and the symbol Sc, for the Latin *Scandia*, meaning Scandinavia, where the element was found. It was found in an ytterbium-containing mineral by Swedish chemist Lars Fredrik Nilson in 1879. Applications include minor aerospace industry components and sports equipment.

scarfing. Cutting surface areas of metal objects, ordinarily by using an oxyfuel gas torch. The operation permits surface imperfections to be cut from ingots, billets, or the edges of plates that are to be beveled for butt welding. See also *chipping*.

scarf joint. A butt joint in which the plane of the joint is inclined with respect to the main axis of the members (Fig. 47).

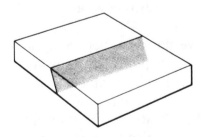

Fig. 47 Example of a scarf butt joint

Scleroscope test. A hardness test in which the loss in kinetic energy of a falling metal tup, absorbed by indentation upon impact of the tup on the metal being tested, is indicated by the height of the rebound. The same as *Shore hardness test*.

scorification. Oxidation, in the presence of fluxes, of molten lead containing precious metals, to partly remove the lead in order to concentrate the precious metals.

scoring. (1) The marring or scratching of a smooth surface, most often caused by sliding contact with a mating member having a hard projection or embedded particle on its surface. (2) Reducing the thickness of a material along a line to intentionally weaken it.

scouring. (1) A wet or dry cleaning process involving mechanical scrubbing. (2) A wet or dry mechanical finishing operation, using fine abrasive and low pressure, carried out by hand or with a cloth or wire wheel to produce *satin* or *butler*-type finishes.

scrap. (1) Products that are discarded because they are defective or otherwise unsuitable for sale. (2) Discarded metallic material, from whatever source, that may be reclaimed through melting and refining.

scratch hardness. The hardness of a metal as determined by the width of a scratch made by drawing a cutting point across the surface under a given pressure.

screen. (1) One of a set of sieves, designated by the size of the openings, used to classify granular aggregates such as sand, ore, or coke by particle size. (2) A perforated sheet placed in the gating system of a mold to separate impurities from the molten metal.

screw dislocation. A type of dislocation—a linear imperfection in a crystalline array of atoms—that corresponds to the axis of a spiral structure in a crystal, characterized by a distortion that joins together normally parallel planes to form a continuous helical ramp (with a pitch of one interplanar distance) winding about the dislocation.

screw press. A press whose slide is operated by a screw rather than by a crank or other means.

screw stock. A free-machining type of alloy in the form of bar, rod, or wire, used to produce screw machine products.

scruff. A mixture of tin oxide and iron-tin alloy formed as dross on a tin-coating bath.

scuffing. A form of *adhesive wear* that produces superficial scratches or a high polish on the rubbing surfaces. It is most often observed on inadequately lubricated parts.

sea coal. Finely ground coal, used as an ingredient in molding sands.

sealing. (1) Closing pores in anodic coatings to render them less absorbent. (2) Plugging leaks in a casting by introducing thermosetting plastics into porous areas and subsequently setting the plastic with heat.

seal weld. Any weld used primarily to obtain tightness and prevent leakage.

seam. On a metal surface, an unwelded fold or lap that appears as a crack, usually resulting from a discontinuity.

seam welding. (1) Arc or resistance welding in which a series of overlapping spot welds is produced with rotating electrodes or rotating work, or both. (2) Making a longitudinal weld in sheet metal or tubing.

season cracking. Cracking resulting from the combined effects of corrosion and internal stress. A term usually applied to *stress-corrosion cracking* of brass, it occurs especially in α brasses such as 70/30.

secant modulus. The slope of the secant drawn from the origin to a specified point on the *stress-strain curve*. See also *modulus of elasticity*.

secondary creep. Time-dependent strain occurring under stress at a minimum and almost constant rate. See also *creep*.

secondary hardening. The hardening phenomenon that occurs during high-temperature tempering of certain steels containing one or more carbide-forming elements (ASTM A941).

secondary metal. Metal recovered from scrap by remelting and refining.

segment die. A die made of parts that can be separated for ready removal of the workpiece. Also known as *split die*.

segregation. Nonuniform distribution of alloying elements, impurities, or microphases in metals and alloys.

seizing. The stopping of a moving part by a mating surface as a result of excessive friction caused by *galling*.

Sejournet process. A direct extrusion process for metals that uses molten glass to insulate the hot billet and to act as a lubricant. Also known as *Ugine-Sejournet process*.

selective flotation. Separating a complex ore into two or more valuable minerals and *gangue* by *flotation*. Also known as *differential flotation*.

selective heating. Intentionally heating only certain portions of a workpiece.

selective leaching. Corrosion in which one element is preferentially removed from an alloy, leaving a residue (often porous) of the elements that are more resistant to the particular environment. See also *decarburization, denickelification, dezincification,* and *graphitic corrosion*.

selective quenching. Quenching only certain portions of an object. See also *quenching*.

selenium. A chemical element having atomic number 34, atomic weight 79, and the symbol Se, for Selene, the Greek goddess of the moon. The metal was isolated by Jôns Jakob Berzelius in 1817 while attempting to purify tellurium metal. Selenium is classed as one of the nine metalloids and is seen as the most useful of all of the metalloids from engineering and metallurgical standpoints. Most high-purity selenium is used for rectifying electric current and photoelectric devices for operating mechanisms. Selenium is added to copper alloys and austenitic stainless steels, in amounts up to 0.5%, to enhance machinability.

self-diffusion. The thermally activated movement of an atom to a new site in a crystal of its own species, as, for example, a copper atom within a crystal of copper.

self-hardening steel. A steel of alloy content sufficient to ensure hardening throughout the section during slow cooling in still air from the austenitizing temperature. Clearly dependent on section size, the usual limit is a ruling section of 2 1/4 in. See preferred term *air-hardening steel.*

semiautomatic plating. *Plating* in which prepared cathodes are mechanically conveyed through the plating baths, with intervening manual transfers.

semiconductor. An electronic conductor whose conductivity is intermediate between that of a metal and an insulator, ranging from approximately 10^5 to 10^{-7} siemens per meter and in which the conductivity increases with increasing temperature over some temperature range.

semicontinuous conveyance furnace. A heating device through which steel objects are intentionally moved in accordance with a predetermined start-stop-start pattern during the thermal processing cycle.

semifinisher. An impression in a forging die that only approximates the finish dimensions of the forging. Semifinishers are often used to extend die life of the finishing impression, ensure proper control of grain flow during forging, and assist in obtaining desired tolerances. Also known as semifinishing impression.

semifinishing. Preliminary operations performed prior to finishing.

semikilled steel. Steel that is incompletely deoxidized and contains sufficient dissolved oxygen to react with the carbon to form carbon monoxide and thus offset solidification shrinkage.

semipermanent mold. A *permanent mold* in which sand cores are used.

Sendzimir mill. A type of cluster mill with small-diameter working rolls and larger-diameter backup rolls, backed up by bearings on a shaft mounted eccentrically so that it can be rotated to increase the pressure between the bearings and backup rolls.

sensitivity. The smallest difference in values that can be detected reliably with a given measuring instrument.

sensitization. In austenitic stainless steels, the precipitation of chromium carbides, usually at grain boundaries, on exposure to temperatures of approximately 550 to 850 °C (1000 to 1550 °F), leaving the grain boundaries depleted of chromium and therefore susceptible to preferential attack by a corroding (oxidizing) medium.

sequence timer. In resistance welding, a device used for controlling the sequence and duration of any or all of the elements of a complete welding cycle except *weld time* or *heat time*.

sequence weld timer. Same as *sequence timer* except that either *weld time* or *heat time*, or both, also are controlled.

sequestering agent. A material that combines with metallic ions to form water-soluble complex compounds.

series welding. Resistance welding in which two or more spot, seam, or projection welds are made simultaneously by a single welding transformer with three or more electrodes forming a series circuit (Fig. 48).

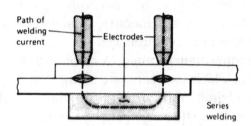

Fig. 48 Series welding

set. The shape of the solidifying surface of a metal, especially copper, with respect to concavity or convexity. May also be known as *pitch*.

set copper. An intermediate copper product containing approximately 3.5% cuprous oxide, obtained at the end of the oxidizing portion of the fire-refining cycle.

settling. (1) The separation of solids from suspension in a fluid of lower density, solely by gravitational effects. (2) A process for removing iron from liquid magnesium alloys by holding the melt at a low temperature after manganese has been added to it.

shadowing. (1) Same as *shielding* in electroplating. (2) Directional deposition of carbon or a metallic film on a plastic replica so as to highlight features to be analyzed by transmission electron microscopy.

shakeout. Removal of castings from a sand mold.

shank. (1) The handle for carrying a small ladle or crucible. (2) The portion of a die, tool, or forging by which it is held. (3) The main body of a lathe tool. If the tool is an inserted type, the shank is the portion that supports the insert. See *single-point tool* (Fig. 50).

shank-type cutter. A cutter having a straight or tapered shank to fit into a machine-tool spindle or adapter.

shaping. Producing flat surfaces using single-point tools. The work is held in a vise or fixture, or is clamped directly to the table. The ram supporting the tool is reciprocated in a linear motion past the work.

shatter cracks. Short, discontinuous internal fissures in ferrous metals attributed to stresses produced by localized transformation and decreased solubility of hydrogen during cooling after hot working. In a fracture surface, shatter cracks appear as bright silvery areas; on an etched surface, they appear as short, discontinuous cracks. Also known as *flakes*.

shaving. (1) As a finishing operation, the accurate removal of a thin layer of a work surface by straightline motion between a cutter and the surface. (2) Trimming parts such as stampings, forgings, and tubes to remove uneven sheared edges or to improve accuracy.

shear. (1) That type of force that causes or tends to cause two contiguous parts of the same body to slide relative to each other in a direction parallel to their plane of contact. (2) A type of cutting tool with which a material in the form of a wire, sheet, plate, or rod is cut between two opposing blades. (3) The type of cutting action produced by *rake* so that the direction of chip flow is other than at right angles to the cutting edge.

shear angle. The angle that the *shear plane*, in metal cutting, makes with the work surface.

shear fracture. A ductile fracture in which a crystal (or a polycrystalline mass) has separated by sliding or tearing under the action of shear stresses.

shearing strain. The change in angle (expressed in radians) between two lines originally at right angles. Also known as shear strain. See also *strain*.

shear ledges. Lines on a fracture surface that radiate from the fracture origin, and are visible to the unaided eye or at low magnification. Shear ledges result from the intersection and connection of brittle fractures propagating at different levels. Also known as *radial marks*. See also *chevron pattern*.

shear lip. A narrow, slanting ridge along the edge of a fracture surface. The term sometimes also denotes a narrow, often crescent-shaped, fibrous region at the edge of a fracture that is otherwise of the cleavage type, even though this fibrous region is in the same plane as the rest of the fracture surface.

shear modulus. A measure of the rigidity of metal. The ratio of stress, below the proportional limit, to the corresponding strain. Specifically, the modulus obtained in torsion or shear is shear modulus, modulus of rigidity, or modulus of torsion. See *modulus of elasticity*.

shear plane. A confined zone along which shear takes place in metal cutting. It extends from the cutting edge to the work surface.

shear strain. The change in angle (expressed in radians) between two lines originally at right angles. Also known as *shearing strain*. See also *strain*.

shear strength. The stress required to produce fracture in the plane of cross section, the conditions of loading being such that the direction of force and of resistance are parallel and opposite, although their paths are offset a specified minimum amount. The maximum load divided by the original cross-sectional area of a section separated by shear.

shear stress. Force per unit area, often thought of as force acting through a small area within a plane. It can be divided into components, with the stress parallel to the plane called *shear stress*. See also *stress*.

sheet. A flat-rolled metal product of some maximum thickness and minimum width arbitrarily dependent on the type of metal. It is thinner than *plate* and has a width-to-thickness ratio greater than about 50.

sheet separation. In spot, seam, or projection welding, the gap that exists between faying surfaces surrounding the weld, after the joint has been welded.

shelf roughness. Roughness on upward-facing surfaces where undissolved solids have settled on parts during a plating operation.

shell. (1) A hollow structure or vessel. (2) An article formed by deep drawing.

(3) The metal sleeve remaining when a billet is extruded with a dummy block of somewhat smaller diameter. (4) In shell molding, a hard layer of sand and thermosetting plastic or resin formed over a pattern and used as the mold wall. (5) A tubular casting used in making seamless drawn tube. (6) A pierced forging.

shell core. A shell-molded sand core.

shell hardening. A surface-hardening process in which a suitable steel workpiece, when heated through and quench hardened, develops a martensitic layer or shell that closely follows the contour of the piece and surrounds a core of essentially pearlitic transformation product. This result is accomplished by a proper balance among section size, steel hardenability, and severity of quench.

shell molding. Forming a mold from thermosetting resin-bonded sand mixtures brought in contact with preheated (150 to 260 °C, or 300 to 500 °F) metal patterns, resulting in a firm shell with a cavity corresponding to the outline of the pattern. Also known as *Croning process*.

shielded metal arc welding (SMAW). Arc welding in which metals are fused together by heating them with an arc between a *covered electrode* and the work. Decomposition of the covering on the consumable electrode provides shielding gas, and the electrode itself provides the filler metal. Pressure is not applied to the joint.

shielding. (1) A material barrier that prevents radiation or a flowing fluid from impinging on an object or a portion of an object. (2) Placing an object in an electrolytic bath so as to alter the current distribution on the cathode. A nonconductor is called a shield; a conductor is called a *robber*, a thief, or a guard.

shift. A casting imperfection caused by mismatch of cope and drag or of cores and molds.

shim. A thin piece of material used between two surfaces to obtain a proper fit, adjustment, or alignment.

shimmy die. The name sometimes used for the die that oscillates to trim parts in a *flat edge trimmer*, a machine for trimming notched edges on shells.

shoe. (1) A metal block used in a variety of bending operations to form or support the part being processed. (2) An anvil cap or *sow block*.

Shore hardness test. A hardness test in which the loss in kinetic energy of a falling metal tup, absorbed by indentation upon impact of the tup on the metal being tested, is indicated by the height of the rebound. Also known as *Scleroscope test*.

short circuiting transfer. In consumable-electrode arc welding, a type of metal transfer similar to globular transfer, but in which the drops are so large that the arc is short circuited momentarily during the transfer of each drop to the weld pool. Compare with *spray transfer* and *globular transfer*.

shorts. The product that is retained on a specified screen in the screening of a crushed or ground material. See also *plus sieve*.

short transverse. Transverse literally means "across," usually signifying a direction or plane perpendicular to the direction of working. In rolled plate or sheet, *short transverse* is the direction through the thickness. See *transverse*.

shot. Small spherical particles of metal.

shotblasting. *Blasting* with metal *shot*; usually used to remove deposits or mill scale more rapidly or more effectively than can be done by sand blasting.

shot peening. Cold working the surface of a metal by metal shot impingement.

shotting. The production of *shot* by pouring molten metal in finely divided streams. Solidified spherical particles are formed during the descent and are cooled in a tank of water.

shrinkage. (1) Liquid shrinkage: the reduction in the volume of liquid metal as it cools to the liquidus. (2) Solidification shrinkage: the reduction in the volume of metal from the beginning to the end of solidification. (3) Solid shrinkage: the reduction in the volume of metal from the solidus to room temperature. The same as *casting shrinkage*.

shrinkage cavity. A void left in cast metal as a result of solidification shrinkage. See *casting shrinkage*.

shrinkage cracks. Hot tears associated with shrinkage cavities.

shrinkage rule. A measuring ruler with graduations expanded to compensate for the change in the dimensions of the solidified casting as it cools in the mold.

shrink fit. An *interference fit* produced by heating the outside member of mating parts to a temperature practical for easy assembly. Usually the inside member is kept at or near room temperature. Sometimes the inside member is cooled to increase ease of assembly.

shrink forming. Forming of metal wherein the inner fibers of a cross section undergo a reduction in a localized area by the application of heat, cold upset, or mechanically induced pressures.

shut height. For a press, the distance from the top of the bed to the bottom of the slide with the stroke down and adjustment up. In general it is the maximum die height that can be accommodated for normal operation, taking the *bolster* plate into consideration.

side cutting-edge angle. Defined by Fig. 50 see *single-point tool*.

side milling. Milling with cutters having peripheral and side teeth. They are usually profile sharpened but may be form relieved.

side rake. In a single-point turning tool, the angle between the tool face and a reference plane, corresponding to *radial rake* in milling. It lies in a plane perpendicular to the tool base and parallel to the rotational axis of the work. See *single-point tool* (Fig. 50).

sieve analysis. A method of determining *particle-size distribution*, usually expressed as the weight percentage retained on each of a series of standard sieves of decreasing mesh size, and the percentage passed by the sieve of the finest size. Also known as *sieve classification*.

sieve classification. A method of determining *particle-size distribution*; usually expressed as the weight percentage retained on each of a series of standard sieves of decreasing mesh size, and the percentage passed by the sieve of the finest size. Also known as *sieve analysis*.

sieve fraction. The portion of a powder sample that passes through a sieve of specified number and is retained by some finer sieve of specified number.

sigma (σ) phase. A hard, brittle, nonmagnetic intermediate phase with a tetragonal crystal structure, containing 30 atoms per unit cell, space group *P4/mnm*, occurring in many binary and ternary alloys of the transition elements. The composition of this phase in the various systems is not the same, and the phase usually exhibits a wide range in homogeneity. Alloying with a third transition element usually enlarges the field homogeneity and extends it deep into the ternary section.

silica flour. A sand additive, containing approximately 99.5% silica, commonly produced by pulverizing quartz sand in large ball mills to a mesh size of 80 to 325.

silicon. A chemical element having atomic number 14, atomic weight 28, and the symbol Si, from the Latin *silex*, meaning flint. The element was isolated from flint by Swedish chemist Jöns Jakob Berzelius in 1824. It is classed as one of the metalloids and is the second most common element after oxygen.

Silicon is used as an alloying element in aluminum, copper, magnesium, and cast iron. It is a deoxidizer in the making of steel and an alloy in valve steels. Very low carbon steels with 2–4% Si have special hysteresis properties that make them useful for transformer cores and other electrical and electronic uses. Silicon is a principal component of most semiconductor devices and microchips.

siliconizing. Diffusing silicon into solid metal, usually low-carbon steels, at an elevated temperature in order to improve corrosion or wear resistance.

silky fracture. A metal fracture in which the broken metal surface has a fine texture, usually dull in appearance. Characteristic of tough and strong metals. Contrast with *crystalline fracture* and *granular fracture*.

silver. A chemical element having atomic number 47, atomic weight 108, and the symbol Ag, for the Latin *argentum*, which came from the early Sanskrit word *argunus*, meaning shining brightly. The English word *silver* stems from the German *silber* and Old English *siolfor* and *seolfor*. Silver, discovered around 5000 B.C., was one of the seven metals of antiquity, along with gold, copper, tin, lead, iron, and mercury.

The oldest use of silver, jewelry, continues to consume approximately one-quarter of the world's production. Its exceptional workability and beauty have combined to make silver a desirable metal for personal adornment, hollowware, and other decorative products. In the Olympics it is still the metal used for second-place medals. Silver coins have been a commercial medium of exchange since 600 B.C. In 1965, the U.S. Treasury and the Canadian Mint removed silver from circulating coinage, but silver continues to be popular for medals and commemorative medallions.

The largest commercial use of silver is in the electrical industry because it is the best electrical conductor and has excellent corrosion resistance. Silver is used for motor contacts, switches and relays for electrical appliances, and in automobiles.

Silver oxide-zinc batteries are used where energy delivered per unit weight and long-term reliability are of prime importance. Silver-zinc miniature batteries are used in watches and calculators, while full-scale batteries are used in submarines, aircraft, and space vehicles. Silver-zinc batteries are used for portable tools and power packs for commercial TV cameras. Silver soldering and brazing are other major uses.

Photographic and x-ray film uses approximately one-quarter of the silver supply, in the form of silver halides. Silver is used for mirrors. It is superior to all other materials for bearings, such as the main shafts of railroad diesel engines. Silver thiosulfate is now used by florists to retard the wilting of cut flowers. Silver amalgams have been used for tooth restoration for over 150 years. Colloidal silver has healing properties and kills bacteria.

silver soldering. The nonpreferred term used to denote brazing with a silver-base filler metal. See preferred terms *furnace brazing, induction brazing,* and *torch brazing.*

single-bevel groove weld. A groove weld in which the joint edge of one member is beveled from one side (Fig. 49).

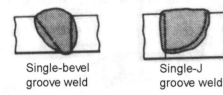

Single-bevel groove weld   Single-J groove weld

Fig. 49 Examples of single-bevel and single-J groove welds

single-impulse welding. Spot, projection, or upset welding by a single impulse of current. Where alternating current is used, an impulse may be any fraction or number of cycles.

single-J groove weld. A groove weld in which the joint edge of one member is prepared in the form of a J, from one side (Fig. 49).

single-point tool. See illustration of nomenclature in Fig. 50.

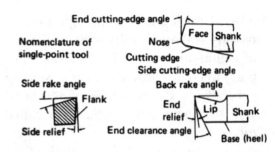

Fig. 50 Single-point tool nomenclature

single relief angle. Defined by Fig. 50. See *single-point tool.*

single-stand mill. A rolling mill of such design that the product contacts only two rolls at a given moment. Contrast with *tandem mill.*

single-U groove weld. A groove weld in which each joint edge is prepared in the form of a J or half-U from one side (Fig. 51).

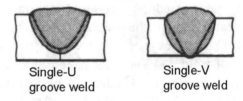

Single-U groove weld   Single-V groove weld

Fig. 51 Examples of single-U and single-V groove welds

single-V groove weld. A groove weld in which each member is beveled from the same side (Fig. 51).

single welded joint. In arc and gas welding, any joint welded from one side only.

sinkhead. A reservoir of molten metal connected to a casting to provide additional metal to the casting, required as the result of shrinkage before and during solidification. Also known as *riser.*

sinking. Drawing tubing through a die or passing it through rolls without the use of an interior tool (such as a mandrel or plug) to control inside diameter; sinking generally produces a tube of increased wall thickness and length. Also called *tube sinking*.

sinter. To heat a mass of fine particles for a prolonged time below the melting point, usually to cause agglomeration.

sintering. The bonding of adjacent surfaces in a mass of particles by molecular or atomic attraction on heating at high temperatures below the melting temperature of any constituent in the material. Sintering strengthens a powder mass and normally produces densification and, in powdered metals, recrystallization. See also *liquid phase sintering*.

size effect. The effect of the dimensions of a piece of metal on its mechanical and other properties and on manufacturing variables such as forging reduction and heat treatment. In general, the mechanical properties are lower for a larger size.

size of weld. (1) The joint penetration in a groove weld. (2) The lengths of the nominal legs of a fillet weld (Fig. 52).

sizing. (1) Secondary forming or squeezing operations required to square up, set down, flatten, or otherwise correct surfaces to produce specified dimensions and tolerances. See also *restriking*. (2) Some burnishing, broaching, drawing, and shaving operations are also called sizing. (3) A finishing operation for correcting ovality in tubing. (4) The final pressing of a sintered powder metallurgy part to obtain a desired dimension.

skelp. The starting stock for making welded pipe or tubing; most often it is strip stock of suitable width, thickness, and edge configuration.

skim gate. A gating arrangement designed to prevent the passage of slag and other undesirable materials into a casting.

skimmer. A tool for removing scum, slag, and dross from the surface of molten metal.

skin. A thin outside metal layer, not formed by bonding as in cladding or electroplating, that differs in composition, structure, or other characteristics from the main mass of metal.

skin lamination. In flat-rolled metals, a surface rupture resulting from the exposure of a subsurface lamination by rolling.

skin pass. A temper pass during light cold rolling of steel sheet, done to improve flatness, minimize the tendency toward formation of stretcher strains and flutes, and obtain the desired texture

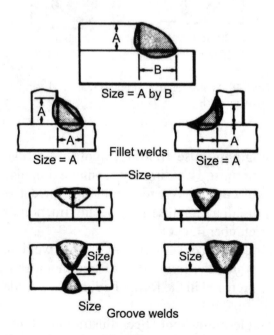

Fig. 52 Size of weld in fillet and groove welds

and mechanical properties. See *temper rolling*.

skiving. (1) The removal of a material in thin layers or chips with a high degree of shear or slippage, or both, of the cutting tool. (2) A machining operation in which the cut is made with a form tool with its face so angled that the cutting edge progresses from one end of the work to the other as the tool feeds tangentially past the rotating workpiece.

skull. A layer of solidified metal or dross on the walls of a pouring vessel after the metal has been poured.

slab. A piece of metal, intermediate between ingot and plate, of which the width is at least twice the thickness.

slabbing mill. A primary mill that produces slabs.

slab milling. Milling a surface parallel to the axis of the cutter. See preferred term *peripheral milling*.

slack quenching. The incomplete hardening of steel due to quenching from the austenitizing temperature at a rate lower than the critical cooling rate for the particular steel, resulting in the formation of one or more transformation products in addition to martensite.

slag. A nonmetallic product resulting from the mutual dissolution of flux and nonmetallic impurities in smelting, refining, and certain welding operations.

slag inclusion. Slag or dross entrapped in a metal.

slant fracture. A type of fracture appearance, typical of *plane-stress* fractures, in which the plane of metal separation is inclined at an angle (typically approximately 45°) to the axis of applied stress.

slide. The main reciprocating member of a mechanical press, guided in a press frame, to which the punch or upper die is fastened.

sliding fit. A loosely defined fit similar to a *slip fit*.

slime. (1) A material of extremely fine particle size encountered in ore treatment. (2) A mixture of metals and some insoluble compounds that forms on the anode in electrolysis.

slip. *Plastic deformation* by the irreversible shear displacement (translation) of one part of a crystal relative to another in a definite crystallographic direction and usually on a specific crystallographic plane. Sometimes called *glide.*

slip band. A group of parallel slip lines so closely spaced as to appear as a single line when observed under an optical microscope. See also *slip line*.

slip direction. The crystallographic direction in which the translation of slip takes place.

slip fit. A loosely defined *clearance fit* between parts assembled by hand without force, but implying slipping contact.

slip flask. A tapered *flask* that depends on a movable strip of metal to hold the sand in position. After closing the mold, the strip is retracted and the flask can be removed and reused. Molds thus made are usually supported by a *mold jacket* during pouring.

slip-interference theory. The theory involving the resistance to deformation offered by a hard phase dispersed in a ductile matrix.

slip line. A trace of the slip plane on the viewing surface; the trace is (usually) observable only if the surface has been polished before deformation. The usual observation on metal crystals (under a light microscope) is of a cluster of slip lines known as a *slip band.*

slip plane. The crystallographic plane in which *slip* occurs in a crystal.

sliver. An imperfection consisting of a very thin elongated piece of metal attached by only one end to the parent metal into whose surface it has been worked.

slope control. Producing electronically a gradual increase or decrease in the welding current between definite limits and within a selected time interval.

slot furnace. A common batch furnace in which stock is charged and removed through a slot or opening.

slotting. Cutting a narrow aperture or groove with a reciprocating tool in a vertical shaper or with a cutter, broach, or grinding wheel.

slot weld. Similar to *plug weld*, the difference being that the hole is elongated and may extend to the edge of a member without closing (Fig. 53).

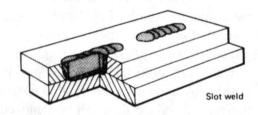

Fig. 53 Slot weld

slug. (1) A short piece of metal to be placed in a die for forging or extrusion. (2) A small piece of material produced by piercing a hole in sheet material.

slugging. The unsound practice of adding a separate piece of material in a joint before or during welding, resulting in a welded joint in which the weld zone is not entirely built up by adding molten filler metal or by melting and recasting base metal, and which therefore does not comply with design, drawing, or specification requirements.

slush casting. A hollow casting usually made of an alloy with a low but wide melting temperature range. After the desired thickness of metal has solidified in the mold, the remaining liquid is poured out.

smelting. Thermal processing wherein chemical reactions take place to produce liquid metal from a beneficiated ore.

smith forging. Manual forging with flat or simple-shaped dies that never completely confine the work. Also known as *open-die forging*, *flat-die forging*, and *hand forging*.

smut. A reaction product sometimes left on the surface of a metal after pickling, electroplating, or etching.

snagging. *Offhand grinding* on castings and forgings to remove surplus metal such as gate and riser pads, fins, and parting lines.

snake. (1) The product formed by the twisting and bending of hot rod prior to its next rolling process. (2) Any crooked surface imperfection in a plate, resembling a snake. (3) A flexible mandrel used in the inside of a shape to prevent flattening or collapse during a bending operation.

snaky edges. Carbonaceous deposits in a wavy pattern along the edges of a sheet or strip. Also known as *carbon edges*.

snap flask. A foundry flask hinged on one corner so that it can be opened and removed from the mold for reuse before the metal is poured.

snap temper. A precautionary interim stress-relieving treatment applied to high-hardenability steels immediately after quenching to prevent cracking caused by a delay in tempering them at the prescribed higher temperature.

S-N diagram. A graphical representation of the relationship of stress (S) to the number of cycles (N) before fracture occurs in fatigue testing.

snowflakes. Short, discontinuous internal fissures in ferrous metals attributed to stresses produced by localized transformation and decreased solubility of hydrogen during cooling after hot working. In a fracture surface, shatter cracks appear as bright silvery areas; on an etched surface, they appear as short, discontinuous cracks. Also known as *flakes* and *shatter cracks*.

snug fit. A loosely defined fit implying the closest clearances that can be assembled manually for firm connection between parts. Similar to a *push fit*.

soak cleaning. *Immersion cleaning* without electrolysis.

soaking. In heat treating of metals, the prolonged holding at a selected temperature to effect homogenization of structure or composition.

sodium. A chemical element having atomic number 11, atomic weight 23, and the symbol Na, for the Latin *natrum*, meaning sodium. The Latin word was derived from the Arabic root originally for sodium, *ntrj*. In most languages except English and Italic languages, the metal is called natrium. Sir Humphry Davy first isolated sodium by the electrolysis of sodium hydroxide in 1807. Sodium is a silvery-white soft metal that can be cut with a knife and molded easily. It reacts with air and violently with water and therefore is generally stored under paraffin in airtight containers. The melting point of sodium is 97.7 °C (207.9 °F). It

has been used as a heat-transfer medium in heavy-duty internal combustion engine exhaust valves.

soft soldering. A form of soldering characterized by having a filler metal whose melting point is below approximately 400 °C (752 °F). See preferred term *soldering*.

soft temper. A *temper* of nonferrous alloys and some ferrous alloys corresponding to the condition of minimum hardness and tensile strength produced by *full annealing*. Also known as *dead soft* temper.

soil. Undesirable material on a surface that is not an integral part of the surface. Oil, grease, and dirt can be soils; a decarburized skin and excess *hard chromium* are not soils. Loose scale is soil; hard scale may be an integral part of the surface and, hence, not soil.

solderability. The ease with which a surface is wetted by molten solder.

solder embrittlement. The reduction in mechanical properties of a metal as a result of local penetration of solder along grain boundaries.

soldering. A group of processes that join metals by heating them to a suitable temperature below the solidus of the base metals and applying a filler metal having a liquidus not exceeding 450 °C (840 °F). Molten filler metal is distributed between the closely fitted surfaces of the joint by capillary action.

solder short. In soldering, an unintended solder connection between two or more conductors, either securely or by mere contact. Also known as a *crossed joint* or *bridging* (5).

solid cutters. Cutters made of a single piece of material rather than a composite of two or more materials.

solidification. The change in state from liquid to solid on cooling through the melting temperature or melting range.

solidification shrinkage. The reduction in volume of metal from the beginning to the end of solidification. See also *casting shrinkage*.

solid shrinkage. The reduction in volume of metal from the solidus to room temperature. See also *casting shrinkage*.

solid solution. A single, solid, homogeneous crystalline phase containing two or more chemical species.

solid-state welding. A group of welding processes that join metals at temperatures essentially below the melting points of the base materials, without the addition of a brazing or soldering filler metal. Pressure may or may not be applied to the joint.

solidus. In a *constitution diagram* or *equilibrium diagram*, the locus of points representing the temperatures at which various compositions finish freezing on cooling or begin to melt on heating. See also *liquidus*.

soluble oil. Specially prepared oil whose water emulsion is used as a cutting or grinding fluid.

solute. The component of either a liquid or solid solution that is present to a lesser or minor extent; the component that is dissolved in the *solvent*.

solution heat treatment. Heating an alloy to a suitable temperature, holding at that temperature long enough to cause one or more constituents to enter into *solid solution*, and then cooling rapidly enough to hold these constituents in solution.

solution potential. *Electrode potential* where half-cell reaction involves only the metal electrode and its ion.

solvent. The component of either a liquid or solid solution that is present to a greater or major extent; the component that dissolves the *solute*.

solvus. In a *constitution diagram* or *equilibrium diagram*, the locus of points representing the temperatures at which the various compositions of the solid phases coexist with other solid phases—that is, the limits of solid solubility.

sonic testing. An inspection method that uses sound waves (in the audible frequency range, approximately 20 to 20,000 Hz) to induce a response from a part or test specimen. Sometimes, but inadvisably, used as a synonym for *ultrasonic testing*.

sorbite. (obsolete) A fine mixture of *ferrite* and *cementite* produced either by regulating the rate of cooling of steel or by tempering steel after hardening. The first type is very fine *pearlite* that is difficult to resolve under the microscope; the second type is tempered *martensite*.

Sorby, Henry Clifton. 1826–1908. A geologist born in Woodbourne (near Sheffield), England, whose microscopic studies of thin slices of rock earned him the title "the father of microscopic petrography" and led to the study of metallography.

sow block. In forging, a removable block of metal set into the hammer anvil to protect the anvil from shock and wear and occasionally to hold insert dies. Also called an anvil cap or a shoe.

space lattice. A regular, periodic array of points (lattice points) in space that represents the locations of atoms of the same kind in a perfect crystal. The concept may be extended, where appropriate, to crystalline compounds and other substances, in which case the lattice points often represent locations of groups of atoms of identical composition, arrangement, and orientation.

spacer strip. A metal strip or bar inserted in the root of a joint prepared for groove welding, to serve as a backing and to maintain root opening throughout the course of the welding operation (Fig. 54).

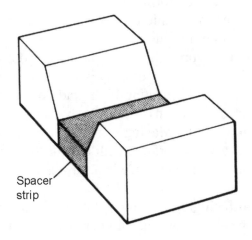

Spacer strip

Fig. 54 A spacer bar (strip) for groove welding

spade drill. A rotary end-cutting tool constructed from a flat piece of material, provided with suitable cutting lips at the cutting end. See preferred term *flat drill*.

spalling. The cracking and flaking of particles out of a surface.

spangle. The characteristic crystalline form in which a hot dipped zinc coating solidifies on steel strip.

spatter. The metal particles expelled during arc or gas welding. They do not form part of the weld.

spatter loss. The metal lost due to *spatter*.

specific energy. In cutting or grinding, the energy expended or the work done in removing a unit volume of material.

specific power. The net amount of power required during machining to remove a unit volume of metal in unit time. Also known as *unit power*.

specimen. A test object, often of standard dimensions and/or configuration, that is used for destructive or nondestructive testing. One or more specimens may be cut from each unit of a *sample*.

speed of travel. In welding, the speed with which a weld is made along its longitudinal axis, usually measured in inches per minute or spots per minute.

speiss. Metallic arsenides and antimonides that result from smelting metal ores such as those of cobalt.

spelter. Crude zinc obtained in smelting zinc ores.

spelter solder. A brazing filler metal of approximately equal parts of copper and zinc.

spheroidite. An aggregate of iron or alloy carbides of essentially spherical shape dispersed throughout a matrix of *ferrite*.

spheroidizing. Heating and cooling to produce a spheroidal or globular form of carbide in steel. Spheroidizing methods frequently used are:

1. Prolonged holding at a temperature just below Ae_1.
2. Heating and cooling alternately between temperatures that are just above and just below Ae_1.
3. Heating to a temperature above Ae_1 or Ae_3 and then cooling very slowly in the furnace or holding at a temperature just below Ae_1.
4. Cooling at a suitable rate from the minimum temperature at which all carbide is dissolved, to prevent the re-formation of a carbide network, and then reheating in accordance with method 1 or 2 above. (Applicable to hypereutectoid steel containing a carbide network.)

spherulitic graphite cast iron. A *cast iron* that has been treated while molten with an element such as magnesium or cerium to induce the formation of free graphite as nodules of spherulites, which imparts a measurable degree of ductility to the cast metal. Also known as *nodular cast iron*, SG iron, and *ductile cast iron*.

spider die. A multiple-section extrusion die capable of producing tubing or intricate hollow shapes without the use of a separate mandrel. Metal is extruded in separate streams through holes in each section and is re-welded by extrusion pressure before it leaves the die. Also known as *porthole die*. Compare with *bridge die*.

spiegeleisen (spiegel). A pig iron containing 15–30% Mn and 4.5–6.5% C.

spindle. (1) The shaft of a machine tool on which a cutter or grinding wheel may be mounted. (2) The metal shaft to which a mounted wheel is cemented.

spinning. Forming a seamless hollow metal part by forcing a rotating blank to conform to a shaped mandrel that rotates concentrically with the blank. In the typical application, a flat-rolled metal blank is forced against the mandrel by a blunt, rounded tool; however, other stock (notably, welded or seamless tubing) can be formed, and sometimes the working end of the tool is a roller.

spinodal structure. A fine homogeneous mixture of two phases that form by the growth of composition waves in a solid solution during suitable heat treatment. The phases of a spinodal structure differ in composition from each other and from the parent phase, but have the same crystal structure as the parent phase.

spline. Any of a series of longitudinal, straight projections on a shaft that fit into slots on a mating part to transfer rotation to or from the shaft.

split die. A die made of parts that can be separated for ready removal of the workpiece. The same as *segment die*.

sponge. A form of metal characterized by a porous condition that is the result of the decomposition or reduction of a compound without fusion. The term is applied to forms of iron, titanium, zirconium, uranium, plutonium, and the platinum-group metals.

sponge iron. Either porous or powdered iron produced directly without fusion, such as by heating high-grade ore with charcoal, or an oxide with a reducing gas.

spot drilling. Making an initial indentation in a work surface, with a drill, to serve as a centering guide in a subsequent machining operation.

spotfacing. Machining a flat seat for a bolt head, nut, or other similar element at the end of and at a right angle to the axis of a previously made hole.

spotting. Fitting one part of a die to another by applying an oil color to the surface of the finished part and bringing it against the surface of the intended mating part, where the high spots are marked by the transferred color.

spotting out. Delayed, uneven staining of metal by entrapment of chemicals during the finishing operation.

spot welding. Welding of lapped parts in which fusion is confined to a relatively small circular area. It is generally resistance welding, but may also be gas tungsten arc, gas metal arc, or submerged arc welding.

spray metallizing. Forming a metallic coating by atomized spraying with molten metal or by *vacuum deposition*. See also *metallizing* (1).

spray quenching. The rapid cooling of a metal (often steel) from a suitable elevated temperature in a spray of liquid.

spray transfer. In consumable-electrode arc welding, a type of metal transfer in which the molten filler metal is propelled across the arc as fine droplets. Compare with *globular transfer* and *short circuiting transfer*.

springback. (1) The elastic recovery of metal after cold forming. (2) The degree to which metal tends to return to its original shape or contour after undergoing a forming operation. (3) In flash, upset, or pressure welding, the deflection in the welding machine caused by the upset pressure.

spring temper. A *temper* of nonferrous alloys and some ferrous alloys characterized by values of tensile strength and hardness approximately two-thirds of the way from those of *full hard* to those of *extra spring* temper.

sprue. (1) The mold channel that connects the *pouring basin* with the runner or, in the absence of a pouring basin, directly into which molten metal is poured. Sometimes referred to as downsprue or downgate. (2) Sometimes used to mean all gates, risers, runners, and similar scrap that are removed from castings after shakeout.

square drilling. Making square holes by means of a specially constructed drill made to rotate and also to oscillate so as to follow accurately the periphery of a square guide bushing or template.

square groove weld. A groove weld in which the abutting surfaces are square (Fig. 55).

Fig. 55 Examples of square groove welds

squaring shear. A machining tool, used for cutting sheet metal or plate, consisting essentially of a fixed cutting knife (usually mounted on the rear of the bed) and another cutting knife mounted on the front of a reciprocally moving crosshead, which is guided vertically in side housings. Corner angles are typically 90°.

squeeze time. In resistance welding, the time between the initial applications of pressure and current.

stabilizing treatment. (1) Before finishing to final dimensions, repeatedly heating a ferrous or nonferrous part to or slightly above its normal operating temperature and then cooling to room temperature to ensure dimensional stability in service. (2) Transforming retained austenite in quenched hardenable steels, usually by *cold treatment*. (3) Heating a solution-treated stabilized grade of austenitic stainless steel to 870 to 900 °C (1600 to 1650 °F) to precipitate all carbon as TiC, NbC, or TaC so that *sensitization* is avoided on subsequent exposure to elevated temperature.

stack cutting. *Oxyfuel gas cutting* of stacked metal plates arranged so that all are severed by a single cut.

stack molding. A molding method that makes use of both faces of a mold section, one face acting as the drag and the other as the cope. Sections, when assembled to other similar sections, form several tiers of mold cavities, all castings being poured together through a common sprue.

stack welding. Resistance *spot welding* of stacked plates, all being joined simultaneously.

staggered-intermittent fillet welding. Making a line of intermittent fillet welds on each side of a joint so that the increments on one side are not opposite those on the other. Contrast with *chain-intermittent fillet welding*.

staggered-tooth cutters. Milling cutters with alternate flutes of oppositely directed helixes.

stainless steel. Any of several steels containing 12–30% Cr as the principal alloying element; they typically exhibit *passivity* in aqueous environments. See *Technical Note 12*.

staking. Fastening two parts together permanently by recessing one part within the other and then causing plastic flow at the joint.

stalagmometer. An apparatus for determining surface tension. The mass of a drop of liquid is measured by weighing

---

TECHNICAL NOTE 12

Stainless Steels

STAINLESS STEELS are iron-base alloys containing at least 10.5% Cr that achieve their stainless characteristics through the formation of an invisible and adherent chromium-rich oxide surface film. This oxide forms and heals itself in the presence of oxygen. Other elements added to improve particular characteristics include nickel, molybdenum, copper, titanium, aluminum, silicon, niobium, nitrogen, sulfur, and selenium. Carbon is normally present in amounts ranging from less than 0.03% to more than 1.0% in certain grades.

Wrought stainless steels are commonly divided into five groups: martensitic stainless steels, ferritic stainless steels, austenitic stainless steels, duplex (ferritic-austenitic) stainless steels, and precipitation-hardening stainless steels.

Martensitic stainless steels are essentially alloys of chromium and carbon that possess a body-centered tetragonal (bct) crystal structure (martensitic) in the hardened condition. They are ferromagnetic, hardenable by heat treatments, and generally resistant to corrosion only in relatively mild environments. Chromium content is generally in the range of 10.5–18%, and carbon content may exceed 1.2%. Additions of nitrogen, nickel, and molybdenum in combination with somewhat lower carbon levels produce steels with improved toughness and corrosion resistance. Sulfur or selenium is added to some allows to improve machinability.

Ferritic stainless steels are essentially iron-chromium (10.5–30% Cr) alloys with body-centered cubic (bcc) crystal structures. Some grades may contain molybdenum, silicon, aluminum, titanium, and niobium to confer particular characteristics. Ferritic alloys are ferromagnetic and have good ductility and formability, but their high-temperature strengths are relatively poor compared with the austenitic grades. Ferritic stainless steels are, however, highly resistant to chloride stress-corrosion cracking.

TECHNICAL NOTE 12 (*continued*)

Compositions of representative standard stainless steels

Type	UNS designation	Composition (a), %							
		C	Mn	Si	Cr	Ni	P	S	Other
Austenitic types									
201	S20100	0.15	5.5–7.5	1.00	16.0–18.0	3.5–5.5	0.06	0.03	0.25 N
205	S20500	0.12–0.25	14.0–15.5	1.00	16.5–18.0	1.0–1.75	0.06	0.03	0.32–0.40 N
302	S30200	0.15	2.00	1.00	17.0–19.0	8.0–10.0	0.045	0.03	...
304	S30400	0.08	2.00	1.00	18.0–20.0	8.0–10.5	0.045	0.03	...
304N	S30451	0.08	2.00	1.00	18.0–20.0	8.0–10.5	0.045	0.03	0.10–0.16 N
310	S31000	0.25	2.00	1.50	24.0–26.0	19.0–22.0	0.045	0.03	...
316	S31600	0.08	2.00	1.00	16.0–18.0	10.0–14.0	0.045	0.03	2.0–3.0 Mo
316LN	S31653	0.03	2.00	1.00	16.0–18.0	10.0–14.0	0.045	0.03	2.0–3.0 Mo; 0.10–0.16 N
321	S32100	0.08	2.00	1.00	17.0–19.0	9.0–12.0	0.045	0.03	5 x %C min Ti
347	S34700	0.08	2.00	1.00	17.0–19.0	9.0–13.0	0.045	0.03	10 x %C min Nb
348	S34800	0.08	2.00	1.00	17.0–19.0	9.0–13.0	0.045	0.03	0.2 Co;10 x %C min Nb; 0.10 Ta
Ferritic types									
405	S40500	0.08	1.00	1.00	11.5–14.5	...	0.04	0.03	0.10–0.30 Al
409	S40900	0.08	1.00	1.00	10.5–11.75	0.50	0.045	0.045	6 x %C min–0.75 max Ti
430	S43000	0.12	1.00	1.00	16.0–18.0	...	0.04	0.03	...
434	S43400	0.12	1.00	1.00	16.0–18.0	...	0.04	0.03	0.75–1.25 Mo
439	S43035	0.07	1.00	1.00	17.0–19.0	0.50	0.04	0.03	0.15 Al; 12 x %C–1.10 Ti
442	S44200	0.20	1.00	1.00	18.0–23.0	...	0.04	0.03	...
446	S44600	0.20	1.50	1.00	23.0–27.0	...	0.04	0.03	0.25 N
Duplex (ferritic-austenitic) types									
329	S32900	0.20	1.00	0.75	23.0–28.0	2.50–5.00	0.040	0.030	1.00–2.00 Mo
...	S31803	0.03	2.00	1.00	21.0–23.0	4.50–6.50	0.030	0.020	...
Martensitic types									
410	S41000	0.15	1.00	1.00	11.5–13.5	...	0.04	0.03	...
414	S41000	0.15	1.00	1.00	11.5–13.5	1.25–2.50	0.04	0.03	...
416Se	S41623	0.15	1.25	1.00	12.0–14.0	...	0.06	0.06	0.15 min Se
420	S42000	0.15 min	1.00	1.00	12.0–14.0	...	0.04	0.03	...
431	S43100	0.20	1.00	1.00	15.0–17.0	1.25–2.50	0.04	0.03	...
440A	S44002	0.60–0.75	1.00	1.00	16.0–18.0	...	0.04	0.03	0.75 Mo
Precipitation-hardening types									
PH 13-8 Mo	S13800	0.05	0.20	0.10	12.25–13.25	7.5–8.5	0.01	0.008	2.0–2.5 Mo; 0.90–1.35 Al; 0.01 N
15-5 PH	S15500	0.07	1.00	1.00	14.0–15.5	3.5–5.5	0.04	0.03	2.5–4.5 Cu; 0.15–0.45 Nb
17-4 PH	S17400	0.07	1.00	1.00	15.5–17.5	3.0–5.0	0.04	0.03	3.0–5.0 Cu; 0.15–0.45 Nb

(a) Single values are maximum values unless otherwise indicated.

The austenitic stainless steels are the most commonly used stainless steels. These materials have a face-centered cubic (fcc) structure attained through the liberal use of austenitizing elements such as nickel, manganese, and nitrogen. These steels are essentially nonmagnetic in the annealed condition and can be hardened only by cold working. They usually possess excellent cryogenic properties and good high-temperature strength. Chromium content generally varies 16–26%; nickel, up to approximately 35%; and manganese, up to 15%. Molybdenum, copper, silicon, aluminum, titanium, and niobium may be added to confer certain characteristics, such as enhanced corrosion or oxidation resistance. Sulfur or selenium may be added to improve machinability.

Duplex stainless steels have a mixed structure of bcc ferrite and fcc austenite. The exact amount of each phase is a function of composition and heat treatment. Most alloys are designed to contain about equal amounts of each phase in the annealed condition. The principal alloying elements are chromium (21–30%) and nickel (3.5–7.5%), but molybdenum (up to 4%), nitrogen, copper, silicon, and tungsten may be added to control structural balance and to impart certain corrosion-resistance characteristics.

Precipitation-hardening (PH) stainless steels are chromium-nickel grades that contain precipitation-hardening elements such as copper and aluminum. These grades may have austenitic, semiaustenitic, or martensitic crystal structures. All are hardened by a final aging treatment to produce very fine precipitates from a supersaturated solid solution.

The selection of stainless steels is usually based on corrosion resistance and mechanical properties. Stainless steels are firmly established as materials for cooking utensils, fasteners, cutlery, flatware, decorative architectural hardware, and equipment for use in chemical plants, pulp and paper mills, dairy and food- and beverage-processing plants, health and sanitation applications, petroleum and petrochemical plants, textile plants, the pharmaceutical and transportation industries, and the power industry (fossil fuel and nuclear power plants).

Selected References

- S.D. Washko and G. Aggen, Wrought Stainless Steels, *Metals Handbook*, 10th ed., Vol 1, ASM International, 1990, p 841–907
- S. Lampman, Elevated-Temperature Properties of Stainless Steels, *Metals Handbook*, 10th ed., Vol 1, ASM International, 1990, p 930–949
- R.M. Davison, T. DeBold, and M.J. Johnson, Corrosion of Stainless Steels, *Metals Handbook*, 9th ed., Vol 13, ASM International, 1987, p 547–565
- G.F. Vander Voort, Metallography and Microstructures of Wrought Stainless Steels, *Metals Handbook*, 9th ed., Vol 9, ASM International, 1985, p 279–296

a known number of drops or by counting the number of drops obtained from a given volume of the liquid.

stamping. A general term covering almost all sheet metal press operations. It includes blanking, shearing, hot or cold forming, drawing, bending, and coining.

stand. A piece of rolling mill equipment containing one set of work rolls. In the usual sense, any pass of a continuous, looping, or cross-country hot rolling mill.

standard electrode potential. The reversible *electrode potential* when all reactants and products are at unit activity.

standard gold. A legally adopted alloy for the coinage of gold. In the United States this alloy contains 10% Cu.

stardusting. An extremely fine form of roughness on the surface of a metal deposit.

starting sheet. A thin sheet of metal used as the cathode in electrolytic refining.

state of strain. A complete description of the deformation within a homogeneously deformed volume or at a point. The description requires, in general, the knowledge of six independent components of *strain*.

state of stress. A complete description of the stresses within a homogeneously stressed volume or at a point. The description requires, in general, the knowledge of six independent components of *stress*.

static fatigue. A term sometimes used to identify a form of hydrogen embrittlement in which a metal appears to fracture spontaneously under a steady stress less than the yield stress. There almost always is a delay between the application of stress (or exposure of the stressed metal to hydrogen) and the onset of cracking. More properly referred to as hydrogen-induced delayed cracking.

steadite. A hard structural constituent of cast iron that consists of a binary eutectic of *ferrite* (containing some phosphorus in solution) and iron phosphide (Fe_3P). The composition of the eutectic is 10.2% P, 89.8% Fe, and the melting temperature is 1050 °C (1920 °F).

Stead's brittleness. A condition of brittleness that causes transcrystalline fracture in the coarse grain structure that results from prolonged annealing of thin sheets of low-carbon steel previously rolled at a temperature below approximately 705 °C (1300 °F). The fracture usually occurs at approximately 45° to the direction of rolling.

steadyrest. In cutting or grinding, a stationary support for a long workpiece.

Steckel mill. A cold reducing mill having two working rolls and two backup rolls, none of which is driven. The strip is drawn through the mill by a power reel in one direction as far as the strip will allow and then reversed by a second power reel, and so on until the desired thickness is attained.

steel. An iron-base alloy, malleable in some temperature ranges as initially cast, containing manganese, usually carbon, and often other alloying elements. In carbon steel and low-alloy steel, the maximum carbon is approximately 2.0%; in high-alloy steel, approximately 2.5%. The dividing line between low-alloy and high-alloy steels is generally regarded as being at approximately 5% metallic alloying elements.

Steel is differentiated from two general classes of "irons": the cast irons, on the high-carbon side; and the relatively pure irons such as ingot iron, carbonyl iron, and electrolytic iron, on the low-carbon side. In some steels containing extremely low carbon, the manganese content is the principal differentiating factor, steel usually containing at least 0.25% and ingot iron considerably less.

Stellite. Stellite is the trademarked name of a high-cobalt chromium alloy of the Deloro Stellite Company. Elwood Haynes of Kokomo, Indiana, was the inventor of Stellite in 1910. Its first application was a lathe tool bit able to outlast high-speed steels. There are now dozens of Stellite alloys that are used for their abrasion resistance, heat resistance, and corrosion resistance. They are used for saw teeth and hard-surfacing alloys.

step aging. The aging of metals at two or more temperatures, by steps, without cooling to room temperature after each step. See also *aging*, and compare with *interrupted aging* and *progressive aging*.

stepped extrusion. The process whereby a single product has one or more abrupt changes in cross section, produced by stopping extrusion to change dies. Often, such an extrusion is made in a complex die having a die section that can be freed from the main die and allowed to ride out with the product when extrusion is resumed. See *extrusion.*

stereoradiography. A technique for producing paired radiographs that may be viewed with a stereoscope to exhibit a shadowgraph in three dimensions with various sections in perspective and spatial relation.

sterling silver. A silver alloy containing at least 92.5% Ag, the remainder being unspecified but usually copper.

stick electrode. A shop term for *covered electrode.*

sticker breaks. Arc-shaped *coil breaks*, typically located near the center of sheet or strip.

stiffness. The ability of a metal or shape to resist elastic deflection. For identical shapes, the stiffness is proportional to the modulus of elasticity. For a given material, the stiffness increases with increasing moment of inertia, which is computed from cross-sectional dimensions.

stock. A general term for solid starting material that is formed, forged, or machined to make parts.

stoking. (obsolete) Presintering, or sintering, in such a way that powder metallurgy compacts are advanced through the furnace at a fixed rate by manual or mechanical means. The preferred term is continuous *sintering*.

stopoff. A material applied to prevent the flow of soldering or brazing filler metal into unwanted areas. See also *resist.*

stopper rod. A device in a bottom-pour ladle for controlling the flow of metal through the nozzle into a mold. The stopper rod consists of a steel rod, protective refractory sleeves, and a graphite stopper head.

stopping off. (1) Applying a *resist.* (2) Depositing a metal (copper, for example) in localized areas to prevent carburization, decarburization, or nitriding in those areas. (3) Filling in a portion of a mold cavity to keep out molten metal.

stored-energy welding. Resistance welding with electrical energy accumulated electrostatically, electromagnetically, or electrochemically at a relatively low rate and made available at the higher rate required in welding.

straddle milling. Face milling a work piece on both sides at once using two cutters spaced as required.

straight polarity. Direct-current arc welding circuit arrangement in which the electrode is connected to the negative terminal. Contrast with *reverse polarity*.

strain. A measure of the relative change in the size or shape of a body. Linear strain is the change per unit length of a linear dimension. True strain (or natural strain) is the natural logarithm of the ratio of the length at the moment of observation to the original gage length. Conventional strain is the linear strain over the original gage length. Shearing strain (or shear strain) is the change in angle (expressed in radians) between two lines originally at right angles. When the term *strain* is used alone, it typically refers to linear strain in the direction of applied stress. See also *state of strain*.

strain-age embrittlement. A loss in *ductility* accompanied by an increase in hardness and strength that occurs when low-carbon steel (especially rimmed or capped steel) is aged following *plastic deformation*. The degree of *embrittlement* is a function of aging time and temperature, occurring in a matter of minutes at approximately 200 °C (400 °F) but requiring several hours to a year at room temperature.

strain aging. A change in the properties of certain metals and alloys that occurs at ambient or slightly elevated temperatures after cold working. See *aging*.

strain energy. (1) The work done in deforming a body. (2) The work done in deforming a body within the elastic limit of the material. It is more properly termed *elastic strain energy* and can be recovered as work rather than heat.

strain hardening. An increase in the hardness and strength of metals caused by plastic deformation at temperatures below the recrystallization range.

strain-hardening exponent. A measure of rate of strain hardening. The constant n in the expression:

$$\sigma = \sigma_0 \delta^n$$

where σ is true stress, σ_0 is true stress at true strain, and δ is true strain.

strain rate. The time rate of straining for the usual tensile test. Strain as measured directly on the specimen gage length is used for determining strain rate. Because strain is dimensionless, the units of strain rate are reciprocal time.

strain-rate sensitivity. Qualitatively, the increase in stress (s) needed to cause a certain increase in plastic strain rate ($\dot{\varepsilon}$) at a given level of plastic strain (ε) and a given temperature (T).

$$\text{Strain-rate sensitivity} = m = \left(\frac{\Delta \log s}{\Delta \log \dot{\varepsilon}} \right)_{\varepsilon, T}$$

strain rods. (1) Rods sometimes used on gapframe metal forming presses to lessen the frame deflection. (2) Rods used to measure elastic strains, and thus stresses, in frames of metal forming presses.

strain state. A complete description of the deformation within a homogeneously deformed volume or at a point. The description requires, in general, the knowledge of six independent components of *strain*. Also known as *state of strain*.

strand casting. A generic term describing *continuous casting* of one or more elongated shapes such as billets, blooms, or slabs; if two or more strands are cast simultaneously, they are often of identical cross section.

stray current. Current flowing in electro-deposition by way of an unplanned and undesired bipolar electrode that may be the tank itself or a poorly connected electrode.

stress. Force per unit area, often thought of as force acting through a small area within a plane. It can be divided into components, normal and parallel to the plane, called normal stress and shear stress, respectively. True stress denotes the stress where force and area are measured at the same time. Conventional stress, as applied to tension and compression tests, is force divided by original area. Nominal stress is the stress computed by simple elasticity formulas, ignoring stress raisers and disregarding plastic flow; in a notch bend test, for example, it is bending moment divided by minimum section modulus. See also *state of stress*.

stress amplitude. One-half the algebraic difference between the maximum and minimum stresses in one cycle of a repetitively varying stress.

stress concentration factor (K_t). A multiplying factor for applied stress that allows for the presence of a structural discontinuity such as a notch or hole; K_t equals the ratio of the greatest stress in the region of the discontinuity to the nominal stress for the entire section.

stress-corrosion cracking (SCC). Failure by cracking under the combined action of corrosion and either external (applied) stress or internal (residual) stress. Cracking may be either intergranular or transgranular, depending on the metal and the corrosive medium. See also *season cracking*.

stress-intensity factor. A scaling factor, usually denoted by the symbol K, used in *linear elastic fracture mechanics* to describe the intensification of applied stress at the tip of a crack of known size and shape. At the onset of rapid crack propagation in any structure containing a crack, the factor is called the critical stress-intensity factor, or the *fracture toughness*. Various subscripts are used to denote different loading conditions or fracture toughnesses. The most common subscripts and their meanings are:

K_c. Plane-stress fracture toughness. The value of stress intensity at which crack propagation becomes rapid in sections thinner than those in which plane-strain conditions prevail.

K_I. Stress-intensity factor for a loading condition that displaces the crack faces in a direction normal to the crack plane (also known as the opening mode of deformation).

K_{Ic}. Plane-strain fracture toughness. The minimum value of K_c for any given material and condition, which is attained when rapid crack propagation in the opening mode is governed by plane-strain conditions.

K_{Id}. Dynamic fracture toughness. The fracture toughness determined under dynamic loading conditions; it is used as an approximation of K_{Ic} for very tough materials.

K_{ISCC}. Threshold stress intensity factor for stress-corrosion cracking (SCC). A value of stress intensity characteristic of a specific combination of material, material condition, and corrosive environment above which SCC propagation occurs and below which the material is immune to SCC.

stress raisers. Changes in contour or discontinuities in structure that cause local increases in stress.

stress range. The algebraic difference between the maximum and minimum stress in one cycle of a repetitively varying stress.

stress ratio. In fatigue, the ratio of the minimum stress to the maximum stress in one cycle, considering tensile stresses as positive and compressive stresses as negative.

stress relieving. Heating to a suitable temperature, holding long enough to reduce residual stresses, and then cooling slowly enough to minimize the development of new residual stresses.

stress-rupture test. A method of evaluating elevated-temperature durability in which a tension-test specimen is stressed under constant load until it breaks. Data recorded commonly include: initial stress, time to rupture, initial extension, creep extension, and reduction of area at fracture. Also known as *creep-rupture test.*

stress state. A complete description of the stresses within a homogeneously stressed volume or at a point. The description requires, in general, the knowledge of six independent components of *stress*. Also known as *state of stress.*

stretcher leveling. Leveling a piece of metal (that is, removing warp and distortion) by gripping it at both ends and subjecting it to a stress higher than its yield strength. Sometimes known as *patent leveling.*

stretcher straightening. Straightening rod, tubing, or shapes by gripping the stock at both ends and applying tension. The products are elongated a definite amount to remove warpage.

stretcher strains. Elongated markings that appear on the surfaces of some materials when deformed just past the yield point. These markings lie approximately parallel to the direction of maximum shear stress and are the result of localized yielding. Also known as *Lüders lines*, Lüders bands, *Hartmann lines*, and *Piobert lines.*

stretch former. (1) A machine used to perform *stretch forming* operations. (2) A device adaptable to a conventional press for accomplishing stretch forming.

stretch forming. The shaping of a sheet or part, usually of uniform cross section, by first applying suitable tension or stretch and then wrapping the sheet or part around a die of the desired shape. The same as *wrap forming.*

stretch wipe forming. A method of curving bars, tubes, or rolled or extruded sections, in which the stock is bent so that it conforms to a fixed form block. Stock is clamped to the form block, then bent by applying force through a wiper block, shoe, or roll that is moved along the periphery of the form block. Also known as *wiper forming* and compression forming. Contrast with *draw forming.*

striation. A fatigue fracture feature, often observed in electron micrographs, that indicates the position of the crack front after each succeeding cycle of stress. The distance between striations indicates the advance of the crack front across that crystal during one stress cycle, and a line normal to the striations indicates the direction of local crack propagation.

strike. (1) A thin electrodeposited film of metal to be overlaid with other plated coatings. (2) A plating solution of high covering power and low efficiency designed to electroplate a thin, adherent film of metal.

striking. Electrodepositing, under special conditions, a very thin film of metal that will facilitate further plating with another metal or with the same metal under different conditions.

striking surface. Those areas on the faces of a set of metal forming dies that are designed to meet when the upper and lower dies are brought together. The striking surface helps to protect impressions from impact shock and aids in maintaining longer die life. Also called beating area.

stringer. In wrought materials, an elongated configuration of microconstituents or foreign material aligned in the direction of working. Commonly, the term is associated with elongated oxide or sulfide inclusions in steel.

stringer bead. A continuous weld bead made without appreciable transverse oscillation. Contrast with *weave bead*.

strip. A flat-rolled metal product of some maximum thickness and width arbitrarily dependent on the type of metal. It is narrower than *sheet*.

stripper punch. A punch that serves as the top or bottom of a metal forming die cavity and later moves farther into the die to eject the part or compact. See also *ejector rod* and *knockout* (1).

stripping. Removing a coating from a metal surface.

strontium. A chemical element having atomic number 38, atomic weight 88, and the symbol Sr, from the mineral strontianite, in which strontium occurs naturally. Both strontium and strontianite are named after Strontian, a Scottish village near where the mineral was discovered. Strontium was discovered in 1787 by Scottish physician Adair Crawford, identified in 1791 by Thomas Charles Hope at Edinburgh, and isolated in 1808 by Sir Humphry Davy. It is similar to calcium in its uses as well as physical and chemical properties, but its higher cost does not justify extraction on a commercial scale.

structural shape. A piece of metal of any of several designs accepted as standard by the structural branch of the iron and steel industries.

stud arc welding. An *arc welding* process that produces coalescence of metals by heating them with an arc between a metal stud, or similar part, and another part. When the surfaces to be joined are properly heated, they are brought together under pressure. Partial shielding may be obtained by the use of a ceramic ferrule surrounding the stud. Shielding gas or flux may or may not be used.

sub-boundary structure. A network of low-angle boundaries, usually with misorientations less than 1° within the main crystals of a microstructure. The same as *substructure*.

subcritical annealing. A process anneal performed on ferrous alloys at a temperature below Ac_1.

subgrain. A portion of a crystal or *grain*, with an orientation slightly different from the orientation of neighboring portions of the same crystal. Generally, neighboring subgrains are separated by low-angle boundaries such as *tilt boundaries* and *twist boundaries*.

submerged arc welding. Arc welding in which the arc, between a bare metal electrode and the work, is shielded by a blanket of granular, fusible material overlying the joint. Pressure is not applied to the joint, and filler metal is obtained from the consumable electrode (and sometimes from a supplementary welding rod).

subsieve analysis. Size distribution of particles that will pass through a 44 μm (No. 325) standard sieve, as determined by specified methods.

subsieve fraction. The portion of a powdered sample that will pass through a 44 μm (No. 325) standard sieve.

substitutional solid solution. A *solid solution* in which the solute atoms are located at some of the lattice points of the solvent, the distribution being random. Contrast with *interstitial solid solution*.

substrate. The layer of metal underlying a coating, regardless of whether that layer is the basis metal.

substructure. A network of low-angle boundaries, usually with misorientations less than one degree within the main crystals of a microstructure. Also known as *sub-boundary structure*.

subsurface corrosion. The formation of isolated particles of corrosion products beneath a metal surface. This results from the preferential reactions of certain alloy constituents to inward diffusion of oxygen, nitrogen, or sulfur.

sulfur dome. An inverted container, holding a high concentration of sulfur dioxide gas, used in die casting to cover a pot of molten magnesium to prevent burning.

sulfur print. A macrographic method of examining for distribution of sulfide inclusions by placing a sheet of wet acidified photographic paper in contact with the polished steel surface to be examined.

superalloy. A type of *heat-resistant alloy*, which is an alloy developed for very-high-temperature service where relatively high stresses (tensile, thermal, vibratory, or shock) are encountered, and where oxidation resistance is frequently required. See *Technical Note 13*.

TECHNICAL NOTE 13

Superalloys

SUPERALLOYS are heat-resistant alloys based on nickel (Ni), Iron-nickel (Fe-Ni), or cobalt (Co) that exhibit a combination of mechanical strength and resistance to surface degradation that is unmatched by other metallic alloys. Superalloys are primarily used in gas turbines, coal conversion plants, and chemical process industries, and for other specialized applications requiring high heat and corrosion resistance.

Superalloys consist of a face-centered cubic (fcc) austenitic gamma (γ) phase matrix plus a variety of secondary phases. The principal secondary phases are carbides (Mc, $M_{23}C_6$, M_6C, and M_7C_3) in all superalloy types and gamma prime (γ') fcc ordered $Ni_3(Al, Ti)$ intermetallic compound in Ni- and Fe-Ni-base alloys. In alloys containing niobium and tantalum, the primary strengthening phase is gamma double prime (γ''), a body-centered tetragonal phase. Superalloys derive their strength from solid-solution hardeners and precipitating phases. Carbides may provide limited strengthening directly (e.g., through dispersion hardening)

Compositions of selected superalloys

Alloy	Cr	Ni	Co	Mo	W	Nb	Ti	Al	Fe	C	Other
Fe-Ni-base											
19-9DL	19.0	9.0	...	1.25	1.25	0.4	0.3	...	66.8	0.30	1.10 Mn; 0.6 Si
Incoloy 800	21.0	32.5	...	...	...	...	0.38	0.38	45.7	0.05	0.8 Mn; 0.5 Si
A-286	15.0	26.0	...	1.25	...	...	2.0	0.2	55.2	0.04	0.005 B; 0.3 V
V-57	14.8	27.0	...	1.25	...	...	3.0	0.25	48.6	0.08 max	0.01 B; 0.5 max V
Incoloy 901	12.5	42.5	...	6.0	...	...	2.7	...	36.2	0.10 max	...
Inconel 718	19.0	52.5	...	3.0	...	5.1	0.9	0.5	18.5	0.08 max	0.15 max Cu
Hastelloy X	22.0	49.0	1.5 max	9.0	0.6	...	...	2.0	15.8	0.15	...
Ni-base											
Waspaloy	19.5	57.0	13.5	4.3	...	...	3.0	1.4	2.0 max	0.07	0.006 B; 0.09 Zr
M252	19.0	56.5	10.0	10.0	...	...	2.6	1.0	<0.75	0.15	0.005 B
Udimet 500	19.0	48.0	19.0	4.0	...	...	3.0	3.0	4.0 max	0.08	0.005 B
Udimet 700	15.0	53.0	18.5	5.0	...	...	3.4	4.3	<1.0	0.07	0.03 B
Astroloy	15.0	56.5	15.0	5.25	...	...	3.5	4.4	<0.3	0.06	0.03 B; 0.06 Zr
René 80	14.0	60.0	9.5	4.0	4.0	...	5.0	3.0	...	0.17	0.015 B; 0.03 Zr
IN-100	10.0	60.0	15.0	3.0	...	...	4.7	5.5	<0.6	0.15	1.0 V; 0.06 Zr; 0.015 B
René 95	14.0	61.0	8.0	3.5	3.5	3.5	2.5	3.5	<0.3	0.16	0.01 B; 0.05 Zr
MAR-M 247	8.25	59.0	10.0	0.7	10.0	...	1.0	5.5	<0.5	0.15	0.015 B; 0.05 Zr; 1.5 Hf; 3.0 Ta
IN MA-754	20.0	78.5	...	...	...	...	0.5	0.3	...	...	0.6 Y$_2$O$_3$
IN MA-6000E	15.0	68.5	...	2.0	4.0	...	2.5	4.5	...	0.05	1.1 Y$_2$O$_3$; 2.0 Ta; 0.01 B; 0.15 Zr
Co-base											
Haynes 25 (L-605)	20.0	10.0	50.0	...	15.0	...	...	...	3.0	0.10	1.5 Mn
Haynes 188	22.0	22.0	37.0	...	14.5	...	...	...	3.0 max	0.10	0.90 La
S-816	20.0	20.0	42.0	4.0	4.0	4.0	...	...	4.0	0.38	...
X-40	22.0	10.0	57.5	...	7.5	...	...	...	1.5	0.50	0.5 Mn; 0.5 Si
WI-52	21.0	...	63.5	...	11.0	...	...	...	2.0	0.45	2.0 Nb + Ta
MAR-M 302	21.5	...	58.0	...	10.0	...	...	...	0.5	0.85	9.0 Ta; 0.005 B; 0.2 Zr
MAR-M 509	23.5	10.0	54.5	...	7.0	...	0.2	...	...	0.6	0.5 Zr; 3.5 Ta
J-1570	20.0	28.0	46.0	...	...	...	4.0	...	2.0	0.2	...

TECHNICAL NOTE 13 (*continued*)

or, more commonly, indirectly (e.g., by stabilizing grain boundaries against excessive shear). In addition to those elements that produce solid-solution hardening and promote carbide and γ' formation, other elements (e.g., boron, zirconium, hafnium, and cerium) are added to enhance mechanical and/or chemical properties.

The three types of superalloys (Fe-Ni-, Ni-, and Co-base) are further subdivided into wrought, cast, and powder metallurgy alloys. Cast alloys can be further broken down into polycrystalline, directionally solidified, and single-crystal superalloys. The most important class of Fe-Ni-base superalloys includes those alloys that are strengthened by intermetallic-compound preceipitation in an fcc matrix. The most common precipitate is γ', typified by alloys A-286 and Incoloy 901, but some alloys precipitate γ", typified by Inconel 718. Another class of cast Fe-Ni-base superalloys is hardened by carbides, nitrides, and carbonitrides: some tungsten and molybdenum may be added to produce solid-solution hardening. Other Fe-Ni-base alloys are modified stainless steels primarily strengthened by solid-solution hardening.

The most important class of Ni-base superalloys is strengthened by intermetallic-compound precipitation in an fcc matrix. The strengthening precipitate is γ', typified by Waspaloy and Udimet 700. Another class is represented by Hastelloy X, which is essentially solid-solution strengthened, but which also derives some strengthening from carbide precipitation produced through a working-plus-aging schedule. A third class includes oxide dispersion strengthened (ODS) alloys, which are strengthened by dispersions of inert particles such as yttria.

Cobalt-base superalloys are strengthened by solid-solution alloying and carbide precipitation. Unlike the Fe-Ni- and Ni-base alloys, no intermetallic phase has been found that will strengthen Co-base alloys to the same degree that γ' or γ" strengthens the other superalloys.

Selected References

· N.S. Stoloff, Wrought and P/M Superalloys, *Metals Handbook*, 10th ed., Vol 1, ASM International, 1990, p 950–980
· G.L. Erickson, Polycrystalline Cast Superalloys, *Metals Handbook*, 10th ed., Vol 1, ASM International, 1990, p 981–994
· K. Harris, G.L. Erickson, and R.E. Schwer, Directionally Solidified and Single-Crystal Superalloys, *Metals Handbook*, 10th ed., Vol 1, 1990, p 995–1006
· G.F. Vander Voort and H.M. James, Metallography and Microstructures of Wrought Heat-Resistant Alloys, *Metals Handbook*, 9th ed., Vol 9 ASM International, 1985, p 305–379

superconductivity. A property of some metals characterized by the abrupt and large increase in electrical conductivity exhibited at temperatures near absolute zero.

superconductors. See *Technical Note 14*.

supercooling. The cooling below the temperature at which an equilibrium phase transformation can take place, without actually obtaining the transformation. The same as *undercooling*.

superficial Rockwell hardness test. A form of the *Rockwell hardness test* using relatively light loads that produce minimum penetration by the indenter. Used for determining surface hardness or hardness of thin sections or small parts, or where a large hardness impression might be harmful.

superfines. The portion of a metal powder that is composed of particles smaller than a specified size, usually 10 μm.

superfinishing. A form of *honing* in which the abrasive stones are spring supported.

superheating. (1) Heating above the temperature at which an equilibrium phase

---

TECHNICAL NOTE 14

Superconductors

SUPERCONDUCTORS are materials that exhibit a complete disappearance of electrical resistance on lowering the temperature below a critical temperature (T_c). For all superconductors presently known, the critical temperatures are well below room temperature, and they are usually attained by cooling with liquefied gas (helium or nitrogen), either at or below atmospheric pressure. A superconducting material must also exhibit perfect diamagnetism, that is, complete exclusion of an applied magnetic field from the bulk of the superconductor. Superconductivity permits electric power generators and transmission lines to have capacities many times greater than recently possible. It also allows the development of levitated transit systems capable of high speeds, and provides an economically feasible way of producing the large magnetic fields required for the confinement of ionized gases in controlled thermonuclear fusion.

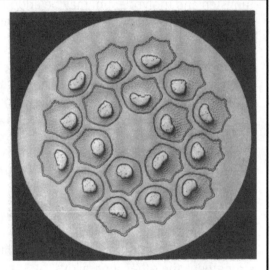

Cross section of a multifilamentary Nb_3Sn superconductor wire. Original magnification: 75x

Superconductivity is observed in a broad range of materials. These include more than half of the metallic elements and a wide range of compounds and alloys. To date, however, the materials that have received the most attention are

---

TECHNICAL NOTE 14 (*continued*)

Approximate superconducting properties of selected superconducting materials

Material	Type	Critical temperature, T_c at 0 T	Thermodynamic critical field, T, at			Magnetic penetration depth (λ), nm	Coherence length (ξ), nm	Crtical current density (J_c)kA · mm^{-2}
			$\mu_0 H_c$	$\mu_0 H_{c1}$	$\mu_0 H_{c2}$			
Pb	I	7.3	0.0803(a)	...	...	40	83	...
Nb	II	9.3	0.37	0.25	0.41	30	40	...
Nb45–50-Ti	II	8.9–9.3	0.16	0.009	10.5–11.0	500	10	3 (at 5 T)
Nb$_3$Sn	II	18	0.46	0.034	19–25	200	6	10 (at 5 T)
Nb$_3$Ge	II	23	0.16	0.004	36–41	650	4	10 (at 5 T)
NbN	II	16–18	0.16	0.004	20–35	600	5	10 (at 0 T)
PbMo$_6$S$_8$	II	14–15	0.4	0.005	40–55	240	4	0.8 (at 5 T)
YBa$_2$Cu$_3$O$_7$	II	92	0.03	0.05(b) 0.01(c)	60(b) >200(c)	150(b) 1000(c)	15(b) 2-3(c)	1 (at 77 K, 0 T)(d)

Parameters at 4.2 K

(a) Thermodynamic critical field at 0 K. (b) Measured with field parallel to the c-axis. (c) Measured with field parallel to the a–b plane. (d) Epitaxial thin film, current in the a–b plane.

TECHNICAL NOTE 14 (*continued*)

niobium-titanium superconductors (the most widely used superconductor), A15 compounds (in which class the important intermetallic Nb_3Sn lies), ternary molybdenum chalcogenides, and high-temperature ceramic superconductors. The chalcogenides and ceramics, however, are only in the research stage.

Niobium-titanium superconductors are actually composite wires that consist of Nb-Ti filaments (<10 μm in diameter) embedded in an oxygen-free, high-purity (99.99%) copper matrix. Commercially pure aluminum (alloy 1100, 99.0% Al) and copper-nickel alloys (typically in concentrations of 90:10 or 70:30) matrices have also been utilized. The filament alloy most widely used is Nb-46.5Ti. Binary Nb-Ti compositions in the range of 45–50% Ti exhibit T_c values of 9.0–9.3 K. Composite conductors containing as few as one to as many as 25,000 filaments have been processed by advanced extrusion and wire-drawing techniques. The primary applications for Nb-Ti superconductors are magnets for use in magnetic resonance imaging (MRI) devices used in hospitals and high-energy physics pulsed accelerator-magnet applications. An example of the latter application is the proposed Superconducting Supercollider for studying the elementary particles of which all matter is composed and the forces through which matter interacts.

A15 superconductors are brittle intermetallic A_3B compounds with a body-centered cubic (bcc) crystal structure. Of the 76 known A15 compounds, 46 are known to be superconducting. Because of its ease of fabrication, Nb_3Sn is the most commercially important A15 compound. Like Nb-Ti superconductors, Nb_3Sn is also assembled into multifilamentary wires. Applications for Nb_3Sn-base superconductors include large commercial magnets, power generators and power transmission lines, and devices for magnetically confining high-energy plasma for thermonuclear fusion.

The ternary molybdenum chalcogenides represent a vast class of materials whose general formula is $M_xMo_6X_8$, where M is a cation and X a chalcogen (sulfur, selenium, or tellurium). Most of the research on these materials has centered around $PbMo_6S_8$ and $SnMo_6S_8$, the former having a T_c of 14–15 K.

High-temperature superconductors (T_c values exceeding 90 K) are ceramic oxides in wire, tape, or thin-film form. The systems being studied include Y-Ba-Cu-O (most notably $YBa_2Cu_3O_7$), Bi-Sr-Ca-Cu-O, and Tl-Ba-Ca-Cu-O.

Selected References

· Superconducting Materials, T.S. Kreilick, Ed., *Metals Handbook*, 10th ed., Vol 2, ASM International, 1990, p 1027–1089
· R.B. Poeppel et al., High-Temperature Superconductors, *Engineered Materials Handbook*, Vol 4, ASM International, 1991, p 1156–1160

transformation should occur without actually obtaining the transformation. (2) Heating molten metal above the normal casting temperature so as to obtain more complete refining or greater fluidity.

superlattice. A lattice arrangement in which solute and solvent atoms of a solid solution occupy different preferred sites in the array. See also *ordering*. Contrast with *disordering*.

superplasticity. The ability of certain metals to undergo unusually large amounts of plastic deformation before local necking occurs (Table 12).

supersonic. Having a speed greater than that of sound. Not the same as ultrasonic; see *ultrasonic frequency*.

support pins. Rods or pins of precise length used to support the overhang of irregularly shaped punches in metal forming presses.

support plate. A plate that supports the draw ring or draw plate in a sheet metal forming press. It also serves as a spacer.

surface checking. Numerous, very fine cracks in a coating or at the surface of a metal part. Checks may appear during processing or during service and most often are associated with thermal treatment or thermal cycling. Also known as *checks*, check marks, checking and *heat checks*.

surface finish. (1) The condition of a surface as a result of a final treatment. (2) Measured surface profile characteristics, the preferred term being *roughness*.

surface grinding. Producing a plane surface by *grinding*.

surface hardening. A generic term covering several processes applicable to a suitable ferrous alloy that produces, by quench hardening only, a surface layer that is harder or more wear resistant than the core. There is no significant alteration of the chemical composition of the surface layer. The processes commonly used are *induction hardening, flame hardening,* and *shell hardening*. Use of the applicable specific process name is preferred.

surface roughness. Relatively finely spaced surface irregularities, the heights, widths, and directions of which establish the predominant surface pattern. Also known as *roughness*.

surface tension. Interfacial tension between two phases of which one is a gas.

surfacing. The deposition of filler metal on a metal surface by welding, spraying, or braze welding, to obtain certain desired properties or dimensions. See also *hardfacing*.

surfacing weld. A type of weld composed of one or more stringer or weave beads deposited on an unbroken surface to obtain desired properties or dimensions.

swaging. Tapering bar, rod, wire, or tubing by forging, hammering, or squeezing; reducing a section by progressively tapering lengthwise until the entire section attains the smaller dimensions of the taper.

swarf. An intimate mixture of grinding chips and fine particles of abrasive and bond resulting from a grinding operation.

sweat. Exudation of a low-melting phase during solidification. Also known as sweatback. For tin bronzes, it is called *tin sweat*.

sweating. A *soldering* technique in which two or more parts are precoated (tinned), then reheated and joined without adding more solder. Also known as sweat soldering.

sweating out. Bringing to the surface small globules of one of the low-melting

Table 12 Superplastic properties of several aluminum and titanium alloys

Alloy	Test temperature		Strain rate, s^{-1}	Strain rate sensitivity, m	Elongation, %
	°C	°F			
Aluminum					
Statically recrystallized					
Al-33Cu	400–500	752–930	8×10^{-4}	0.8	400-1000
Al-4.5Zn-4.5Ca	550	1020	8×10^{-3}	0.5	600
Al-6 to 10Zn-1.5Mg-0.2Zr	550	1020	10^{-3}	0.9	1500
Al-5.6Zn-2Mg-1.5Cu-0.2Cr	516	961	2×10^{-4}	0.8-0.9	800-1200
Dynamically recrystallized					
Al-6Cu-0.5Zr (Supral 100)	450	840	10^{-3}	0.3	1000
Al-6Cu-0.35Mg-0.14Si (Supral 220)	450	840	10^{-3}	0.3	900
Al-4Cu-3Li-0.5Zr	450	840	5×10^{-3}	0.5	900
Al-3Cu-2Li-1Mg-0.2Zr	500	930	1.3×10^{-3}	0.4	878
Titanium					
α/β					
Ti-6Al-4V	840–870	1545–1600	1.3×10^{-4} to 10^{-3}	0.75	750-1170
Ti-6Al-5V	850	1560	8×10^{-4}	0.70	700-1100
Ti-6Al-2Sn-4Zr-2Mo	900	1650	2×10^{-4}	0.67	538
Ti-4.5Al-5Mo-1.5Cr	871	1600	2×10^{-4}	0.63-0.81	>510
Ti-6Al-4V-2Ni	815	1499	2×10^{-4}	0.85	720
Ti-6Al-4V-2Co	815	1499	2×10^{-4}	0.53	670
Ti-6Al-4V-2Fe	815	1499	2×10^{-4}	0.54	650
Ti-5Al-2.5Sn	1000	1830	2×10^{-4}	0.49	420
β and near β					
Ti-15V-3Cr-3Sn-3Al	815	1499	2×10^{-4}	0.5	229
Ti-13Cr-11V-3Al	800	1470	...	...	<150
Ti-8Mn	750	1380	...	0.43	150
Ti-15Mo	800	1470	...	0.60	100
α					
CP Ti	850	1560	1.7×10^{-4}	...	115

constituents of an alloy during heat treatment, such as lead out of bronze.

sweep. A form or template used for shaping sand molds or cores by hand.

sweeps. Floor and table sweepings containing precious metal particles.

sweet roast. A *roasting* process for the complete elimination of sulfur. Also known as *dead roast*.

swing forging machine. Equipment for continuously hot reducing ingots, blooms, or billets to square flats, rounds, or rectangles by the crank-driven oscillating action of paired dies.

swing frame grinder. A grinding machine suspended by a chain at the center point so that it may be turned and swung in any direction for the grinding of billets, large castings, or other heavy work. The principal use is removing surface imperfections and roughness.

synchronous timing. In spot, seam, or projection welding, a method of regulating the welding transformer primary current so that all of the following conditions will prevail: (a) The first half-cycle is initiated at the proper time in relation to the voltage to ensure a balanced current wave; (b) each succeeding half-cycle is essentially identical to the first; and (c) the last half-cycle is of opposite polarity to the first.

syntectic. An isothermal reversible reaction in which a solid phase, on absorption of heat, is converted to two conjugate liquid phases.

synthetic cold rolled sheet. A hot rolled pickled sheet given a sufficient final temper pass to impart a surface approximating that of cold rolled steel.

Dictionary of Metals
H.M. Cobb, editor

tacking. Making *tack welds*.

tack welds. Small, scattered welds made to hold parts of a weldment in proper alignment while the final welds are being made.

taconite. A siliceous iron formation from which certain iron ores of the Lake Superior region are derived; consists chiefly of fine-grain silica mixed with magnetite and hematite.

tailings. The discarded portion of a crushed ore, separated during concentration.

tandem die. A *progressive die* consisting of two or more parts in a single holder, used with a separate lower die to perform more than one operation (such as piercing and blanking) on a part at two or more stations. Also known as *follow die*.

tandem mill. A rolling mill consisting of two or more stands arranged so that the metal being processed travels in a straight line from stand to stand. In continuous rolling, the various stands are synchronized so that the strip can be rolled in all stands simultaneously. Contrast with *single-stand mill*.

tandem welding. Arc welding in which two or more electrodes are in a plane parallel to the line of travel.

tangent bending. Forming one or more identical bends having parallel axes by wiping sheet metal around one or more radius dies in a single operation. The sheet, which may have side flanges, is clamped against the radius die and made to conform to the radius die by pressure from a rocker-plate die that moves along the periphery of the radius die.

tangent modulus. The slope of the stress-strain curve at a specified point. See also *modulus of elasticity*.

tank voltage. The total voltage between the anode and cathode of a plating bath or electrolytic cell during electrolysis. It is equal to the sum of: (a) the equilibrium reaction potential, (b) the IR drop, and (c) the electrode potentials.

tantalum. A chemical element having atomic number 73, atomic weight 181, and the symbol Ta, for Tantalus of Greek mythology. The metal was isolated and named by Swedish chemist Anders Gustav Ekeburg in 1802. Tantalum, a silvery metal, is supplied as sheets, plates, rod, wire, and foil. It is used in aircraft engine parts, electrical devices, and for hearing aid and pacemaker capacitors. The metal is also used in capacitors in automobile computer circuitry to regulate engine controls, air bag deployment, antilock braking systems, and global positioning systems.

tap. A cylindrical or conical thread-cutting tool with one or more cutting elements having threads of a desired form on the periphery. By a combination of rotary and axial motions, the leading end cuts an internal thread, the tool deriving its principal support from the thread being produced.

tap density. The apparent density of a metal powder, obtained when the volume receptacle is tapped or vibrated during loading under specified conditions.

tapping. (1) Opening the outlet of a melting furnace to remove molten metal. (2) Removing molten metal from a furnace. (3) Cutting internal threads with a *tap*.

tarnish. The surface discoloration of a metal caused by formation of a thin film of corrosion product.

Taylor process. A process for making extremely fine metal wire by inserting a piece of larger-diameter wire into a glass tube and stretching the two together at high temperature.

technetium. A chemical element having atomic number 43, atomic weight 99, and the symbol Tc, from the Greek *technetos*, meaning artificial. The metal was synthesized in 1937 by Emelio Segre and Carlo Perrier.

technical cohesive strength. Fracture stress in a notch tensile test. Often used instead of merely "cohesive strength" to avoid confusion among the several definitions of cohesive strength.

tee joint. A weld joint in which the members are oriented in the form of a T (Fig. 56).

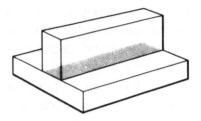

Fig. 56 Example of a tee weld joint

teeming. Pouring molten metal from a ladle into ingot molds. The term applies particularly to the specific operation of pouring either iron or steel into ingot molds.

tellurium. A chemical element having atomic number 52, atomic weight 128, and the symbol Te, for Tellus, the Roman goddess of the earth. Tellurium was discovered by Hungarian-Romanian mining inspector Franz Muller von Reichenstein in 1782, but was isolated and named by German chemist Martin Heinrich Klaproth in 1798 after he continued the work of Muller von Reichenstein.

Tellurium is one of the metalloids. In quantities of less than 1%, it improves the machinability of copper and can be used to improve the machinability of steel, having the same effect as selenium. Tellurium improves the mechanical properties and corrosion resistance of lead. It has been added to lead- and tin-base bearing alloys for the same reasons. Cast irons with less than 0.05% Te inhibit graphite formation.

temper. (1) In heat treatment, to reheat hardened steel or hardened cast iron to some temperature below the eutectoid temperature for the purpose of decreasing hardness and increasing toughness. The process also is sometimes applied to normalized steel. (2) In tool steels, temper is sometimes used, but unadvisedly, to denote carbon content. (3) In nonferrous alloys and in some ferrous alloys (steels that cannot be hardened by heat treatment), the hardness and strength produced by mechanical or thermal treatment, or both, and characterized by a certain structure, mechanical properties, or reduction in area during cold working. (4) To moisten with water the sand for casting molds.

temper brittleness. Brittleness that results when certain steels are held within, or are cooled slowly through, a certain range of temperatures below the transformation range. This brittleness is manifested as an upward shift in ductile-to-brittle transition temperature but only rarely produces a low value of reduction in area in a smooth-bar tension test of the embrittled material.

temper carbon. Fine, apparently amorphous carbon particles formed in white cast iron and certain steels during prolonged annealing. Also known as *annealing carbon*.

temper color. A thin, tightly adhering oxide skin (only a few molecules thick) that forms when steel is tempered at a low temperature, or for a short time, in air or a mildly oxidizing atmosphere. The color, which ranges from straw to blue depending on the thickness of the oxide skin, varies with both tempering time and temperature.

tempering. In heat treatment, reheating a quench-hardened or normalized steel object to a temperature below Ac_1 and then cooling it at any desired rate.

temper rolling. The light cold rolling of sheet steel. This operation is performed to improve flatness, to minimize the tendency toward formation of stretcher strains and flutes, and to obtain the desired texture and mechanical properties.

temper time. In resistance welding, that part of the postweld interval during which the current is suitable for tempering or heat treatment.

tensile strength. In tensile testing, the ratio of maximum load to original cross-sectional area. Also called *ultimate strength*. Compare with *yield strength*.

terbium. A chemical element having atomic number 65, atomic weight 159, and the symbol Tb, named for the Swedish town of Ytterby, where the element was found by Swedish chemist Carl Gustaf Mosander in 1843. Terbium is one of the 17 rare earth metals.

terminal phase. A solid solution having a restricted range of compositions, one end of the range being a pure component of an alloy system.

ternary alloy. An alloy that contains three principal elements.

terne. An alloy of lead containing 3–15% Sn, used as a hot dip coating for steel sheet or plate. Terne coatings, which are smooth and dull in appearance, give the steel better corrosion resistance and enhance its ability to be formed, soldered, or painted. The name is from the French *terne*, meaning tarnished.

tertiary creep. Time-dependent *strain* occurring under *stress* at an accelerating rate. See *creep*.

texture. In a polycrystalline aggregate, the state of the distribution of crystal orientations. In the usual sense, it is synonymous with *preferred orientation.*

thallium. A chemical element having atomic number 81, atomic weight 204, and the symbol Tl. It was named for the Greek goddess Thalles, meaning green, because of its green spectral line found by English chemist Sir William Crookes, who discovered the element spectroscopically in 1861. There are no alloys based on thallium; it is little used because of the toxic nature of the metal and its compounds.

thermal analysis. A method for determining transformations in a metal by noting the temperatures at which thermal arrests occur. These arrests are manifested by changes in slope of the plotted or mechanically traced heating and cooling curves. When such data are secured under nearly equilibrium conditions of heating and cooling, the method is commonly used for determining certain critical temperatures required for the construction of *equilibrium diagrams.*

thermal electromotive force. The *electromotive force* generated in a circuit containing two dissimilar metals when one junction is at a temperature different from that of the other. See also *thermocouple.*

thermal fatigue. Fracture resulting from the presence of temperature gradients that vary with time in such a manner as to produce cyclic stresses in a structure.

thermal shock. The development of a steep temperature gradient and accompanying high stresses within a structure.

thermal spraying. A group of coating or welding processes in which finely divided metallic or nonmetallic materials are deposited in a molten or semimolten condition to form a coating. The coating material may be in the form of powder, ceramic rod, wire, or molten materials. See also *flame spraying* and *plasma spraying.*

thermal stresses. Stresses in a metal resulting from nonuniform temperature distribution.

thermit reactions. Strongly exothermic self-propagating reactions such as that where finely divided aluminum reacts with a metal oxide. A mixture of aluminum and iron oxide produces sufficient heat to weld steel, the filler metal being produced in the reaction.

thermit welding. Welding with heat produced by the reaction of aluminum with a metal oxide. Filler metal, if used, is obtained from reduction of an appropriate oxide.

thermochemical treatment. A heat treatment for steels carried out in a medium suitably chosen to produce a change in the chemical composition of the steel object by exchange with the medium.

thermocouple. A device for measuring temperatures, consisting of lengths of two dissimilar metals or alloys that are electrically joined at one end and connected to a voltage-measuring instrument at the other end. When one junction is hotter than the other, a *thermal electromotive force* is produced that is roughly proportional to the difference in temperature between the hot and cold junctions.

thermomechanical working. A general term covering a variety of metal forming processes combining controlled thermal and deformation treatments to obtain synergistic effects, such as improvement in strength without loss of toughness. Same as thermal-mechanical treatment.

thief. An extra cathode or cathode extension that reduces the current density on what otherwise would be a high-current-density area on work being electroplated. In electroplating, the same as *robber*.

Thomas converter. A Bessemer converter having a basic bottom and lining, usually dolomite, and employing a basic slag.

thorium. A chemical element having atomic number 90, atomic weight 232, and the symbol Th. It was named after Thor, the Scandinavian god of war, by Swedish chemist Jöns Jakob Berzelius in 1815. The use of the metal is restricted because it is radioactive. Most of the metal is used in nuclear engineering applications. It is also used in x-ray equipment and as a radioactive source for nondestructive flaw detection testing.

three-point bending. The bending of a piece of metal, or a structural member, in which the object is placed across two supports and force is applied between and in opposition to them. See also *V-bend die*.

three-quarters hard. A *temper* of nonferrous alloys and some ferrous alloys characterized by values of tensile strength and hardness approximately midway between those of *half hard* and *full hard* tempers.

throat depth. On a resistance-welding machine, the distance from the centerline of the electrodes or platens to the nearest point of interference for flat work.

throat of a fillet weld. (1) (theoretical) The distance from the beginning of the root of the joint perpendicular to the hypotenuse of the largest right triangle that can be inscribed within the fillet weld cross section. (2) (actual) The shortest distance from the root of a fillet weld to its face. (3) (effective) The minimum distance from the root of the weld to its face, minus any reinforcement. See *convex fillet weld* (Fig. 16).

through weld. A nonpreferred term sometimes used to indicate a weld of substantial length made by melting through one member of a lap or tee joint and into the other member.

throwing power. The ability of a plating solution to produce a uniform metal distribution on an irregularly shaped *cathode*. Compare with *covering power*.

thulium. A chemical element having atomic number 69, atomic weight 169, and the symbol Tm, for the Latin *Thule*, an ancient name for Scandinavia. The metal was identified by Swedish chemist Per Theodore Cleve in 1879 and isolated by Charles James in 1911. It is the least abundant of the rare earth metals but is used as a radiation source in portable x-ray devices and in solid-state lasers.

tiger stripes. Continuous bright lines on sheet or strip in the rolling direction.

tight fit. A loosely defined fit of slight negative allowance the assembly of which requires a light press or driving force.

TIG welding. Tungsten inert-gas welding; see preferred term, *gas tungsten arc welding*.

tilt boundary. A subgrain boundary consisting of an array of edge *dislocations*.

tilt mold. A casting mold, usually a book mold, that rotates from a horizontal to a vertical position during pouring, which reduces agitation and thus the formation and entrapment of oxides.

tilt mold ingot. An ingot made in a *tilt mold*.

time quenching. Interrupted quenching in which the time in the quenching medium is controlled. See *quenching*.

tin. A chemical element having atomic number 50, atomic weight 119, and the symbol Sn, from the Latin *stannum*. The English word *tin* is from the Anglo-Saxon. Tin is one of the seven metals of antiquity, along with gold, silver, lead, iron, copper, and mercury. It may have been discovered around 3200 B.C. by the Egyptians. When tin was alloyed with copper, it created bronze—a hard metal used for knives, spear heads, utensils, and jewelry—giving birth to the Bronze Age. Pure tin was also used for utensils. See *Technical Note 15*.

TECHNICAL NOTE 15

Tin and Tin Alloys

TIN, which is a soft, brilliant white, low-melting-point material, was one of the first metals known to man. Throughout ancient history, various cultures recognized the virtues of tin in coatings, alloys, and compounds, and the use of the metal increased with advancing technology. Today, tin is an important metal in industry even though the annual tonnage used is relatively small compared with many other metals. One reason for this fact is that, in most applications, only very small amounts of tin are used at a time.

Tin is produced from both primary and secondary sources. Secondary tin is produced from recycled materials. Primary tin originates from the mineral cassiterite, a naturally occurring oxide of tin. These ores are smelted and refined to produce high-purity tin, which is cast into ingots weighing 12 to 25 kg (26 to 56 lb) or bars in weights of 1 kg (2.2 lb).

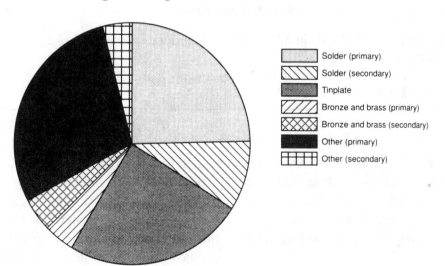

Solder (primary)
Solder (secondary)
Tinplate
Bronze and brass (primary)
Bronze and brass (secondary)
Other (primary)
Other (secondary)

Relative consumption of tin in the United States by application. 1988 data. Source: U.S. Bureau of Mines.

TECHNICAL NOTE 15 (*continued*)

Because of its low strength (yield strength of only 11 MPa, or 1.6 ksi, at room temperature), the pure metal is not regarded as a structural material and is rarely used in monolithic form. Rather, tin is most frequently used as a coating for other metals. The largest single application of tin worldwide is in the manufacture of tinplate (steel sheet coated with tin), which accounts for approximately 40% of total world tin consumption. Since 1940, the traditional hot dip method of making tinplate has been largely replaced by electrodeposition of tin on continuous strips of rolled steel. Electrolytic tinplate can be produced with either equal or unequal amounts of tin on the two surfaces of the steel base metal. Coating thicknesses range from approximately 0.1 μm up to 0.60 mm (0.004 mil to 0.02 in.). More than 90% of tinplate is used for containers (tin cans). Tin coatings are also deposited on nonferrous alloys, primarily copper, and copper-base alloys. These include tin-cadmium, tin-cobalt, tin-copper, tin-lead (see *terne*), tin-nickel, and tin-zinc alloys.

Tin also finds wide use in alloys, the most important of which are tin-base soft solders, bearing alloys (see the table accompanying the term *Babbitt metal*), and copper-base bronzes. Solders (primarily tin-lead solders) account for the largest use of tin in the United States. Tin-base solders are used to join food can seams, electronic and electrical components, and plumbing fixtures. Other applications for tin include jewelry and servingware (see *pewter* and *white metal*), organ pipes, *type metal*, and *fusible alloys*. Tin is also an alloying element in battery grid alloys, cast irons, dental amalgams, titanium alloys, and zirconium alloys.

Selected References

- W.B. Hampshire, Tin and Tin Alloys, *Metals Handbook*, 10th ed., Vol 2, ASM International, 1990, p 517–526
- D.J. Maykuth and W.B. Hampshire, Corrosion of Tin and Tin Alloys, *Metals Handbook*, 9th ed., Vol 13, 1987, p 770–783

---

tin cry. A creaking noise that may occur when tin or tin alloys are deformed. It is caused by the twinning of grains.

tinning. Coating metal with a very thin layer of molten solder or brazing filler metal.

tin pest. A polymorphic modification of tin that causes it to crumble into a powder known as gray tin. It is generally accepted that the maximum rate of transformation occurs at approximately −40 °C (−40 °F), but transformation can occur at as high as approximately 13 °C (55 °F).

tin sweat. Exudation of a low-melting phase during solidification in tin bronzes. See also *sweat*.

tin tossing. Oxidizing impurities in molten tin by pouring it from one vessel to

---

another in air, forming a dross that is mechanically separable.

TIR. Abbreviation for *total indicator reading*.

titanium. A chemical element with atomic number 22, atomic weight 48, and symbol Ti, for the mythological Greek Titans, first sons of the Earth. In 1791, William Gregor found the metal in ilmenite (iron titanate). Gregor named the metal *menachanite* because his discovery was in the Manaccan Valley in Cornwall. In 1795, Martin Heinrich Klaproth confirmed that it was an element and renamed it titanium. See *Technical Note 16*.

TECHNICAL NOTE 16

Titanium and Titanium Alloys

TITANIUM is the fourth most abundant structural metal in the crust of the earth after aluminum, iron, and magnesium. The development of its alloys and processing technologies started only in the late 1940s; thus, titanium metallurgy just missed being a factor in World War II. The difficulty in extracting titanium from ores, its high reactivity in the molten state, its forging complexity, its machining difficulty, and its sensitivity to segregation and inclusions necessitated the development of special processing techniques. These techniques have contributed to the high cost of titanium raw materials, alloys, and final products. On the other hand, the low density of titanium alloys (approximately 60% of the density of steel) provides high structural efficiencies based on a wide range of mechanical properties, coupled with an excellent resistance to aggressive environments. These alloys have contributed to the quality and durability of military aircraft, helicopters, and turbofan jet engines as well as to the increased reliability of heat exchangers and surgical implants.

Titanium metal passes through three major steps during processing from ore to finished product: (1) reduction of titanium ore (rutile, or TiO_2) to a porous form of titanium metal called "sponge," (2) melting of sponge to form an ingot, and (3) remelting and casting into finished shape, or primary fabrication, in which ingots are converted into general mill products followed by secondary fabrication of finished shapes from mill products. Powder metallurgy processing is also commonly used.

Titanium exists in two crystallographic forms. At room temperature, unalloyed (commercially pure) titanium has a hexagonal close-packed (hcp) crystal structure referred to as alpha (α) phase. At 883 °C (1621 °F), this transforms to a body-centered cubic (bcc) structure known as beta (β) phase. The manipulation of these crystallographic variations through alloying additions and thermomechanical processing is the basis for the development of a wide range of alloys and properties. These phases also provide a convenient way to categorize titanium alloys. Depending on their microstructure, titanium alloys fall into one of four classes: α, near- α, α-β, or β. These classes denote the general type of microstructure after processing.

TECHNICAL NOTE 16 (continued)

Summary of commercial and semicommercial grades and alloys of titanium

Designation	Tensile strength (min)		0.2% yield strength (min)		Impurity limits, wt%					Nominal composition, wt%				
	MPa	ksi	MPa	ksi	N (max)	C (max)	H (max)	Fe (max)	O (max)	Al	Sn	Zr	Mo	Others
Unalloyed grades														
ASTM Grade 1	240	35	170	25	0.03	0.10	0.015	0.20	0.18	...	...	...	...	...
ASTM Grade 2	340	50	280	40	0.03	0.10	0.015	0.30	0.25	...	...	...	...	...
ASTM Grade 3	450	65	380	55	0.05	0.10	0.015	0.30	0.35	...	...	...	...	...
ASTM Grade 4	550	80	480	70	0.05	0.10	0.015	0.50	0.40	...	...	...	...	...
ASTM Grade 7	340	50	280	40	0.03	0.10	0.015	0.30	0.25	...	...	...	...	0.2 Pd
Alpha and near-alpha alloys														
Ti-0.3Mo-0.8Ni	480	70	380	55	0.03	0.10	0.015	0.30	0.25	...	...	...	0.3	0.8 Ni
Ti-5Al-2.5Sn	790	115	760	110	0.05	0.08	0.02	0.50	0.20	5	2.5	...	...	...
Ti-5Al-2.5Sn-ELI	690	100	620	90	0.07	0.08	0.0125	0.25	0.12	5	2.5	...	...	...
Ti-8Al-1Mo-1V	900	130	830	120	0.05	0.08	0.015	0.30	0.12	8	...	...	1	1 V
Ti-6Al-2Sn-4Zr-2Mo	900	130	830	120	0.05	0.05	0.0125	0.25	0.15	6	2	4	2	...
Ti-6Al-2Nb-1Ta-0.8Mo	790	115	690	100	0.02	0.03	0.0125	0.12	0.10	6	1	...	1	2 Nb, 1 Ta
Ti-2.25Al-11Sn-5Zr-1Mo	1000	145	900	130	0.04	0.04	0.008	0.12	0.17	2.25	11.0	5.0	1.0	0.2 Si
Ti-5Al-5Sn-2Zr-2Mo(a)	900	130	830	120	0.03	0.05	0.0125	0.15	0.13	5	5	2	2	0.25 Si
Alpha-beta alloys														
Ti-6Al-4V(b)	900	130	830	120	0.05	0.10	0.0125	0.30	0.20	6.0	...	...	...	4.0 V
Ti-6Al-4V-ELI(b)	830	120	760	110	0.05	0.08	0.0125	0.25	0.13	6.0	...	...	...	4.0 V
Ti-6Al-6V-2Sn(b)	1030	150	970	140	0.04	0.05	0.015	1.0	0.20	6.0	2.0	...	...	0.75 Cu, 6.0 V
Ti-8Mn(b)	860	125	760	110	0.05	0.08	0.015	0.50	0.20	...	...	...	...	8.0 Mn
Ti-7Al-4Mo(b)	1030	150	970	140	0.05	0.10	0.013	0.30	0.20	7.0	...	...	4.0	...
Ti-6Al-2Sn-4Zr-6Mo(c)	1170	170	1100	160	0.04	0.04	0.0125	0.15	0.15	6.0	2.0	4.0	6.0	...
Ti-5Al-2Sn-2Zr-4Mo-4Cr(a)(c)	1125	163	1055	153	0.04	0.05	0.0125	0.30	0.13	5.0	2.0	2.0	4.0	4.0 Cr

(continued)

(a) Semicommercial alloy; mechanical properties and composition limits subject to negotiation with suppliers. (b) Mechanical properties given for annealed condition; may be solution treated and aged to increase strength. (c) Mechanical properties given for solution treated and aged condition; alloy not normally applied in annealed condition. Properties may be sensitive to section size and processing. (d) Primarily a tubing alloy; may be cold drawn to increase strength.

TECHNICAL NOTE 16 (continued)

Summary of commercial and semicommercial grades and alloys of titanium (continued)

Designation	Tensile strength (min)		0.2% yield strength (min)		Impurity limits, wt%					Nominal composition, wt%				
	MPa	ksi	MPa	ksi	N (max)	C (max)	H (max)	Fe (max)	O (max)	Al	Sn	Zr	Mo	Others
Alpha-beta alloys (continued)														
Ti-6Al-2Sn-2Zr-2Mo-2Cr(a)(b)	1030	150	970	140	0.03	0.05	0.0125	0.25	0.14	5.7	2.0	2.0	2.0	2.0 Cr, 0.25 Si
Ti-10V-2Fe-3Al(a)(c)	1170	170	1100	160	0.05	0.05	0.015	2.5	0.16	3.0	...	...	...	10.0 V
Ti-3Al-2.5V(d)	620	90	520	75	0.015	0.05	0.015	0.30	0.12	3.0	...	...	...	2.5 V
Beta alloys														
Ti-13V-11Cr-3Al(c)	1170	170	1100	160	0.05	0.05	0.025	0.35	0.17	3.0	...	...	...	11.0 Cr, 13.0 V
Ti-8Mo-8V-2Fe-3Al(a)(c)	1170	170	1100	160	0.05	0.05	0.015	2.5	0.17	3.0	...	...	8.0	8.0 V
Ti-3Al-8V-6Cr-4Mo-4Zr(a)(b)	900	130	830	120	0.03	0.05	0.020	0.25	0.12	3.0	...	4.0	4.0	6.0 Cr, 8.0 V
Ti-11.5Mo-6Zr-4.5Sn(b)	690	100	620	90	0.05	0.10	0.020	0.35	0.18	...	4.5	6.0	11.5	...

(a) Semicommercial alloy; mechanical properties and composition limits subject to negotiation with suppliers. (b) Mechanical properties given for annealed condition; may be solution treated and aged to increase strength. (c) Mechanical properties given for solution treated and aged condition; alloy not normally applied in annealed condition. Properties may be sensitive to section size and processing. (d) Primarily a tubing alloy; may be cold drawn to increase strength.

TECHNICAL NOTE 16 (*continued*)

Aerospace applications—including use in both structural (airframe) components and jet engines—account for the largest share of titanium alloy use, because titanium saves weight in highly loaded components that operate at low to moderately elevated temperatures. Many titanium alloys have been custom designed to have optimum tensile, compressive, and/or creep strength at selected temperatures and, at the same time, to have sufficient workability to be fabricated into products suitable for specific applications. During the life of the titanium industry, alloy Ti-6Al-4V has been consistently responsible for about 45% of total industry applications. Titanium is also often used in applications that exploit its corrosion resistance. These applications include chemical processing, the pulp and paper industry, marine applications, energy production and storage, and biomedical applications that take advantage of titanium's inertness in the human body for use in surgical implants and prosthetic devices. Other materials of interest are titanium-niobium alloys used for superconductors (see *superconductors*, Technical Note 14) and titanium-nickel alloys that exhibit a shape memory effect. Titanium-matrix composites and titanium-base intermetallics are also under development.

Selected References

- S. Lampman, Wrought Titanium and Titanium Alloys, *Metals Handbook*, 10th ed., Vol 2, ASM International, 1990, p 592–633
- D. Eylon, J.B. Newman, and J.K. Thorne, Titanium and Titanium Alloy Castings, *Metals Handbook*, 10th ed., Vol 2, ASM International, 1990, p 634–646
- D. Eylon and F.H. Froes, Titanium P/M Products, *Metals Handbook*, 10th ed., Vol 2, ASM International, 1990, p 647–660
- R.W. Schutz and D.E. Thomas, Corrosion of Titanium and Titanium Alloys, *Metals Handbook*, 9th ed., Vol 13, ASM International, 1987, p 669–706

TIV. Abbreviation for *total indicator variation*.

toe crack. A base-metal crack at the *toe of weld*.

toe of weld. The junction between the face of a weld and the base metal. See *fillet weld* (Fig. 27).

toggle press. A mechanical press in which the slide is actuated by one or more toggle links or mechanisms.

tolerance. The specified permissible deviation from a specified nominal dimension, or the permissible variation in size or other quality characteristic of a part.

tolerance limits. The boundaries that define the range of permissible variation in size or other *quality characteristic* of a part.

tonghold. The portion of a forging billet, usually on one end, that is gripped by the operator's tongs. It is removed from the part at the end of the forging operation. Common to drop hammer and press-type forging.

tool steel. Any of a class of carbon and alloy steels commonly used to make tools. Tool steels are characterized by high hardness and resistance to abrasion, often accompanied by high toughness and resistance to softening at elevated temperature. These attributes are generally attained with high carbon and alloy contents. See *Technical Note 17*.

TECHNICAL NOTE 17

Tool Steels

A TOOL STEEL is any steel used to make tools for cutting, forming, or otherwise shaping a material into a final part or component. These complex alloy steels, which contain relatively large amounts of tungsten, molybdenum, vanadium, manganese, and chromium, make it possible to meet increasingly severe service demands. In service, most tools are subjected to extremely high loads that are applied rapidly. The tools must withstand these loads a great number of times without breaking and without undergoing excessive wear or deformation. In many applications, tool steels must provide this capability under conditions that develop high temperatures in the tool. Most tool steels are wrought products, but precision castings can be used in some applications. The powder metallurgy process is also used in making tool steels. It provides, first, a more uniform carbide size and distribution in large sections and, second, special compositions that are difficult or impossible to produce in wrought or cast alloys.

Tool steels are classified according to their composition, application, or method of quenching. Each group is identified by a capital letter; individual tool steel types are assigned code numbers (see the accompanying table). High-speed steels are tool materials developed largely for use in high-speed metal cutting applications. There are two classifications of high-speed steels: molybdenum high-speed steels, or group M, which contain from 0.75–1.52% C and 4.50–11.00% Mo, and tungsten high-speed steels, or group T, which have similar carbon contents but high (11.75–21.00%) tungsten contents. Group M steels constitute more than 95% of all high-speed steel produced in the United States.

Hot work steels (group H) have been developed to withstand the combinations of heat, pressure, and abrasion associated with punching, shearing, or forming of metals at high temperatures. Group H steels usually have medium carbon contents (0.35–0.45%) and chromium, tungsten, molybdenum, and vanadium contents of 6–25%. H steels are divided into chromium hot work steels, tungsten hot work steels, and molybdenum hot work steels.

Cold work tool steels are restricted in application to those uses that do not involve prolonged or repeated heating above 205 to 260 °C (400 to 500 °F).

Composition limits of selected types of wrought tool steels

Designation		Composition(a), %								
AISI	UNS	C	Mn	Si	Cr	Ni	Mo	W	V	Co
Molybdenum high-speed steels										
M1	T11301	0.78–0.88	0.15–0.40	0.20–0.50	3.50–4.00	0.30 max	8.20–9.20	1.40–2.10	1.00–1.35	...
M2	T11302	0.78–0.88; 0.095–1.05	0.15–0.40	0.20–0.45	3.75–4.50	0.30 max	4.50–5.50	5.50–6.75	1.75–2.20	...
M4	T11304	1.25–1.40	0.15–0.40	0.20–0.45	3.75–4.75	0.30 max	4.25–5.50	5.25–6.50	3.75–4.50	...
M35	T11335	0.82–0.88	0.15–0.40	0.20–0.45	3.75–4.50	0.30 max	4.50–5.50	5.50–6.75	1.75–2.20	4.50–5.50
M42	T11342	1.05–1.15	0.15–0.40	0.15–0.65	3.50–4.25	0.30 max	9.00–10.00	1.15–1.85	0.95–1.35	7.75–8.75
M62	T11362	1.25–1.35	0.15–0.40	0.15–0.40	3.50–4.00	0.30 max	10.00–11.00	5.75–6.50	1.80–2.10	...
Tungsten high-speed steels										
T1	T12001	0.65–0.80	0.10–0.40	0.20–0.40	3.75–4.50	0.30 max	...	17.25–18.75	0.90–1.30	...
T15	T12015	1.50–1.60	0.15–0.40	0.15–0.40	3.75–5.00	0.30 max	1.00 max	11.75–13.00	4.50–5.25	4.75–5.25
Chromium hot work steels										
H11	T20811	0.33–0.43	0.20–0.50	0.80–1.20	4.75–5.50	0.30 max	1.10–1.60	...	0.30–0.60	...
H19	T20819	0.32–0.45	0.20–0.50	0.20–0.50	4.00–4.75	0.30 max	0.30–0.55	3.75–4.50	1.75–2.20	4.00–4.50
Tungsten hot work steels										
H21	T20821	0.26–0.36	0.15–0.40	0.15–0.50	3.00–3.75	0.30 max	...	8.50–10.00	0.30–0.60	...
H23	T20823	0.25–0.35	0.15–0.40	0.15–0.60	11.00–12.75	0.30 max	...	11.00–12.75	0.75–1.25	...
H26	T20826	0.45–0.55(b)	0.15–0.40	0.15–0.40	3.75–4.50	0.30 max	...	17.25–19.00	0.75–1.25	...
Molybdenum hot work steels										
H42	T20842	0.55–0.70(b)	0.15–0.40	...	3.75–4.50	0.30 max	4.50–5.50	5.50–6.75	1.75–2.20	...
Air-hardening, medium-alloy, cold work steels										
A2	T30102	0.95–1.05	1.00 max	0.50 max	4.75–5.50	0.30 max	0.90–1.40	...	0.15–0.50	...
A6	T30106	0.65–0.75	1.80–2.50	0.50 max	0.90–1.20	0.30 max	0.90–1.40	...	...	...
A10	T30110	1.25–1.50(c)	1.60–2.10	1.00–1.50	...	1.55–2.05	1.25–1.75	...	...	...

(continued)

(a) All steels except group W contain 0.25 max Cu, 0.03 max P, and 0.03 max S; group W steels contain 0.20 max Cu, 0.025 max P, and 0.025 max S. Where specified, sulfur may be increased to 0.06–0.15% to improve machinability of group A, D, H, M, and T steels. (b) Available in several carbon ranges. (c) Contains free graphite in the microstructure. (d) Optional. (e) Specified carbon ranges are designated by suffix numbers.

TECHNICAL NOTE 17 (*continued*)

Composition limits of selected types of wrought tool steels (*continued*)

Designation		Composition(a), %								
AISI	UNS	C	Mn	Si	Cr	Ni	Mo	W	V	Co
High-carbon, high-chromium, cold work steels										
D2	T30402	1.40–1.60	0.60 max	0.60 max	11.00–13.00	0.30 max	0.70–1.20	...	1.10 max	...
D3	T30403	2.00–2.35	0.60 max	0.60 max	11.00–13.50	0.30 max	...	1.00 max	1.00 max	...
Oil-hardening cold work steels										
O1	T31501	0.85–1.00	1.00–1.40	0.50 max	0.40–0.60	0.30 max	...	0.40–0.60	0.30 max	...
O2	T31502	0.85–0.95	1.40–1.80	0.50 max	0.50 max	0.30 max	0.30 max	...	0.30 max	...
O6	T31506	1.25–1.55(c)	0.30–1.10	0.55–1.50	0.30 max	0.30 max	0.20–0.30	...	...	...
Shock-resisting steels										
S1	T41901	0.40–0.55	0.10–0.40	0.15–1.20	1.00–1.80	0.30 max	0.50 max	1.50–3.00	0.15–0.30	...
S2	T41902	0.40–0.55	0.30–0.50	0.90–1.20	...	0.30 max	0.30–0.60	...	0.50 max	...
S7	T41907	0.45–0.55	0.20–0.90	0.20–1.00	3.00–3.50	...	1.30–1.80	...	0.20–0.30(d)	...
Low-alloy special-purpose tool steels										
L2	T61202	0.45–1.00(b)	0.10–0.90	0.50 max	0.70–1.20	...	0.25 max	...	0.10–0.30	...
L6	T61206	0.65–0.75	0.25–0.80	0.50 max	0.60–1.20	1.25–2.00	0.50 max	...	0.20–0.30(d)	...
Low-carbon mold Steels										
P2	T51602	0.10 max	0.10–0.40	0.10–0.40	0.75–1.25	0.10–0.50	0.15–0.40	...	...	...
P5	T51605	0.10 max	0.20–0.60	0.40 max	2.00–2.50	0.35 max	...	...	...	...
P20	T51620	.28–0.40	0.60–1.00	0.20–0.80	1.40–2.00	...	0.30–0.55	...	...	...
Water-hardening tool steels										
W1	T72301	0.70–1.50(e)	0.10–0.40	0.10–0.40	0.15 max	0.20 max	0.10 max	0.15 max	0.10 max	...
W2	T72302	0.85–1.50(e)	0.10–0.40	0.10–0.40	0.15 max	0.20 max	0.10 max	0.15 max	0.15–0.35	...

(a) All steels except group W contain 0.25 max Cu, 0.03 max P, and 0.03 max S; group W steels contain 0.20 max Cu, 0.025 max P, and 0.025 max S. Where specified, sulfur may be increased to 0.06–0.15% to improve machinability of group A, D, H, M, and T steels. (b) Available in several carbon ranges. (c) Contains free graphite in the microstructure. (d) Optional. (e) Specified carbon ranges are designated by suffix numbers.

TECHNICAL NOTE 17 (*continued*)

There are three categories of cold work steel: air-hardening steels, or group A; high-carbon, high-chromium steels, or group D; and oil-hardening steels, or group O.

Shock-resisting, or group S, steels contain manganese, silicon, chromium, tungsten, and molybdenum, in various combinations; carbon content is approximately 1.50%. Group S steels are used primarily for chisels, rivet sets, punches, and other applications requiring high toughness and resistance to shock loading.

The low-alloy special-purpose, or group L, tool steels contain small amounts of chromium, vanadium, nickel, and molybdenum. Group L steels generally are used for machine parts and other special applications requiring good strength and toughness.

Mold steels, or group P, contain chromium and nickel as principal alloying elements. Because of their low resistance to softening at elevated temperatures, group P steels are used almost exclusively in low-temperature die casting dies and in molds for injection or compression molding of plastics.

Water-hardening, or group W, tool steels contain carbon as the principal alloying element (0.70–1.50% C). Group W steels, which also have low resistance to softening at elevated temperatures, are suitable for cold heading, coining, and embossing tools, woodworking tools, metal cutting tools, and wear-resistance machine tool components.

Selected References

- A.M. Bayer, Wrought tool Steels, *Metals Handbook*, 10th ed., Vol 1, ASM International, 1990, p 757–779
- K.E. Pinnow and W. Stasko, P/M Tool Steels, *Metals Handbook*, 10th ed., Vol 1, 1990, p 780–792
- G.A. Roberts and R.A. Gary, *Tool Steels*, 4th ed., American Society for Metals, 1980

tooth. (1) A projection on a multipoint tool (such as on a saw, milling cutter, or file) designed to produce cutting. (2) A projection on the periphery of a wheel or segment thereof (as on a gear, spline, or sprocket, for example) designed to engage another mechanism and thereby transmit force or motion, or both. A similar projection on a flat member such as a rack.

tooth point. On a face mill, the chamfered cutting edge of the blade, to which a flat is sometimes added to produce a shaving effect and to improve finish. See *face mill* (Fig. 26).

top-and-bottom process. A process for separating copper and nickel, in which their molten sulfides are separated into two liquid layers by the addition of sodium sulfide. The lower layer holds most of the nickel.

torch. A gas burner used to solder, braze, weld, or cut metals. For brazing or welding, it has two gas feed lines: one for fuel, such as acetylene or hydrogen, the other for oxygen. For cutting, there may be an additional feed line for oxygen. See *oxygen cutting*.

torch brazing. Brazing in which the heat is supplied by a fuel gas flame emanating from a *torch*.

torsion. A twisting action resulting in shear stresses and strains.

torsional moment. In a body being twisted, the algebraic sum of the couples or the moments of the external forces about the axis of twist, or both.

total carbon. The sum of the free and combined carbon (including carbon in solution) in a ferrous alloy.

total cyanide. Cyanide content of an electroplating bath (including both simple and complex ions).

total indicator reading (TIR). The difference between the maximum and minimum indicator readings during a checking cycle. See preferred term *total indicator variation*.

total indicator variation (TIV). The difference between the maximum and minimum indicator readings during a checking cycle.

toughness. The ability of a metal to absorb energy and deform plastically before fracturing. It usually is measured by the energy absorbed in a notch impact test, but the area under the stress-strain curve in tensile testing is also a measure of toughness.

tough pitch copper. Copper containing from 0.02–0.04% O, obtained by refining copper in a reverberatory furnace.

tracer milling. The duplication of a three-dimensional form by means of a cutter controlled by a tracer that is directed by a master form.

traffic mark. (1) The process in which hard particles or protuberances are forced against and moved along a solid surface. (2) The roughening or scratching of a surface due to abrasive wear. (3) The process of grinding or wearing away through the use of abrasives.

tramp alloys. Residual alloying elements that are introduced into steel when unidentified alloy steel is present in the scrap charge to a steelmaking furnace.

transcrystalline. Within or across the crystals or grains of a metal. Also known as *transgranularand instracrystalline*.

transference. The movement of ions through the electrolyte associated with the passage of the electric current. Also called *transport* or *migration*.

transference number. The proportion of total electroplating current carried by ions of a given kind. The same as *transport number*.

transformation-induced plasticity. A phenomenon, occurring chiefly in certain highly alloyed steels that have been heat treated to produce metastable *austenite* or metastable austenite plus *martensite*, whereby, on subsequent deformation, part of the austenite undergoes strain-induced transformation to martensite. Steels capable of transforming in this manner, commonly referred to as TRIP steels, are highly plastic after heat treatment, but exhibit a very high rate of strain hardening and thus have high tensile and yield strengths after plastic deformation at temperatures between approximately 20 and 500 °C (70 and 930 °F). Cooling to −195 °C (−320 °F) may or may not be required to complete the transformation to martensite. Tempering is usually done following transformation.

transformation ranges. Those ranges of temperature within which a phase forms during heating and transforms during cooling. The two ranges are distinct, sometimes overlapping but never coinciding. The limiting temperatures of the ranges depend on the composition of the alloy and on the rate of change of temperature, particularly during cooling. See also *transformation temperature.*

transformation temperature. The temperature at which a change in phase occurs. This term is sometimes used to denote the limiting temperature of a transformation range. The following symbols are used for irons and steels:

Ac_{cm}. In hypereutectoid steel, the temperature at which solution of cementite in austenite is completed during heating.

Ac_1. The temperature at which austenite begins to form during heating.

Ac_3. The temperature at which transformation of ferrite to austenite is completed during heating.

Ac_4. The temperature at which austenite transforms to delta ferrite during heating.

Ae_{cm}, Ae_1, Ae_3, Ae_4. The temperatures of phase changes at equilibrium.

Ar_{cm}. In hypereutectoid steel, the temperature at which precipitation of cementite starts during cooling.

Ar_1. The temperature at which transformation of austenite to ferrite or to ferrite plus cementite is completed during cooling.

Ar_3. The temperature at which austenite begins to transform to ferrite during cooling.

Ar_4. The temperature at which delta ferrite transforms to austenite during cooling.

Ar'. The temperature at which transformation of austenite to pearlite starts during cooling.

M_f. The temperature at which transformation of austenite to martensite is completed during cooling.

M_s (or Ar''). The temperature at which transformation of austenite to martensite starts during cooling.

NOTE: All of these changes, except formation of martensite, occur at lower temperatures during cooling than during heating, and depend on the rate of change of temperature.

transgranular. Within or across the crystals or grains of a metal. Also known as *transcrystalline* and *intracrystalline.*

transitional fit. A fit that may have either clearance or interference resulting from specified tolerances on hole and shaft.

transition lattice. An unstable crystallographic configuration that forms as an intermediate step in a solid-state reaction such as precipitation from solid solution or eutectoid decomposition.

transition metal. A metal in which the available electron energy levels are occupied in such a way that the *d*-band contains less than its maximum number of ten electrons per atom, for example, iron, cobalt, nickel, and tungsten. The distinctive properties of the transition metals result from the incompletely filled *d*-levels.

transition point. At a stated pressure, the temperature (or at a stated temperature, the pressure) at which two solid phases exist in equilibrium—that is, an allotropic transformation temperature (or pressure).

transition temperature. (1) An arbitrarily defined temperature that lies within the temperature range in which metal fracture characteristics (as usually determined by tests of notched specimens) change rapidly, such as from primarily fibrous (shear) to primarily crystalline (cleavage) fracture. Commonly used definitions are: transition temperature for 50% cleavage fracture; 10-ft-lb transition temperature; and transition temperature for half maximum energy. (2) Sometimes used to denote an arbitrarily defined temperature within a range in which the ductility changes rapidly with temperature.

transport. The movement of ions through the electrolyte associated with the passage of the electric current. Also known as *transference or migration*.

transport number. The proportion of total electroplating current carried by ions of a given kind. Also known as *transference number*.

transverse. Literally, "across," usually signifying a direction or plane perpendicular to the direction of working. In rolled plate or sheet, the direction across the width often is called long transverse, and the direction through the thickness, short transverse.

transverse rolling machine. Equipment for producing complex preforms or finished forgings from round billets inserted transversely between two or three rolls that rotate in the same direction and drive the billet. The rolls, carrying replaceable die segments with appropriate impressions, make several revolutions for each rotation of the workpiece.

trees. Visible projections of electrodeposited metal formed at sites of high current density.

trepanning. A type of boring where an annular cut is made into a solid material with the coincidental formation of a plug or solid cylinder.

triaxiality. In a *triaxial stress* state, the ratio of the smallest to the largest principal stress, all stresses being tensile.

triaxial stress. A state of stress in which none of the three *principal stresses* is zero.

tribology. The science and technology concerned with the design, friction, lubrication, and wear of contacting surfaces that move relative to each other (as in bearings, cams, or gears, for example).

trimmer blades. The portion of the *trimmers* through which a forging is pushed to shear off the flash.

trimmer punch. The upper portion of the *trimmers*, which comes in contact with a forging and pushes it through the *trimmer blades*. The lower end of the trimmer punch is generally shaped to fit the surface of the forging against which it pushes.

trimmers. The combination of the *trimmer punch*, *trimmer blades*, and perhaps the *trimming shoe* used to remove the flash from the forging.

trimming. (1) In drawing, shearing the irregular edge of the drawn part. (2) In forging or die casting, removing any parting-line flash and gates from the part by shearing. (3) In casting, the removal of gates, risers, and fins.

trimming shoe. The holder used to support *trimmers*. Sometimes called the trimming chair.

triple-action press. A mechanical or hydraulic press having three slides with three motions properly synchronized for triple-action drawing, redrawing, and forming. Usually, two slides—the blankholder slide and the plunger—are located above, and a lower slide is located within the bed of the press. See also *hydraulic press* and *mechanical press*.

triple point. A point on a phase diagram where three phases of a substance coexist in equilibrium.

tripoli. Friable and dustlike silica used as an abrasive.

TRIP steel. A commercial steel product exhibiting *transformation-induced plasticity*.

trommel. A revolving cylindrical screen used in grading coarsely crushed ore.

troostite. (obsolete) A previously unresolvable, rapidly etching, fine aggregate of carbide and ferrite produced either by tempering martensite at low temperature or by quenching a steel at a rate slower than the critical cooling rate. Preferred terminology for the first product is tempered *martensite*; for the latter, fine *pearlite*.

true current density. Current density at a point or on a small area. See preferred term *local current density*.

true rake. The angle between a plane containing a tooth face and the axial plane through the tooth point as measured in the direction of chip flow through the tooth point. Thus, it is the rake resulting from both cutter configuration and direction of chip flow. See preferred term *effective rake*.

true strain. A measure of the relative change in the size or shape of a body. True strain (or natural strain) is the natural logarithm of the ratio of the length at the moment of observation to the original gage length. See *strain*.

true stress. Force per unit area, often thought of as force acting through a small area within a plane. True stress denotes the stress where force and area are measured at the same time. See *stress*.

truing. The removal of the outside layer of abrasive grains on a grinding wheel to restore its face to running true or to alter the cutting face for grinding of special contours.

tuballoy. A code name for "enriched uranium" used at the Oak Ridge National Laboratory.

tube reducing. Reducing both the diameter and wall thickness of tubing with a mandrel and a pair of rolls with tapered grooves. The Rockrite process uses a fixed tapered mandrel, and the rolls reciprocate along the tubing with corresponding reversal in rotation. Roll reliefs at the initial and final diameters permit, respectively, the advance and rotation of the tubing. The Pilger process uses a uniform rod (broach), which reciprocates with the tubing. The fixed rolls rotate continuously. During the gap in each revolution, the tubing is advanced and rotated and then, on roll contact, reduced and partially returned.

tube sinking. Drawing tubing through a die or passing it through rolls without the use of an interior tool (such as a mandrel or plug) to control inside diameter; sinking generally produces a tube of increased wall thickness and length.

tube stock. A semifinished tube suitable for subsequent reduction and finishing.

tubular products, steel. The general term used for all hollow carbon and low-alloy steel products used as conveyors of liquids and as structural members. Although usually produced in cylindrical form, these products are often processed to make square, oval, rectangular, and other symmetrical shapes. Table 13 provides a classification of steel tubular products.

tumbling. Rotating workpieces, usually castings or forgings, in a barrel partly filled with metal slugs or abrasives, to remove sand, scale, or fins. It may be done

Table 13   Classification of steel tubular products

Product	Typical use	Production processes	Outside diameter(a), in.	Typical grades	Usual finished status
Oil country goods					
Casing	To line oil and gas wells to prevent collapse of the hole	Seamless, electric resistance welding	4.5–20	H-40, J-55, K-55 C-75, L-80, N-80, C-90, G-95, P-110, Q-125	As-rolled Normalized or quench and temper
Tubing	To convey oil or gas from the producing strata to the earth's surface	Seamless, continuous welding, electric resistance welding	1.050–4.5	All others H-40, J-55, N-80, P-105	Quench and temper As-rolled Normalize or quench and temper
Drill pipe	Rotary stem for drill bits	Seamless	2.375–6.625	E	Normalize and temper, or quench and temper
				X-95, G-105, S-135	Quench and temper
Line pipe	To convey oil, gas, or water	Seamless, electric resistance welding, continuous welding, double submerged are welding	0.125(nom)–80	All grades B, X42, X46, X52, X60, X65, X70	As-rolled As-rolled Control rolled
Standard pipe	Plumbing, electrical conduit, low pressure conveyance of fluids, and nonstringent structural applications	Seamless, electric resistance welding, continuous welding, double submerged arc welding	0.125(nom)–80	All grades	As-rolled
Mechanical tubing	Variety of round, hollow mechanical parts, such as automotive axles, bearing races, and hydraulic pistons	Seamless, electric resistance welding	0.375–10.75	Carbon and alloy	Hot rolled or cold drawn
Pressure tubing	Boiler tubes, condenser tubes, heat exchanger tubes, and refrigeration tubes	Seamless, electric resistance welding	0.5–10.75	Carbon and alloy	Hot rolled or cold drawn

Note: Because steel tubular products manufactured in the United States are customarily produced to standard inch and fractional inch sizes, tubular product sizes are given only in inches in this article. 1 in. = 25.4 mm or 2.54 cm. (a) nom: nominal dry, or with an aqueous solution added to the contents of the barrel. Sometimes called rumbling or rattling.

tungsten. A chemical element having atomic number 74, atomic weight 184, and the symbol W, for its European name of German origin, *wolfram*. Tungsten is a metal with unique properties that led to its use in cutting and forming other metals and in important high-temperature applications. It has the highest melting point of any metal (3410 °C, or 6170 °F) and the lowest vapor pressure.

Tungsten has one of the highest tensile strengths, reaching 600,000 psi in wire form. Its corrosion resistance is one of the highest, and its density is exceeded only by members of the platinum group and rhenium. Tungsten is elastic and ductile. Its compound with carbon, tungsten carbide, is the hardest known metallic substance.

The iron-manganese-tungstate mineral wolfram was first described in scientific literature in 1574 and was originally believed to be a mineral of tin, with which it is commonly associated. The name *wolfram* probably makes reference to the wolflike characteristic of tungsten described by early miners as devouring tin and causing low recoveries in the tin smelting operation. The word *tungsten*, an adaptation of the Swedish *tung sten* (heavy stone), was applied to the calcium tungstate mineral subsequently known as scheelite. Scheelite was named after Swedish chemist Karl Wilhelm Scheele, who in 1781 discovered that a new acid, tungstic acid, could be made from scheelite (at the time named tungsten). Wolfram was first identified as a mineral

of tungsten in 1783 by Spanish brothers and chemists Juan José and Fausto de Elhuyar, who are also credited with the first production of metallic tungsten from wolframite in that same year.

The element did not find industrial applications until early in the 20th century, but it has since become one of the most important of industrial metals. The first major application of tungsten was in incandescent lamp filaments, an application that remained important for many years, until fluorescent light became common. By 1941, 95% of the total tungsten production went into steel production. Products include ferroalloys, carbides, hardfacing alloys, Co-Cr-W-Mo high-temperature alloys, electrical equipment, and chemicals.

tungsten inert gas (TIG) welding. A fusion welding process in which metals are joined by heating them with an electric arc between a nonconsumable tungsten electrode and the work. Shielding is obtained from a gas or gas mixture. Pressure may or may not be applied to the joint, and filler metal may or may not be added. See preferred term *gas tungsten arc welding*.

Turk's-head rolls. Four undriven working rolls, arranged in a square or rectangular pattern, through which metal strip, wire, or tubing is drawn to form square or rectangular sections.

turning. Removing material by forcing a cutting tool (often a *single-point tool*) against the surface of a rotating workpiece. The tool may or may not be moved toward or along the axis of rotation while it cuts away material.

tuyere. An opening in the shell and refractory lining of a furnace, through which air is forced (Fig. 57).

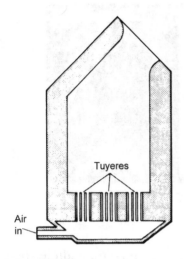

Fig. 57 Schematic showing tuyeres

twin. Two portions of a crystal having a definite crystallographic relationship; one may be regarded as the parent, the other as the twin. The orientation of the twin is either a mirror image of the orientation of the parent across a twinning plane or an orientation that can be derived by rotating the twin portion about a twinning axis. See also *annealing twin* and *mechanical twin*.

twist boundary. A subgrain boundary consisting of an array of screw *dislocations*.

two-high mill. A type of rolling mill in which only two rolls, the working rolls, are contained in a single housing. Compare with *four-high mill* and *cluster mill*.

type metal. Any of a series of alloys containing 54–95% Pb, 2–28% Sb, and 2–20% Sn, used to make printing type.

Dictionary of Metals
H.M. Cobb, editor

U-bend die. A die, commonly used in press-brake forming, machined horizontally with a square or rectangular cross-sectional opening that provides two edges over which metal is drawn into a channel shape.

Ugine-Sejournet process. A direct extrusion process for metals that uses molten glass to insulate the hot billet and to act as a lubricant. Also known as *Sejournet process.*

ultimate strength. The maximum conventional stress (tensile, compressive, or shear) that a material can withstand without fracture.

ultrahigh-strength steels. Structural steel with a minimum yield strength of 1380 MPa (200 ksi). Such steels include medium-carbon low-alloy steels, medium-alloy air-hardening steels, and high fracture toughness steels. Table 14 lists compositions of commercial ultrahigh-strength steels.

ultrasonic beam. A beam of acoustical radiation with a frequency higher than the frequency range for audible sound—that is, above approximately 20 kHz.

ultrasonic cleaning. Immersion cleaning aided by ultrasonic waves that cause microagitation.

ultrasonic frequency. A frequency, associated with elastic waves, that is greater than the highest audible frequency, generally regarded as being higher than 20 kHz.

ultrasonic machining. A form of abrasive machining in which a tool vibrating at ultrasonic frequency causes a grit-loaded slurry to impinge on the surface of a workpiece and thereby remove material.

ultrasonic testing. A nondestructive test applied to sound-conductive materials having elastic properties for the purpose of locating inhomogeneities or structural discontinuities within a material by means of an *ultrasonic beam.*

ultrasonic welding. A solid-state process in which materials are welded by locally applying high-frequency vibratory energy to a joint held together under pressure.

Table 14 Compositions of ultrahigh-strength steels

Designation or trade name	Composition(a), wt%							
	C	Mn	Si	Cr	Ni	Mo	V	Co
Medium-carbon low-alloy steels								
4130	0.28–0.33	0.40–0.60	0.20–0.35	0.80–1.10	...	0.15–0.25	...	...
4140	0.38–0.43	0.75–1.00	0.20–0.35	0.80–1.10	...	0.15–0.25	...	...
4340	0.38–0.43	0.60–0.80	0.20–0.35	0.70–0.90	1.65–2.00	0.20–0.30	...	...
AMS 6434	0.31–0.38	0.60–0.80	0.20–0.35	0.65–0.90	1.65–2.00	0.30–0.40	0.17–0.23	...
300M	0.40–0.46	0.65–0.90	1.45–1.80	0.70–0.95	1.65–2.00	0.30–0.45	0.05 min	...
D-6a	0.42–0.48	0.60–0.90	0.15–0.30	0.90–1.20	0.40–0.70	0.90–1.10	0.05–0.10	...
6150	0.48–0.53	0.70–0.90	0.20–0.35	0.80–1.10	...	...	0.15–0.25	...
8640	0.38–0.43	0.75–1.00	0.20–0.35	0.40–0.60	0.40–0.70	0.15–0.25	...	...
Medium-alloy air-hardening steels								
H11 mod	0.37–0.43	0.20–0.40	0.80–1.00	4.75–5.25	...	1.20–1.40	0.40–0.60	...
H13	0.32–0.45	0.20–0.50	0.80–1.20	4.75–5.50	...	1.10–1.75	0.80–1.20	
High fracture toughness steels								
AF1410(b)	0.13–0.17	0.10 max	0.10 max	1.80–2.20	9.50–10.50	0.90–1.10	....	13.50–14.50
HP 9-4-30(c)	0.29–0.34	0.10–0.35	0.20 max	0.90–1.10	7.0–8.0	0.90–1.10	0.06–0.12	4.25–4.75

(a) P and S contents may vary with steelmaking practice. Usually, these steels contain no more than 0.035 P and 0.040 S. (b) AF1410 is specified to have 0.008 P and 0.005 S composition. Ranges utilized by some producers are narrower. (c) HP 9-4-30 is specified to have 0.10 max P and 0.10 max S. Ranges utilized by some producers are narrower.

underbead crack. A subsurface crack in the base metal near a weld.

undercooling. The cooling below the temperature at which an equilibrium phase transformation can take place, without actually obtaining the transformation. The same as *supercooling.*

undercut. (1) In weldments, a groove melted into the base metal adjacent to the toe of a weld and left unfilled by weld metal (Fig. 58). (2) For castings or forgings, the same as *back draft:* A reverse taper on a casting pattern or a forging die that prevents the pattern or forged stock from being removed from the cavity.

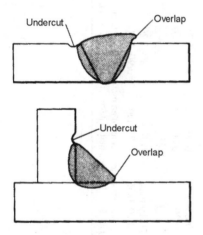

Fig. 58 Undercut and overlap in a fillet and groove weld

underdraft. A condition wherein a metal curves downward on leaving a set of rolls because of a higher speed in the upper roll.

underfill. A portion of a forging that has insufficient metal to give it the true shape of the impression.

understressing. Applying a cyclic stress lower than the *endurance limit.* This may improve fatigue life if the member is later cyclically stressed at levels above the endurance limit.

uniaxial stress. A state of stress in which two of the three principal stresses are zero.

uniform strain. The strain occurring prior to the beginning of localization of strain (necking); the strain to maximum load in the tension test.

unit cell. In crystallography, the fundamental building block of a space lattice. Space lattices are constructed by stacking identical unit cells—that is, parallelepipeds of identical size, shape, and orientation, each having a lattice point at every corner—face to face in perfect three-dimensional alignment.

unit die. A *die block* that contains several cavity inserts for making different kinds of castings.

unit power. The net amount of power required during machining to remove a unit volume of metal in unit time.

universal forging mill. A combination of four hydraulic presses arranged in one plane equipped with billet manipulators and automatic controls, used for radial or draw forging.

universal mill. A rolling mill in which rolls with a vertical axis roll the edges of the metal stock between some of the passes through the horizontal rolls.

UNS. A Unified Numbering System for Metals and Alloys established in the United States in 1974 to provide an organized system for numbering virtually all metals and alloys in commercial use in the United States. The system was developed jointly by the American Society for Testing and Materials (ASTM) and the Society of Automotive Engineers (SAE), with a committee comprised of representatives from most of the organizations involved with the numbering of metals, including the Aluminum Association, American Welding Society, Copper Development Association, Department of Defense, International Nickel Company, and Steel Founders Society of America.

The UNS System, which consists of a prefix letter and five digits, attempts to incorporate to the extent possible the existing numbers of the various organizations. AA 2024 becomes A92024 in the UNS System; CDA 272 becomes C27200; and AISI 304L becomes S30403. The metals are classified into 18 groups, as shown in Table 15. The system has provided numbers for alloys without previous numbers and provides for the numbering of new alloys. The UNS has become a popular system for listings in standards and for cataloging in metal directories in a manner comparable to the European EN numbering system.

upset. (1) The localized increase in cross-sectional area of a workpiece or weldment resulting from the application of pressure during mechanical fabrication or welding. (2) That portion of a welding cycle during which the cross-sectional area is increased by the application of pressure.

upset forging. A forging obtained by *upset* of a suitable length of bar, billet, or bloom.

upsetter. A horizontal mechanical press used to make parts from bar stock or tubing by *upset forging,* piercing, bending, or otherwise forming in dies. Also known as a *header.*

upsetting. Working metal so that the cross-sectional area of a portion or all of the stock is increased. See also *heading.*

Table 15 Unified Number System (UNS) metals and alloys
Applies to plain carbon and low-alloy steels and cast steel and to a limited extent to high-alloy and/or work-hardened steel

UNS Series	Metal
A00001 to A99999	Aluminum and aluminum alloys
C00001 to C99999	Copper and copper alloys
D00001 to D99999	Specified mechanical property steels
E00001 to E99999	Rare earth and rare earth-like metals and alloys
F00001 to F99999	Cast irons
G00001 to G99999	AISI and SAE carbon and alloy steels (except tool steels)
H00001 to H99999	AISI and SAE H-steels
J00001 to J99999	Cast steels (except tool steels)
K00001 to K99999	Miscellaneous steels and ferrous alloys
L00001 to L99999	Low-melting metals and alloys
M00001 to M99999	Miscellaneous nonferrous metals and alloys
N00001 to N99999	Nickel and nickel alloys
P00001 to P99999	Precious metals and alloys
R00001 to R99999	Reactive and refractory metals and alloys
S00001 to S99999	Heat and corrosion resistant (stainless) steels
T00001 to T99999	Tool steels, wrought and cast
W00001 to W99999	Welding filler metals
Z00001 to Z99999	Zinc and zinc alloys

upset welding. A resistance welding process in which the weld is produced, simultaneously over the entire area of abutting surfaces or progressively along a joint, by applying mechanical force (pressure) to the joint, then causing electrical current to flow across the joint to heat the abutting surfaces. Pressure is maintained throughout the heating period. See also *open-gap upset welding.*

upslope time. In resistance welding, the time associated with current increase using *slope control.*

uranium. A chemical element having atomic number 92, atomic weight 238, and the symbol U, after the planet Uranus, which was named from the Greek *ouranus,* meaning the sky. Uranium was identified and named by German chemist Martin Heinrich Klaproth in 1789. See *Technical Note 18.*

TECHNICAL NOTE 18

Uranium and Uranium Alloys

URANIUM is a moderately strong and ductile metal that can be cast, formed, and welded by a variety of standard methods. It is used in nonnuclear applications primarily because of its very high density (19.1 g/cm^3; 68% greater than lead). Uranium is frequently selected over other very dense metals because it is easier to cast and/or fabricate than the refractory metal tungsten and much less costly than such precious metals as gold and platinum. Typical nonnuclear applications for uranium and uranium alloys include radiation shields, counterweights, and armor-piercing kinetic energy penetrators.

Natural uranium contains approximately 0.7% of the fissionable isotope U-235 and 99.3% U-238. Ore of this isotopic ratio is processed by mineral beneficiation and chemical procedures to produce uranium hexafluoride (UF_6). Isotopic separation is performed at this stage. This produces both enriched (radioactive) UF_6, which contains more than the natural isotopic abundance of U-235 and is subsequently processed and used for nuclear applications, and depleted UF_6, which typically contains 0.2% U-235. Access to enriched UF_6 is tightly controlled, but depleted material can be purchased for industrial applications. The UF_6 is reduced to uranium tetrafluoride (UF_4) by chemical reduction with hydrogen. The UF_4 is then reduced with magnesium or calcium in a closed vessel at elevated temperature, producing 150 to 500 kg (330 to 1100 lb) ingots of metallic uranium commonly referred to as derbies. These derbies are typically vacuum induction remelted and cast into shapes required for engineering components or for subsequent mechanical working at elevated temperatures.

Uranium is frequently alloyed to improve its corrosion resistance and mechanical properties. Alloying results in substantial decreases in density;

therefore, it is desirable to obtain the necessary properties with small amounts of alloying additions. Principal alloying elements include titanium, niobium, molybdenum, and zirconium. Like unalloyed uranium, uranium alloys are produced by vacuum induction or vacuum arc melting and can be fabricated at elevated temperatures. A wide range of properties can be obtained by post-fabrication heat treatment of uranium alloys.

Nominal compositions and properties of uranium alloys

Alloy	Density, g/cm^3	Hardness	Yield strength, MPa (ksi)	Tensile strength, MPa (ksi)	Elongation(a), %	Reduction of area(a), %
Unalloyed uranium(b)	19.1	93 HRB	295 (43)	700 (101)	22	...
	19.1	94 HRB	270 (39)	720 (104)	31	...
U-0.75Ti(b)	18.6	36 HRC	650 (94)	1310 (190)	31	52
	18.6	42 HRC	965 (140)	1565 (227)	19	29
	18.6	52 HRC	1215 (176)	1660 (241)	<2	<2
U-2.0Mo	18.5	34 HRC	675 (98)	1100 (160)	23	25
U-2.3Nb	18.5	32 HRC	545 (79)	1060 (154)	28	33
U-4.5Nb	17.9	42 HRC	900 (130)	1190 (173)	10	8
U-6.0Nb	17.3	82 HRB	160 (23)	825 (120)	31	34
U-10Mo	16.3	28 HRC	900 (130)	930 (134)	9	30
U-7.5Nb-2.5Zr	16.4	20 HRC	540 (78)	850 (123)	23	50

(a) All based on high-purity alloys with low hydrogen contents. (b) Property values may vary due to heat treating practice.

Selected References

• K.H. Eckelmeyer, Uranium and Uranium Alloys, *Metals Handbook*, 10th ed., Vol 2, ASM International, 1990, p 670–682
• G.M. Ludtka and E.L. Bird, Heat Testing of Uranium and Uranium Alloys, *ASM Handbook*, Vol 4, ASM International, 1991, p 928–938

Dictionary of Metals
H.M. Cobb, editor

Copyright © 2012 ASM International®
All rights reserved
www.asminternational.org

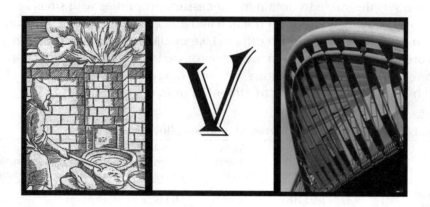

vacancy. A type of lattice imperfection in which an individual atom site is temporarily unoccupied. Diffusion (of other than interstitial solutes) generally is visualized as the shifting of vacancies.

vacuum arc remelting (VAR). A *consumable-electrode remelting* process in which heat is generated by an electric arc between the electrode and the ingot. The process is performed inside a vacuum chamber. Exposure of the droplets of molten metal to the reduced pressure reduces the amount of dissolved gas in the metal.

vacuum deposition. The condensation of thin metal coatings on the cool surface of work in a vacuum.

vacuum fusion. An analytic technique for determining the amount of gases in metals; ordinarily used for hydrogen and oxygen, and sometimes for nitrogen. Applicable to many metals, but not to alkali or alkaline earth metals.

vacuum induction melting (VIM). A process for remelting and refining metals in which the metal is melted inside a vacuum chamber by induction heating. The metal may be melted in a crucible and then poured into a mold. This process may also be operated in a configuration similar to that used in *consumable-electrode remelting,* except that the heat is supplied by an induction heating coil rather than from the passage of electric current through the electrode.

vacuum melting. Melting in a vacuum to prevent contamination from air, as well as to remove gases already dissolved in the metal; the solidification may also be carried out in a vacuum or at low pressure.

vanadium. A chemical element having atomic number 23, atomic weight 51, and the symbol V. The element was identified around 1830 by Swedish chemist Nils Gabriel Sefström, who named it for Vanadis, the Nordic goddess of love and beauty. The element had been observed years earlier, in 1801, by Spanish-born Mexican mineralogist Andres Manuel del Rio y Fernandez, who named the metal erythronium after the colorful flower of the erythronia plant. Fernandez later withdrew his claims because he suspected that the colorful compounds were actually due to chromium.

The largest use of vanadium is as an alloy in steel. Vanadium is added to cast iron to improve the hardenability and fatigue strength. Small additions of vanadium are used for age hardening aluminum casting alloys, for grain refinement, to increase the response to thermal treatment, and to improve fatigue strength.

vapor blasting. Producing a finely polished finish by directing an air-injected chemical emulsion containing fine abrasives against the surface to be finished. Also known as *liquid honing.*

vapor degreasing. The degreasing of work in the vapor over a boiling liquid solvent, the vapor being considerably heavier than air. At least one constituent of the soil must be soluble in the solvent. Modifications of this cleaning process include vapor-spray-vapor, warm liquid-vapor, boiling liquid-warm liquid-vapor, and ultrasonic degreasing.

vapor plating. The deposition of a metal or compound on a heated surface by the reduction or decomposition of a volatile compound at a temperature below the melting points of the deposit and the base material. The reduction is usually accomplished by a gaseous reducing agent such as hydrogen. The decomposition process may involve thermal dissociation or reaction with the base material. Occasionally used to designate deposition on cold surfaces by vacuum evaporation. See also *vacuum deposition.*

V-bend die. A die commonly used in press-brake forming, usually machined with a triangular cross-sectional opening to provide two edges as fulcrums for accomplishing three-point bending.

vector field. The magnetic field that is the result of two or more magnetizing forces impressed on the same area of a magnetizable object. Also known as *resultant field.*

Vegard's law. The rule that states that the lattice parameters of *substitutional solid solutions* vary linearly between the values for the components, with composition expressed in atomic percentage.

veining. A type of sub-boundary structure in a metal that can be delineated because of the presence of a greater-than-average concentration of precipitate or possibly solute atoms.

vent. A small opening in a foundry mold for the escape of gases.

vermicular iron. Cast iron having a graphite shape intermediate between the flake form typical of gray cast iron and the spherical form of fully spherulitic ductile cast iron. Compacted graphite cast iron is produced in a manner similar to that of ductile cast iron but using a technique that inhibits the formation of fully spherulitic graphite nodules. The same as *compacted graphite cast iron* and *CG iron.*

vertical-position welding. Welding in which the axis of the weld is essentially vertical.

vibratory finishing. A process for deburring and surface finishing in which the product and an abrasive mixture are placed in a container and vibrated.

Vickers hardness test. A microindentation hardness test using a 136° diamond pyramid indenter (Vickers) and variable loads, enabling the use of one hardness scale for all ranges of hardness—from very soft lead to tungsten carbide. Also known as the *diamond pyramid hardness test.*

virgin metal. Metal extracted from minerals and free of reclaimed metal scrap. Also known as *primary metal.* Compare with *secondary metal* and *native metal.*

Vitallium. A trademark for an early cobalt-chromium corrosion-resistant alloy used for orthopaedic and dental appliances. The composition was approximately 60% Co, 28–32% Cr, 5–7% Mo, 0.75% Mn, and 0.05% maximum carbon. Vitallium was developed in 1932 by Albert W. Merrick for Austenal Laboratories Inc.

voltage efficiency. The ratio, usually expressed as a percentage, of the equilibrium-reaction potential in a given electrochemical process to the bath voltage.

Wallner lines. A distinct pattern of intersecting sets of parallel lines, usually producing a set of V-shaped lines, sometimes observed in viewing brittle fracture surfaces at high magnifications in an electron microscope. Wallner lines are attributed to interaction between a shock wave and a brittle crack front propagating at high velocity. Wallner lines are sometimes misinterpreted as *fatigue striations.*

wandering sequence. A longitudinal welding sequence wherein the weld-bead increments are deposited at random to minimize distortion. Also known as *random sequence.*

warm working. Plastically deforming metal at a temperature above ambient (room) temperature but below the temperature at which the material undergoes recrystallization.

wash. (1) A coating applied to the face of a mold prior to casting. (2) An imperfection at a cast surface similar to a *cut.*

wash metal. Molten metal used to wash out a furnace, ladle, or other container.

water break. The appearance of a discontinuous film of water on a surface signifying nonuniform wetting and usually associated with a surface contamination.

waviness. A wavelike variation from a perfect surface, generally much larger and wider than the *roughness* caused by tool or grinding marks (Fig. 59).

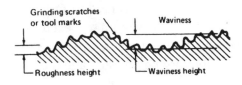

Fig. 59 Schematic showing roughness and waviness

wear pad. In forming, an expendable pad of rubber or rubberlike material of nominal thickness that is placed against the diaphragm to lessen the wear on it. See *diaphragm* (2).

weave bead. A weld bead made with oscillations transverse to the axis of the weld. Contrast with *stringer bead.*

web. (1) For twist drills and reamers, the central portion of the tool body that joins the lands. (2) In forging, the thin section of metal remaining at the bottom of a cavity or depression or at the location of the top and bottom punches. The former type may be removed by piercing or machining; the latter, by the trim punch. (3) A plate or thin portion between stiffening ribs or flanges, as in an I-beam, H-beam, or other similar section.

weight percent. Percentage composition by weight. Contrast with *atomic percent.*

weld. A union made by *welding.*

weldability. A specific or relative measure of the ability of a material to be welded under a given set of conditions. Implicit in this definition is the ability of the completed weldment to fulfill all functions for which the part was designed.

weld bead. A deposit of filler metal from a single welding *pass.*

weld crack. A crack in weld metal. See also *crater crack, root crack, toe crack,* and *underbead crack.*

weld-delay time. In spot, seam, or projection welding, the time during which the current is delayed with respect to starting the forge delay timer in order to synchronize the forging pressure and the welding heat.

welder. A person who makes welds using manual or semiautomatic equipment. Formerly used as a synonym for *welding machine.*

weld gage. A device for checking the shapes and sizes of welds.

welding. (1) Joining two or more pieces of material by applying heat or pressure, or both, with or without filler material, to produce a localized union through fusion or recrystallization across the interface.

The thickness of the filler material is much greater than the capillary dimensions encountered in *brazing.* (2) May also be extended to include brazing and soldering.

welding current. The current flowing through a welding circuit during the making of a weld. In resistance welding, the current used during preweld or postweld intervals is excluded.

welding cycle. The complete series of events involved in making a resistance weld. Also applies to semiautomatic mechanized fusion welds.

welding force. The same as *electrode force*—the force between electrodes—in resistance welding.

welding generator. A generator used for supplying current for welding.

welding ground. The electrical conductor connecting the source of arc welding current to the work. The same as *work lead.* Also called work connection or ground lead.

welding leads. The electrical cables that serve as either the *work lead* or the *electrode lead* of an arc welding circuit.

welding machine. Equipment used to perform the welding operation. For example, *spot welding* machine, *arc welding* machine, and *seam welding* machine.

welding procedure. The detailed methods and practices, including joint preparation and welding procedures, involved in the production of a *weldment.*

welding rod. Welding or brazing filler metal, usually in rod or wire form, but not a *consumable electrode.* A welding rod does not conduct the electrical current to an arc and may be either fed into the weld puddle or preplaced in the joint.

welding sequence. The order in which the various component parts of a weldment or structure are welded.

welding stress. *Residual stress* caused by localized heating and cooling during welding.

welding technique. The details of a welding operation that, within the limitations of a welding procedure, are performed by the *welder.*

welding tip. (1) A torch tip designed for welding. (2) The electrode tip that contacts the work in resistance spot welding.

weld interval. The total heat and cool time in making one multiple-impulse resistance weld.

weld-interval timer. A device used in resistance welding to control heat and cool times and weld interval when making multiple-impulse welds singly or simultaneously.

weld line. The junction of the weld metal and the base metal, or the junction of the base-metal parts when filler metal is not used.

weldment. An assembly whose component parts are joined by welding.

weld metal. That portion of a weld that has been melted during welding.

weld nugget. The weld metal in spot, seam, or projection welding.

weldor. (obsolete) Formerly used to designate a person who makes welds. The preferred term is *welder.*

weld time. In single-impulse and flash welding, the time that the welding current is applied to the work.

weld timer. A device used in resistance welding to control the weld time only.

Wenstrom mill. A rolling mill similar to a universal mill but one in which the edges and sides of a rolled section are acted on simultaneously.

wet blasting. A process for cleaning or finishing by means of a slurry of abrasive in water directed at high velocity against the workpieces.

wetting. A condition in which the interfacial tension between a liquid and a solid is such that the contact angle is 0 to 90°.

wetting agent. A surface-active agent that produces *wetting* by decreasing the *cohesion* within the liquid.

whiskers. Metallic filamentary growths, often microscopic, sometimes formed during electrodeposition and sometimes spontaneously during storage or service, after finishing.

white brass. A copper alloy containing over 50% Zn that is too brittle for common use.

white cast iron. *Cast iron* that shows a white fracture because the carbon is in combined form.

whiteheart malleable. A cast iron made by prolonged annealing of *white cast iron* in which decarburization or graphitization, or both, take place to eliminate some or all of the cementite. The graphite is in the form of temper carbon. If decarburization is the predominant reaction, the product will exhibit a light fracture surface, hence "whiteheart malleable." See *malleable cast iron.*

white metal. (1) A general term covering a group of white-colored metals of relatively low melting points (lead, antimony, bismuth, tin, cadmium, and zinc) and the alloys based on these metals. (2) A copper matte of approximately 77% Cu obtained from smelting of sulfide copper ores.

white rust. Zinc oxide; the powdery product of corrosion of zinc or zinc-coated surfaces.

Widmanstätten structure. A structure characterized by a geometrical pattern resulting from the formation of a new phase along certain crystallographic planes of the parent solid solution. The orientation of the lattice in the new phase is related crystallographically to the orientation of the lattice in the parent phase. This structure was originally observed in meteorites, but is readily produced in many other alloys by appropriate heat treatment.

wildness. A condition that exists when molten metal, during cooling, evolves so much gas that it becomes violently agitated, forcibly ejecting metal from the mold or other container.

Williams riser. An *atmospheric riser:* a riser that uses atmospheric pressure to aid feeding. Essentially a *blind riser* from which a small core or rod protrudes, the function of the core or rod being to provide an open passage so that the molten interior of the riser will not be under a partial vacuum when metal is withdrawn to feed the casting, but will always be under atmospheric pressure.

winning. Recovering a metal from an ore or chemical compound using any suitable hydrometallurgical, pyrometallurgical, or electrometallurgical method.

wiped coat. A hot dipped galvanized coating from which virtually all free zinc is removed by wiping prior to solidification, leaving only a thin zinc-iron alloy layer.

wiped joint. A joint wherein filler metal is applied in liquid form and distributed by mechanical action.

wiper forming. A method of curving bars, tubes, or rolled or extruded sections, in which the stock is bent so that it conforms to a fixed form block. The stock is clamped to the form block, then bent by applying force through a wiper block, shoe, or roll that is moved along the periphery of the form block. Sometimes called compression forming. Contrast with *draw forming.*

wiping effect. The activation of a metal surface by mechanical rubbing or wiping to enhance the formation of conversion coatings, such as phosphate coatings.

wire. (1) A thin, flexible, continuous length of metal, usually of circular cross section, and usually produced by drawing through a die. See also *flat wire.* (2) A length of single metallic electrical conductor; it may be of solid, stranded, or tinsel construction, and may be either bare or insulated.

wire bar. A cast shape, particularly of tough pitch copper, that has a cross section approximately square with tapered ends, designed for hot rolling to rod for subsequent drawing into wire.

wire drawing. Reducing the cross section of wire by pulling it through a die (Fig. 60). See *Taylor process.*

wire rod. Hot rolled coiled stock that is to be cold drawn into wire.

wiring. The formation of a curl along the edge of a shell, tube, or sheet, and insertion of a rod or wire within the curl for stiffening the edge. See also *curling.*

wolfram. The principal ore of tungsten, which is a tungstate of iron and manganese. The chemical symbol for tungsten, W, is taken from wolfram.

wood flour. A pulverized wood product used in the foundry to furnish a reducing atmosphere in the mold, help overcome sand expansion, increase flowability, improve casting finish, and provide easier shakeout.

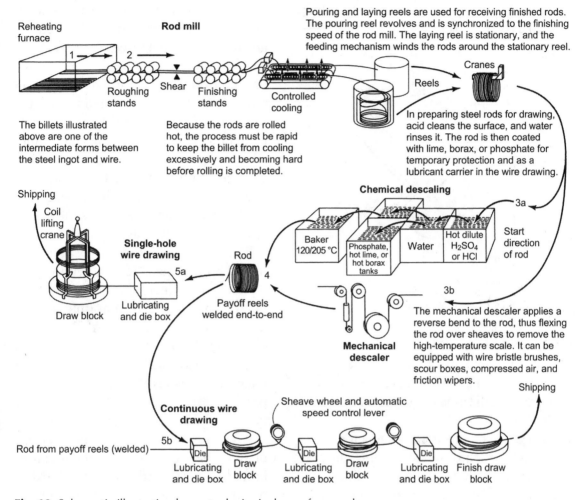

Reheating furnace

Rod mill

Pouring and laying reels are used for receiving finished rods. The pouring reel revolves and is synchronized to the finishing speed of the rod mill. The laying reel is stationary, and the feeding mechanism winds the rods around the stationary reel.

Cranes

Roughing stands

Shear

Finishing stands

Controlled cooling

Reels

The billets illustrated above are one of the intermediate forms between the steel ingot and wire.

Because the rods are rolled hot, the process must be rapid to keep the billet from cooling excessively and becoming hard before rolling is completed.

In preparing steel rods for drawing, acid cleans the surface, and water rinses it. The rod is then coated with lime, borax, or phosphate for temporary protection and as a lubricant carrier in the wire drawing.

Chemical descaling

3a

Shipping

Coil lifting crane

Single-hole wire drawing

Rod

5a

Baker 120/205 °C

Phosphate, hot lime, or hot borax tanks

Water

Hot dilute H_2SO_4 or HCl

Start direction of rod

Draw block

Lubricating and die box

Payoff reels welded end-to-end

4

3b

Mechanical descaler

The mechanical descaler applies a reverse bend to the rod, thus flexing the rod over sheaves to remove the high-temperature scale. It can be equipped with wire bristle brushes, scour boxes, compressed air, and friction wipers.

Shipping

Continuous wire drawing

Sheave wheel and automatic speed control lever

5b

Rod from payoff reels (welded)

Die

Lubricating and die box

Draw block

Die

Lubricating and die box

Draw block

Die

Lubricating and die box

Finish draw block

Fig. 60 Schematic illustrating how steel wire is drawn from rods

Wood's metal. A low-melting-point (70 °C, or 158 °F) fusible alloy used as a low-melting solder, low-temperature casting alloy, high-temperature coupling fluid in heat baths, and in fire sprinkler systems. It is a eutectic composition of 50% Bi, 26.7% Pb, 13.3% Sn, and 10% Cd. It is named after American physicist Robert W. Wood.

woody structure. A macrostructure, found particularly in wrought iron and in extruded rods of aluminum alloys, that shows elongated surfaces of separation when fractured.

Wootz Process. The name of the old process of cementation and crucible melting for steelmaking. It was used in India, and then in Damascus and Toledo for the production of high-quality steel blades. The crucible steel process was lost in the Middle Ages but rediscovered by Benjamin Huntsman in 1740.

work angle. In arc welding, the angle between the electrode and one member of the joint, taken in a plane normal to the weld axis (Fig. 61).

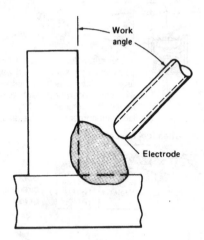

Fig. 61 Weld electrode work angle

work hardening. An increase in the hardness and strength of metals caused by plastic deformation at temperatures below the recrystallization range. Also known as *strain hardening.*

work lead. The electrical conductor connecting the source of arc welding current to the work. Also known as work connection, *welding ground,* or ground lead.

worm. An exudation (sweat) of molten metal forced through the top crust of solidifying metal by gas evolution. See also *zinc worms.*

wrap forming. The shaping of a sheet or part, usually of uniform cross section, by first applying suitable tension or stretch and then wrapping the sheet or part around a die of the desired shape. Also known as *stretch forming.*

wringing fit. A fit of nominally zero allowance.

wrinkling. A wavy condition obtained in deep drawing of sheet metal, in the area of the metal that passes over the draw radius. Wrinkling may also occur in other forming operations when unbalanced compressive forces are set up.

wrought iron. Wrought iron was the "iron" of the Iron Age, a period that began around 1500 B.C. depending on the location. It was one of the seven metals of antiquity, which also included gold, silver, copper, tin, mercury, and lead. Iron generally replaced bronze for the making of weapons and implements.

The metal has been described as: "A ferrous metal aggregated from a solidifying mass of pasty particles of highly refined metallic iron, with which, without subsequent fusion is incorporated a minutely and uniformly-distributed quantity of slag" (ASTM 1930) (Fig. 62). Wrought iron, however, is a virtually unknown and forgotten metal today, for reasons that will become apparent. A typical chemical analysis of the metal is 0.02% C, 0.01 % Mn, 0.1 % P, and 3.0 % slag by weight. The presence of slag fibers gives the metal a tough fibrous structure similar to that of hickory wood, as seen easily by examining pieces that have been nicked and slowly broken (Fig. 63).

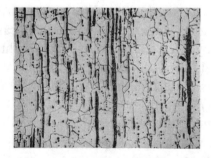

Fig. 62 Photomicrograph of wrought iron showing glass-like siliceous slag fibers embedded in the high purity iron base metal

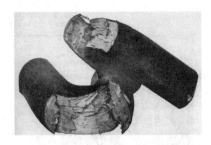

Fig. 63 Wrought iron bars fractured to show characteristic fibrous, hickory-like structure

The strength of wrought iron is approximately 80% that of structural steel, according to the former ASTM A42 specification for wrought iron plate, which requires a minimum tensile strength of 48,000 psi, a yield strength of 27,000 psi, and a minimum elongation of 14% in 8 in.

Wrought iron is malleable and weldable and has good fatigue resistance and good corrosion resistance. The presence of the slag is believed to enhance these properties.

In 1703, the first record of the term *wrought iron* appeared in an Act of Parliament that mentioned "wares of wrought iron." The two names led to some confusion, but metallurgists seemed to prefer using "wrought iron." Applications of wrought iron in colonial times included horseshoes and nails. During the American Revolution, a stout chain strung across the Hudson River at West Point could be raised to stop a British ship. Starting around 1840, iron was the material for railroad rails and bridges until Bessemer steel was invented later in the century. In 1843 the SS *Great Britain*, the first Atlantic liner to be built of wrought iron, and having a screw propeller and sails, was launched. At 88 m (322 ft.), it was the world's largest ship.

Wrought iron was used later for battleship armor plate. The 1000-foot-tall Eiffel Tower erected in 1889 is built of wrought iron. In the 20th century there was widespread use of iron, especially as it became known for its durability in corrosive conditions. Industries specifying wrought iron included building construction, railroads and shipping, petroleum, and public works. About 1929, in order to increase wrought iron production, the A.M. Byers Co. of Pittsburgh built a new mechanized plant that would use the patented Aston Process. The process involved pouring molten refined slag into liquid pig iron to create an 8000-pound sponge ball. The ball was transferred to a 9000-ton hydraulic press in which the excess slag was squeezed out, an operation that was repeated with another ball of iron every five minutes. Forty years later, however, the demand for wrought iron had declined to such an extent that the product was no longer profitable, forcing the plant to close in 1969.

Dictionary of Metals
H.M. Cobb, editor

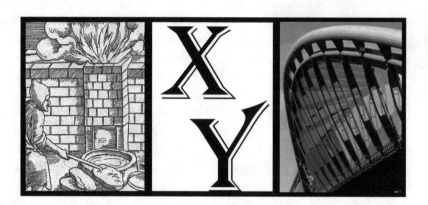

x-ray. A form of electromagnetic radiation, of wavelengths less than approximately 10 nm, emitted as the result of the deceleration of fast-moving electrons (bremsstrahlung, continuous spectrum) or decay of atomic electrons from excited orbital states (*characteristic radiation*); specifically, the radiation produced when an electron beam of sufficient energy impinges on a target of suitable material.

Y-block. A single *keel block,* which is a standard test casting, for steel and other high-shrinkage alloys, consisting of a rectangular bar that resembles the keel of a boat, attached to the bottom of a large riser, or shrinkhead.

yellow brass, 63%. A 63Cu–37Zn zinc alloy, UNS C27400.

yellow brass, 65%. A 65Cu-35Zn zinc alloy, UNS C27000.

yellow brass, 66%. A 66Cu-34Zn zinc alloy, UNS C26800.

yield point. The first stress in a material, usually less than the maximum attainable stress, at which an increase in strain occurs without an increase in stress. Only certain metals exhibit a yield point. If there is a decrease in stress after yielding, a distinction may be made between upper and lower yield points.

yield strength. The stress at which a material exhibits a specified deviation from proportionality of stress and strain. An offset of 0.2% is used for many metals. Compare with *tensile strength.*

Young's modulus. A measure of the rigidity of metal; the ratio of stress, below the proportional limit, to the corresponding strain. Specifically, the modulus obtained in tension or compression. See *modulus of elasticity.*

ytterbium. A chemical element having atomic number 70, atomic weight 173, and the symbol Yb, named for Ytterby, the Swedish village near where the element was discovered by Jean Charles Galissard de Marignac in 1878. Ytterbium metal was first made by Klemm and Bonner in 1937. Ytterbium is one of the 11 rare earth metals.

yttrium. A chemical element having atomic number 39, atomic weight 89, and the symbol Y, for Ytterby, the Swedish village near where the metal was found by Finnish chemist Johan Gadolin in 1794. It is one of the 11 rare earth metals.

Dictionary of Metals
H.M. Cobb, editor

zinc. A chemical element having atomic number 30, atomic weight 65, and the symbol Zn, from the German *zink* or *zinke*. The element was identified as a metal in 1746 by German chemist Andreas Marggraf. Zinc is a bluish-white metal that is rather brittle at room temperature but is soft and ductile at 100 °C (212 °F).

Zinc has had an unusual history because it was widely used without being recognized. The Romans became involved in the business of coloring copper to create a golden metal. That was done by heating copper pieces and calamine powder with charcoal in a closed crucible. The colored copper pieces then were melted and cast. The Romans called the colored metal *aurichalcan* (golden copper). They did not realize that the calamine power was zinc carbonate and that they had produced what we now call *brass*. See *Technical Note 19*.

zincum. An early name for *zinc*.

zinc worms. Surface imperfections, characteristic of high-zinc brass castings, that occur when zinc vapor condenses at the mold/metal interface, where it is oxidized and then becomes entrapped in the solidifying metal.

zirconium. A chemical element having atomic number 40, atomic weight 91, and the symbol Zr, from zircon and the Persian word *zar-gun,* meaning goldlike. In 1789, Martin Heinrich Klaproth discovered the presence of a new metal oxide in the mineral zircon and named the compound zirconia. Jöns Jakob Berzelius is given credit for producing the first metallic zirconium in 1824. In Germany in 1914, D. Lely, Jr. and L. Hamburger were the first to produce a metal sufficiently pure to be ductile. In 1923, it was found that what was thought to be zirconium actually was zirconium alloyed with a second metal. The second metal was the recently-discovered hafnium, which was chemically similar to zirconium and had not been detected.

TECHNICAL NOTE 19

Zinc and Zinc Alloys

ZINC, its alloys, and its chemical compounds represent the fourth most industrially utilized metal (behind iron, aluminum, and copper). Zinc is used in five principal areas of application: in coatings and anodes for corrosion protection of irons and steels, in zinc casting alloys, as an alloying element in copper, aluminum magnesium, and other alloys, in wrought zinc alloys, and in zinc chemicals.

The use of zinc as a coating to protect iron and steel from corrosion is the largest single application for zinc worldwide. Metallic zinc coatings are applied to steels from a molten metal bath (hot dip galvanizing), by electrochemical means (electrogalvanizing), from a spray of molten metal (metallizing), and in the form of zinc powder by chemical/mechanical means (mechanical galvanizing). Zinc coatings are applied to many different types of products, ranging in size from small fasteners to continuous strip to large structural shapes and assemblies. The hot dip galvanizing industry is currently the largest consumer of zinc in the coatings field.

Almost all of the zinc used in zinc casting alloys is employed in die casting compositions. Two alloy groups—hypoeutectic and hypereutectic—make up zinc alloy castings. The hypoeutectic alloys contain approximately 4% Al. Of these, Alloy 3 (Zn-4Al-0.4Mg) is the most commonly used. Alloy 7 (Zn-4Al-0.015Mg) is a modification of this alloy. Alloy 5 (Zn-4Al-1Cu-0.05Mg) is used when higher

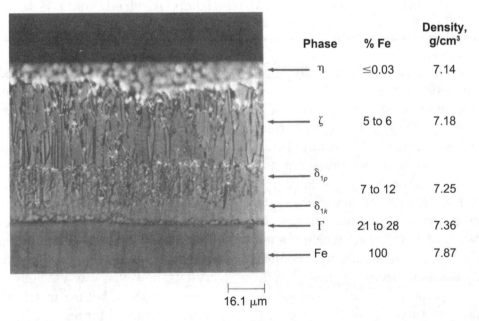

Phase	% Fe	Density, g/cm³
η	≤0.03	7.14
ζ	5 to 6	7.18
δ_{1p}	7 to 12	7.25
δ_{1k}		
Γ	21 to 28	7.36
Fe	100	7.87

16.1 μm

Typical hot dip galvanized coating. Note the gradual transition from layer to layer, which results in a strong bond between base metal and coating.

TECHNICAL NOTE 19 (*continued*)

tensile strength and/or hardness is required. The hypereutectic alloys have higher aluminum contents (>5% Al) and are used for high-performance applications. These alloys include ZA-8 (Zn-8Al-1Cu-0.02Mg), ZA-12 (Zn-11Al-1Cu-0.025Mg), and ZA-27 (Zn-27Al-2Cu-0.015Mg). Zinc castings are used extensively in the transportation industry for parts such as carburetors, fuel pump bodies, wiper parts, speedometer frames, heater components, radio bodies, instrument panels, and body moldings. Zinc castings are also used extensively in electronic and electrical fittings of all kinds as well as for hardware used in the computer industry, in business machines, and in such items as recording machines and cameras.

Among zinc-containing alloys, copper-base alloys such as brasses are the largest zinc consumers. Rolled zinc is the principal form in which wrought products are supplied, although drawn zinc wire for metalizing is showing increasing usage. In the zinc-chemical category, zinc oxide is the major compound utilized.

Selected References

· R.J. Barnhurst, Zinc and Zinc Alloys, *Metals Handbook*, 10th ed., Vol 2, ASM International, 1990, p 527–542
· D.C.H. Nevison, Cast Zinc and Zinc Alloys, *Metals Handbook*, 9th ed., Vol 15, 1988, p 786–797
· D.C.H. Nevison, Corrosion of Zinc, *Metals Handbook*, 9th ed., Vol 13, 1987, p 755–769

The hafnium had been chemically separated in a six-step process. It was found that the amount of hafnium normally associated with zirconium was from 0.5 to 2%. Hafnium was found to be almost twice as dense as zirconium and approximately the density of lead. The greatest and most surprising differences, however, were the thermal neutron absorption cross sections, which varied from 0.5 barn per atom for zirconium to 105 barns per atom for the hafnium. These properties, plus their corrosion resistance and strength, would make both metals vital in the future nuclear industry.

In 1925, A.E. van Arkel and J. de Boer described the thermal decomposition of metal halides as a means of producing pure metal. Zirconium deposited from zirconium iodide on a hot filament was very ductile and could readily be drawn into fine wire or rolled into thin sheets. Some years later, the Foote Mineral Company of Philadelphia started the production of iodide zirconium. The single-crystal zirconium bars were approximately 2.54 cm (1 in.) in diameter and 25.40 cm (10 in.) long. Growing the crystals, however, was a tedious and expensive process, but the metal provided valuable material for research.

By 1944, the potential value of zirconium and hafnium in nuclear reactors was fully realized and plans were on the drawing board for nuclear submarines, but there was no substantial source of zirconium. The Bureau of Mines in Albany, Ore., had just developed the first practical process for producing titanium, using the Kroll process. The process permitted ingots 30.48 cm (12 in.) in diameter by 182.88 cm (72 in.) tall to be melted from a compacted titanium sponge electrode by the vacuum arc consumable-electrode process. A similar method worked for zirconium. As luck would have it, there was an abundant source of zirconia right on the beaches of Oregon. The Atomic Energy Commission geared up for an enormous production of zirconium. The ore would be processed into sponge and melted by four private organizations. Four other companies would be set up to manufacture the nuclear reactors and zirconium-clad fuel elements.

The atomic-powered *Nautilus* submarine was launched in 1955. Nuclear power plants were built—all using zirconium fuel cladding.

zircon sand. A very refractory mineral, composed chiefly of zirconium silicate; it has low thermal expansion and high thermal conductivity.

zone melting. Highly localized melting, usually by induction heating, of a small volume of an otherwise solid metal piece, usually a rod. By moving the induction coil along the rod, the melted zone can be transferred from one end to the other. In a binary mixture where there is a large difference in composition on the *liquidus* and *solidus* lines, high purity can be attained by concentrating one of the constituents in the liquid as it moves along the rod.

APPENDIX 1

Metals History Timeline

6000 B.C. Chalcolithic Period. Copper comes into common use.

ca. 5000 B.C. Gold. Metal No. 1. Gold was one of the seven metals of antiquity. It was and is the most attractive and most desired of the metals. It has been called a noble metal and a precious metal.

4000 B.C. Copper. Metal No. 2. Copper was one of the seven metals of antiquity. It is an excellent electrical conductor and a metal that can be alloyed to produce bronze and brass.

4000 B.C. Silver. Metal No. 3. Silver was one of the seven metals of antiquity. It is a noble metal and a precious metal.

4000 B.C. Iron. Metal No. 4. Iron was the strongest of the seven metals of antiquity. Iron beads were worn in Egypt as early as 4000 B.C. It is the metal that later became known as wrought iron.

3500 B.C. Lead. Metal No. 5. Lead was one of the seven metals of antiquity.

2400 B.C. Beginning of the Bronze Age. Bronze is an alloy chiefly of copper and tin that was suitably hard and strong for making weapons and tools.

2000 B.C. Beginning of the Iron Age.

1750 B.C. Tin. Metal No. 6. Tin was one of the seven metals of antiquity, noted for the formation of bronze when alloyed with copper.

1350 B.C. Hammered Iron. The oldest known article of iron shaped by hammering is a dagger found in Egypt that was made before 1350 B.C. It is believed not to have been made in Egypt but to be of Hittite workmanship.

753 B.C. Founding of Rome.

750 B.C. Mercury. Metal No. 7. Mercury was the last of the seven metals of antiquity.

509 B.C. Beginning of the Roman Republic.

500 B.C. Wootz Process. The name of the old process of cementation and crucible melting for steelmaking. It was used in India and then in Damascus and Toledo for the production of high-quality steel blades. The crucible steel process was lost in the Middle Ages but rediscovered by Benjamin Huntsman in 1740.

27 B.C. Start of the Roman Empire. The Empire is established that will last to the fifth century.

A.D.

73. Pliny the Elder's *Natural History*. Pliny (23–79 A.D.), a Roman whose actual name was Gaius Plinius Secundus, was the author of *Natural History*, a 37-volume work that included discussions on the mining and metallurgy of his time.

1st Century. Calamine Brass. At the beginning of the first century, the Romans were producing what was called *aurichalcum* (golden copper) without realizing that it actually was a brass that they had created by heating pieces of copper with calamine powder (zinc silicate). The zinc vaporized and was absorbed by the copper to create brass. The Romans believed they were coloring the copper.

300. Roman Love of Lead. During the Roman Empire, which lasted from 27 B.C. to the fifth century, there was an extensive use of what the Romans called *plumbum*. It is said that lead was used for roofing, cisterns, water pipes, and ornaments. Romans were accustomed to boiling wine in lead pots or lead-lined copper kettles, claiming that the lead gave the wine a sweetish taste. In the upper-class households it was common to use lead-containing pewter mugs and plates.

ca. 400. Iron Pillar of Delhi. The pillar is the oldest known metal monument. It is an inscribed wrought iron pillar standing 7 m (22 ft) tall, weighing an estimated 5.44 metric tons (6 U.S. tons). The diameter at the base is 420 mm (16.4 in.). It is remarkable that little corrosion has occurred. The pillar is said to have been fashioned by Emperor Chandragupta Vikramaditya (375–413), although other authorities give dates as early as 900 B.C.

5th Century. Fall of Rome.

897. Earliest Reference to Iron. The earliest reference to iron in the *Oxford English Dictionary* is written in Middle English as: "Durh ocet isern ocet" (*isern* was a word for iron).

900–1100. Alchemy. Alchemists believe it possible to turn a base metal such as iron into a precious metal such as gold.

1198. Dartmoor Tin Production. The Pipe Roll (an official record) of Henry II shows production of tin in Dartmoor, Devon, England, at approximately 45 metric tons (50 U.S. tons) annually.

1200. Zinc. Metal No. 8. Zinc is the first metal to be found in the 2000 years since the metals of antiquity were discovered. A Hindu book describes a tinlike material made from calamine that was distilled to produce zinc.

1239. Copper Roof on German Cathedral. A copper roof on a cathedral in Hildesheim remained intact until bombed in 1945, making it the longest lived copper roof.

1250. Arsenic. Metal No. 9. Discovered about 1250 by the German monk Alburtus Magnus, arsenic was first definitely prepared in 1641.

1292. Early Record of Pewter (Peutre). *Pewter* is from the French *peutre*. The *Oxford English Dictionary* by John Reeves contains the following quotation by Britton about criminal offenses: "Qi mauvaise chose vendent pure bone, sicum peutre pur argen ou latoun pur or." (He who sells a bad thing as good, such as pewter as silver, or brass as gold.)

1293. Catalan Forge Iron. The iron workers of Catalan in Spain made a major advance in the manufacture of wrought iron from ore. They developed a hearth-type furnace consisting of a hearth or crucible in which a mixture of ore and fuel was placed. An air blast produced by a trompe or water blower entered the furnace through tuyeres near the bottom. The output of the Catalan furnace is said to have been approximately 63.5 kg (140 lb) of wrought iron in 5 h, which was considerably more than the output of other furnaces.

1337. Cornwall Tin. By 1337, Cornwall became England's center of tin production, recording 590 metric tons (650 U.S. tons) that year.

1340. Sheffield Cutlery. The first recorded mention of Sheffield cutlery is in the inventory of King Edward III, who listed in his will a knife and the beneficiary to whom he was leaving it.

1350. Cast Iron Begins. A furnace known as a *stuckofen* was the earliest blast furnace in which air was blown in at the bottom. The product, which was very brittle, could be cast into various useful shapes. The new metal was called cast iron.

1350. Pewter. In 14th century England, pewter is in widespread use, especially for church ornaments. Pewter gradually starts replacing wooden cups and bowls in the more affluent households. Pewters are alloys of tin, antimony, and, originally, as much as 15% Pb.

1352. Gold Leaf in Egyptian Tombs. Many mummies and their cases in Egyptian tombs, including King Tutankhamun's, were overlaid with gold leaf. The art of beating gold into a sheet 75 millionths of a millimeter (3 millionths of an inch) thick was developed in antiquity. It required neither heat nor mechanical equipment.

1373. Solingen is Chartered. In Westphalia, Germany, on the Wupper River east of Dusseldorf, Solingen is chartered. The town was noted for sword making in the Middle Ages and became known as a cutlery center, a distinction shared with Sheffield in England and Thiers in France.

1374. Zinc. Hindus recognize zinc to be a new metal, the eighth known to man. There is a limited amount of zinc production.

1380. Chaucer's Sheffield Knife. Geoffrey Chaucer (1343–1400), the English poet of the Middle Ages, writes about a Sheffield knife in *The Reeves Tale*, the third story of *The Canterbury Tales*, and can be seen in portraits wearing such a knife.

1436. Gutenberg's Printing Press. Johannes Gutenberg builds what is believed to have been the first European printing press with moveable type cast in a mold. Metal type also was developed in the 15th century; it consisted of lead with small amounts of arsenic, which caused the metal to expand on freezing, producing type with excellent sharpness.

1457. The Mons Meg Gun. This medieval siege gun given to James II of Scotland at Edinburgh Castle is said to have been able to fire a stone ball almost 3.22 km (2 miles). It is made of iron bars welded together and has a bore of 0.5 m (19.5 in.). The length is 406 cm (80 in.).

1473. Worshipful Company of Pewterers. A trade guild is created in England to ensure the quality of pewter. Pewter was an alloy of tin, antimony, and lead.

1489. Strike while the Iron is Hot. This expression apparently has its origin when William Caxton (1415~1422–1492),

English merchant, writer, and printer, wrote: "When the yron is well hoote, hit werketh the better" (*Oxford English Dictionary*).

1546. *De La Pirotechnia.* Vannoccio Biringuccio (1480–c. 1539), an Italian worker in the armory of Siena, traveled through Germany and Italy and wrote about ores, assaying, melting, casting, and fireworks. The book becomes one of the two classics of the period, the other being *De Re Metallica*. The book was translated from the Latin into English in 1942 by the noted American metallurgist, Cyril Stanley Smith, and Martha Teach Gnudi. The translation was published by the American Institute of Mining, Metallurgical and Petroleum (AIME) in a small edition that has been reprinted by Dover Publications.

1556. *De Re Metallica* **(Concerning metals).** Georgius Agricola (1494–1555), a German doctor, spent his free time roaming the mining district collecting information to write a book on mining and metallurgy. The book, which was 20 years in the making, was published one year after Agricola's death. Written in Latin, it became a classical text on the subject for 200 years. In 1912, future U.S. president Herbert Hoover, a mining engineer, and his wife Lou Henry, a Latin scholar, made an English translation of the book and included many of Hoover's footnotes. The translation has been reprinted by Dover Publications.

1560. Antimony. Metal No. 10. Georgius Agricola, a German doctor, reported the technique of forming antimony by roasting stibium or antimony sulphide in an iron pot.

1565. Company of Mineral and Battery Works. The company is one of two mining monopolies created by Queen Elizabeth I of England. The company had the right to make battery ware (items of beaten metal), cast work and wire of latten (brass), iron, and steel; and to mine calamine (nickel carbonate) stone and use it to make latten and other mixed metals.

1568. First British Wireworks. Queen Elizabeth I grants a patent of incorporation to William Humfrey to set up a wireworks in Tintern, Monmouthshire, England. Humfrey hires Christopher Schutz, a German copper maker, along with his entire workshop. The initial product is to be brass in addition to the iron wire needed for making cards (combs) for the British wool industry.

1607. Coke. It is discovered how to convert coal to coke, revolutionizing the iron industry.

1624. Company of Cutlers. An act of Parliament forms the Company of Cutlers, a trade guild of metalworkers of cutlery and other wares of wrought iron and steel in Hallamshire, which includes the entire Sheffield area.

1646. Saugus Iron Works. The Saugus Iron Works on the Saugus River near Boston, Mass., is equipped with a furnace and mill for making iron from bog iron. The colonial furnace was in operation for 22 years, from 1646 to 1668, when the bog became depleted.

1664. First Record of "Cast Iron." The first record of "cast iron" in print occurred in 1664 with the following notation in the *Oxford English Dictionary*: "The pipes should be the best cast iron" (Evelyn Kal. Hort.).

1700. White Copper.. Although *pai-tung* (white copper)—a metal that resembles silver but contains copper—had been in use in China for many centuries, it was not until the 18th century that it was discovered that the copper was alloyed with a new metal called nickel.

1703. Wrought Iron. The earliest known use of the term *wrought iron* was in the records of British Parliament in 1703, according to the *Oxford English Dictionary*, which recorded: "Wares of wrought iron." Prior to that time, the word *iron* was commonly used.

1709. Cast Iron. Abraham Darby of Coalbrookdale, Wales, discovers how to make coke from coal, eliminating the need for charcoal in the production of cast iron. Cast iron becomes one of the major materials of the Industrial Revolution. It is used for many bridges in England during the 19th century.

1725. Tinplate Invented. Tin-plated steel food containers originated in Bohemia (Czech Republic) and appeared in England in 1725, as noted in port books in Gloucestershire, England.

1727. Copper Roof for Philadelphia Church. A copper roof is installed on the famous Christ Church, the oldest known copper-roofed church in America.

1728. John Wilkinson (1728–1808). Wilkinson was a British industrialist, known as the great Staffordshire ironmaster, who found new applications for iron. He devised a boring machine that was essential to the success of James Watt's steam engine. His boring machine (1775) could bore engine cylinders and cannon barrels with extreme accuracy. Wilkinson produced wrought iron on a large scale by using a steam engine to drive a large air pump for its manufacture. In 1787, he produced an iron-hulled barge that was a sensation at the time. He produced all the pipe, cylinders, and ironwork for the Paris water works. He was buried in a self-designed cast iron coffin. His descendants built Wilkinson Sword, Ltd.

1735. Cobalt. Metal No. 11. Cobalt is discovered by Georg Brandt, Swedish chemist and mineralogist.

1737. Iron Mining at Cornwall. John Grubb discovers high-quality magnetite iron ore in Lebanon County, Pa., and opens the Cornwall Iron Mines, naming them after his father's birthplace in Cornwall, England. The mines proved to be the largest iron deposits east of Lake Superior and were once the largest open-pit iron mines in the world. Mining continued for 234 years of uninterrupted production, closing in 1973.

1738. Zinc Process Patented. William Champion patents the zinc distillation process.

1739. New Jersey Zinc Mine. The Sterling Hill Mine in northwestern New Jersey was rich in zinc in seams that were worked to a depth of 777 m (2550 ft) below the surface in tunnels totaling more than 56 km (35 miles). Ten million metric tons (eleven million U.S. tons) of zinc were mined, along with quantities of manganese and iron.

1740. Crucible Steel Made by Huntsman. Benjamin Huntsman, an English clockmaker, reinvents the long-lost art of making crucible steel, a process discovered in India around 500 B.C. Huntsman mixed broken pieces of blister steel and slag in a closed pot that was fired for three hours. High quality, but very expensive, steel was made by this process into the

20th century, when it largely was superseded by the electric furnace process.

1742. The Franklin Stove. Benjamin Franklin invented, but did not patent, what became known as the Franklin Stove. He believed that everyone should have free use of inventions. His stove, which was made of cast iron plates bolted together, generated twice the heat of an open fireplace using one-third the fuel and was far safer than an open fireplace. His and other stove designs created a substantial business for cast iron foundries.

1742. Galvanizing. The first record of galvanizing is described by P.J. Malouin, a French chemist, who coats iron with molten zinc. The term *galvanizing* is coined later, after Italian physician and physicist Luigi Galvani.

1748. Platinum. Metal No. 12. One of the precious metals, platinum is identified by Spanish mathematician Antonio de Ulloa and named *platina*, meaning little silver.

1750. Iron Act. The British enacted a rule intended to stem the development of industries in the American colonies that would compete with British mills. Steel furnaces, slitting mills, and hard wire manufacture were prohibited, but pig iron and bar iron could be exported to England—now with no duty. The rule was never strictly enforced.

1751. Nickel. Metal No. 13. Discovered by Baron Axel Friedrik Cronstedt, Swedish chemist and mineralogist.

1753. Bismuth. Metal No. 14. Identified by Claude Geoffroy, French chemist. It had often been confused with tin.

1755. Magnesium. Metal No. 15. Discovered by Joseph Black, Scottish chemist and physicist. It was first isolated by Sir Humphry Davy in England in 1808.

1770. Hopewell Furnace. Colonel Marcus (Mark) Bird set up an iron furnace in southeastern Pennsylvania approximately 56 km (35 miles) from Philadelphia. A water wheel provided the power to blow air into the furnace to make pig iron that was remelted and cast in molds. One of the earliest applications was stove plates that could be assembled with nuts and bolts to create Hopewell Stoves. During the Revolutionary War, shells and 115 cast iron cannons were made. Hopewell was one of the "iron plantations" where entire families worked and lived. Trees in the forest were cut, from which charcoal was made. Iron ore was hauled 32 km (20 miles) from the Cornwall Mines, and limestone for flux came from nearby quarries. Hopewell was closed in 1883, when it was no longer possible to compete with furnaces using anthracite coal instead of charcoal.

1770. Britannia Metal. Britannia metal, named for the early name of the English island, was a type of pewter noted for its silvery color first produced in Sheffield, England with a composition of approximately 93% Sn, 5% Sb, and 2% Cu. The 3.6 kg (8 lb) hollow Oscar statuettes handed out as Hollywood's Academy Awards are said to be gold-plated Britannia metal.

1772. Phenomenon of Combustion. Antoine Laurent Lavoisier, a French chemist, reveals that the phenomenon of combustion is an oxidation process.

1772. Electrophysiology Experiment. Luigi Galvani (1737–1798), an Italian surgeon, discovers the electrochemical process that occurs between two dissimilar metals in the well-known experiment with a frog. His name was used for the zinc coating process of galvanizing.

1774. Manganese. Metal No. 16. Identified by German-Swedish chemist Carl Wilhelm Scheele and isolated the same year by Swedish chemist Johan Gottlieb Gahn.

1778. Humphry Davy (1778–1829). Davy was an outstanding English scientist who made discoveries in the fields of metallurgy and chemistry. He is perhaps best known for inventing the Davy safety mining lamp. At age 23, he worked at the Royal Institution, where he was assistant lecturer in chemistry, director of the chemical laboratory, and assistant editor of the journals of the Institution. Davy was a pioneer in the field of electrolysis. Using a voltaic cell to electrolyze molten salts, he discovered the alkali metals, sodium and potassium, and also discovered magnesium, boron, and calcium, making him a major discoverer of the metals.

1778. A Chain across the Hudson. During the American Revolutionary War, a great iron chain was placed across the Hudson River at West Point, N.Y., which could be raised to prevent British naval vessels from sailing up the river. A 549 m (1800 ft) wrought iron chain with 61 cm (24 in.) links was constructed at the Sterling Iron Works in Warwick, N.Y. The chain was secured on both banks of the river and could be raised and lowered with pulleys. It is said that the British never attempted to pass the chain.

1779. The Ironbridge. The first metal bridge is built. It is a 30 m (100 ft) span across the River Severn Gorge, near the village of Ironbridge, in Shropshire, England. The bridge was designed by Thomas Pritchard and built by Abraham Darby III using cast iron beams and girders, from the foundry nearby, that form a graceful arch over the river. Because cast iron is brittle, the arch is used so that the major stress is in compression. The bridge is one of the first accomplishments of the Industrial Revolution and is the first of over 50 cast iron bridges that would be built in England over the next 75 years.

In July 1779 the arches of the first bridge ever to be built of iron spanned the River Severn at Coalbrookdale; today that bridge is perhaps the most important industrial monument in Great Britain, and the centerpiece of the world-famous Ironbridge Gorge Museum. The building of the Iron Bridge was important in two respects. First, in the early eighteenth century the Shropshire coalfield and the industries associated with it—especially iron founding—were the cradle of the Industrial Revolution, and the river its economic lifeline; yet there was no bridge across the Severn Gorge, and its ferries were incapable of carrying vehicles. Secondly—and more importantly—the manufacture and application of iron had been developed further in Shropshire than in any other area as a result of Abraham Darby's successful experiment in smelting iron with coke in 1709. Thus in the authors' words, the bridge was "a magnificent declaration of the structural potential of cast iron."

The bridge not only changed the pattern of communication in the area, it attracted thousands of sightseers, from overseas as well as in Britain, for no structure of its time so caught public imagination, and in the decade following its construction "it epitomized not just the progress of engineering science, but a whole optimistic view of the future of industrial society . . ."

(N. Cossons and B. Trinder,
*The Iron Bridge: Symbol of the
Industrial Revolution*)

1780. Pewter in America. Following the American Revolution, the pewter trade flourishes for a hundred years. Pewter pieces are made by casting into bronze or brass molds.

1781. Molybdenum. Metal No. 17. Molybdenum, the first refractory metal, is first isolated by Peter Jakob Hjelm, Swedish chemist. Carl Wilhelm Scheele identified the metal in 1778.

1781. Brass. Zinc is added to liquid copper to make brass, in contrast with the calamine diffusion process in use for hundreds of years.

1781. Tungsten Discovered. Metal No. 18. Swedish chemist Karl Wilhelm Scheele first identified tungsten in a calcium tungstate mineral that later became known as scheelite. The word *tungsten* is an adaptation of the Swedish *tung sten*, meaning heavy stone.

Tungsten also became known in Europe as wolfram, which is a pejorative term based on the German *wolf* plus *ram*, meaning dirt. It is explained that tin miners gave the material this name because it was considered a worthless material. W, therefore, became the chemical symbol for wolfram.

The reason for the more common use of the name *tungsten* is explained in the 1844 entry on columbium, another contentious name.

Tungsten is the second refractory metal. Wolfram was first identified as a mineral of tungsten by brothers Juan and Fausto Elhuyar, Spanish chemists and mineralogists, who are also credited with the first production of metallic tungsten from wolframite in that same year.

1782. First Shot Tower. William Watts builds a tower in Redcliffe, Bristol, England, for making shot by a process he has patented. Lead is melted at the top of a high tower and poured through a copper sieve. The liquid lead solidifies as it falls and by surface tension forms spherical balls that are caught at the floor of the tower in a basin full of water.

1782. Tellurium. Metal No. 19. Discovered by Franz Müller von Reichenstein, Hungarian-Romanian mineralogist and mining inspector.

1783. Hydrogen Reduction Process. A revolutionary step takes place in smelting when hydrogen is used to produce metallic tungsten by the gaseous reduction of its oxide.

1784. Wrought Iron. Englishman Henry Cort refines pig iron in a puddling furnace, a new and more efficient way to produce wrought iron.

1786. British Naval Vessels Coppered. The Admiralty orders naval ships to have their underwater hulls sheathed with copper to prevent the attack of borers and the accumulation of marine growth. Despite its high cost, coppering allowed vessels to stay at sea longer without the need for cleanings. This method of protection was used until the development of modern anti-fouling paint.

1787. Strontium. Metal No. 20. Discovered in 1787 by Adair Crawford, Scottish physician; identified in 1791 by Thomas Charles Hope, Scottish physician and chemist, at Edinburgh; and isolated in 1808 by Sir Humphry Davy, British chemist and inventor.

1789. Uranium. Metal No. 21. Discovered by Martin Heinrich Klaproth, German chemist and mineralogist. The element is isolated in 1841 by Eugène Péligot, French chemist.

1789. Zirconium. Metal No. 22. The first reactive metal is discovered by Martin Heinrich Klaproth, a German chemist and mineralogist.

1790. Phoenix Iron Works. The Phoenix Iron Company is founded at a place that was named for the company: Phoenixville, Pennsylvania. In 1812, the company was purchased by Robert Waln, a New Jersey industrialist, and the company began making pig iron and wrought iron. The company grew to include a huge blast furnace and a puddling furnace for making wrought iron. A foundry and other buildings were sent up, including a row of houses for the workers. In 1855, the company produced 7.62 cm (3 in.) Griffen Ordnance rifles with wrought iron barrels that were said to be more reliable than the cast iron gun tubes made by competitors. The "Phoenix Column," developed during the Civil War, was a hollow cylinder composed of four, six, or eight wrought iron segments that were riveted together to make strong columns for building construction. The company also made wrought iron railroad rails and wrought iron bridge sections that could be shipped and assembled at the bridge site. The Eiffel Tower was built with Phoenix wrought iron sections.

1791. Titanium. Metal No. 23. The second reactive metal is discovered by William Gregor, English vicar, in ilmenite (iron titanate) and identified in 1795 by Martin Heinrich Klaproth.

1793. Lukens Steel Company. Lukens, in Coatesville, Pennsylvania, is the oldest continuously-running steel mill in the United States, in operation now for 218 years. The mill began as The Federal Slitting Mill in 1793 and became the Brandywine Iron Works and Nail Factory in 1810. In 1817, the mill was leased to Dr. Charles Lloyd Lukens. In the following year Lukens made the first boiler plate in the United States. It was made of wrought iron. In 1818, they made the hull for the first iron-hulled vessel in the United State, the *Codorus*. Dr. Lukens died in 1925, leaving the mill to his wife, Rebecca. The inheritance made Rebecca Lukens the first woman in the United States to have an important role in the iron industry. She served as the manager of the mill for 22 years, saving it from bankruptcy by selling boiler plates to England for their steam locomotives.

During 1881, Lukens started making steel and iron. Although a small mill, they excelled in metallurgical skill and innovation, becoming known for making the widest steel plates. In 1903, their steam-driven mill produced plates 3.45 m (136 in.) wide, and by 1919 their mill could produce 5.23 m (206 in.) plates, a record that would stand for 40 years. In 1930, they were the pioneers in producing clad steel plates—metallurgically bonding two plates together. They delivered plates for the hull of the *Nautilus*, the first atomic-powered submarine, in 1955. Production of raw steel reached a peak of 870,000 metric tons (958,000 U.S. tons) in 1974.

Their largest order was a $74 million contract to produce plates for the largest warships: two Nimitz-class nuclear aircraft carriers in 1988.

In 1997, Bethlehem, the second largest steel company in the United States, bought Lukens for $400 million. Bethlehem Steel went bankrupt and was acquired by the International Steel Group (ISG) for $1.5 billion in 2003. In 2004, Mittal Steel entered the picture and paid $4.5 billion for ISG. That was followed in 2006 by a merger between Mittal and Arcelor to create a steel company three times larger than any other.

1794. Yttrium. Metal No. 24 Discovered by Johan Gadolin, Finnish chemist and mineralogist. It is one of the rare earth metals.

1797. Extrusion Process Patented. Joseph Bramah, prolific English inventor, patents a process for hot extruding lead pipe.

1797. Beryllium. Metal No. 25. Discovered by Nicolas-Louis Vauquelin, French mineralogist. It is isolated by Friedrich Wöhler, German chemist, in 1828.

1797. Chromium. Metal No. 26. Discovered by Nicolas-Louis Vauquelin, French mineralogist.

1800. Paul Revere Sets up Rolling Mill. Paul Revere (1735–1818), the famous American rider of 1775, became a master silversmith. About the year 1800, he set up a rolling mill for copper sheet and produced copper sheathing for many ships, including the *Constitution*, and the Massachusetts statehouse. Revere also set up a brass casting foundry.

1801. Vanadium. Metal No. 27. Discovered by Andrés Manuel del Rio y Fernandez, Spanish-born Mexican scientist and naturalist (Mexican College of Mines), vanadium was initially named panchromiun and later erythronium. It was rediscovered about 1830 and named vanadium by Nils Gabriel Sefström, Swedish chemist.

1801. Columbium (Niobium). Metal No. 28. Discovered by Charles Hatchett, English chemist, in a mineral sent from America called columbite. Columbite had been discovered about 1734 by John Winthrop the Younger, the first governor of Connecticut. About 1844, Heinrich Rose in Germany discovered what he thought was a new metal and called it niobium, because it was associated with a tantalum ore, and Niobe was the daughter of Tantalus in Greek mythology.

Americans continued to call the metal columbium and Europeans preferred niobium. The dispute was settled officially in 1949 when the International Union of Pure and Applied Chemistry (IUPAC) ruled that niobium is the official name.

1801. First American Copper Mill. At age 65, Paul Revere sets up a copper rolling mill 24 km (15 miles) from Boston at Canton, Mass.

1801. First Suspension Bridge with Wrought Iron Chains. James Finley (1756–1828) of Fayette County, Pennsylvania., designed and built the world's first suspension bridge supported by wrought iron chains. The Jacob's Creek Bridge, connecting Uniontown and Greensburg, Pennsylvania., had a span of 21.34 m (70 ft) and was 3.81 m (12.5 ft) wide. The bridge stood for 32 years before being demolished.

1802. Tantalum. Metal No. 29. Discovered by Anders Gustav Ekeberg, Swedish chemist.

1803. Cerium. Metal No. 30. Discovered by Jöns Jakob Berzelius and Wilhelm Hisinger, Swedish chemist and Swedish geologist, respectively, at Uppsala University.

1803. Iridium. Metal No. 31. Discovered by Smithson Tennant, English chemist at Cambridge.

1803. Osmium. Metal No. 32. Discovered by Smithson Tennant, English chemist at Cambridge.

1803. Palladium. Metal No. 33. Discovered by William Hyde Wollaston. It is in the platinum group of metals.

1803. Rhodium. Metal No. 34. Discovered by William Hyde Wollaston. It is in the platinum group of metals.

1804. Meetinghouse with Tin Roof. The Arch Street Meetinghouse in Philadelphia is built with a roof of tin-plated shingles.

1804. Henry Nock Becomes Royal Gun Maker. Henry Nock, who had set up a gun maker's shop in London in 1772, becomes royal gun maker to King George III of England.

1805. Belgian Zinc. Jean-Jacques Daniel Dony (1759–1819) set up the first Belgian zinc plant. It used a horizontal retort for the distillation of calamine to make zinc. This was the predecessor of the Societé de la Vieille Monatagne, which became the largest zinc producing company in the world.

1807. Fulton's Steamboat. Robert Fulton's (1765–1815) 45.72 m (150 ft) steamboat, *Clermont*, starts providing passenger service on the Hudson River from New York City to Albany. The *Clermont* was the first commercially successful steamboat in America.

1807. Potassium and Sodium. Metals No. 35 and 36. Discovered by Humphry Davy, British chemist and inventor.

1808. Barium. Metal No. 37. Discovered by Humphry Davy, British chemist and inventor.

1808. Iron Chain Suspension Bridge. James Finley (1756–1828) designed an iron chain suspension bridge that was built over The Falls of the Schuylkill, 8 km (5 miles) upriver from Philadelphia. Part of the superstructure broke in September 1810 when a drove of cattle was crossing. In January 1816 the bridge collapsed from the weight of snow.

1808. Boron. Metal No. 38. Discovered by Louis-Josef Gay Lussac and Louis-Jacques Thénard, French chemists.

1808. Calcium. Metal No. 39. Discovered by Humphry Davy, British chemist and inventor.

1808. Ruthenium. Metal No. 40. Discovered by Jedrzej Andrei Śniadecki, Polish chemist, in 1808, and isolated by Karl Klaus, Russian chemist, in 1844.

1809. Iron Nail Machine. Iron nails were made by machine for the first time by the French Creek Nail Works, which was established near Philadelphia by Benjamin Longstreth. The company was later named the Phoenix Iron Works and the surroundings became the village of Phoenixville.

1810. First Tin Can. Peter Durand receives the first British patent for the tin can, which apparently is the time when the inexact name originated. Of course, it should have been called the "tinned" can because the can was made of steel. It was many years before a convenient way of opening tin cans was discovered.

1810. Patent Chain Bridge. James Finley, an American engineer, publishes "A Description of the Patent Chain Bridge,"

The Port Folio, vol. III, and eventually builds some 40 small and crude suspension bridges.

1810. Merrimack River Chain Bridge. The fifth of James Finley's wrought iron chain bridges is built over the Merrimack River in Newburyport, Massachusetts. With a span of 74 m (244 ft) it may have been the longest chain bridge in America. The Newburyport bridge collapsed in 1827 after 27 years of service, under the weight of a team of oxen and horses.

1814. First Steam Locomotive. George Stevenson builds the world's first steam locomotive in England.

1815. Thorium. Metal No. 41. Discovered by Jöns Jakob Berzelius, Swedish chemist.

1816. The Spider Bridge. The Chain Bridge at the Falls of the Schuylkill had just collapsed early in the year and the nearest bridge, The Collossus, was several miles downstream toward Philadelphia. Josiah White and Erskine Hazard, who operated a wire mill nearby, erected an unusual temporary footbridge across the river using three iron wires from which the floorboards of the bridge were suspended. The total length of the bridge was 121.9 m (400 ft) without support. The floor boards were 60.96 cm (2 ft) long, 7.62 cm (3 in.) wide, and 2.54 cm (1 in.) thick. Eight iron wires on either side served as guide rails. The toll charge was one cent per person until the cost of the bridge, $125, was collected; after that, it was free. A British visitor, Captain Joshua Rowley Watson, saw great potential for military use, and made sketches, calling it "The Spider Bridge at the Falls of the Schuylkill." The bridge

apparently lasted at least until a wooden covered bridge was built nearby.

1817. Cadmium. Metal No. 42. Discovered simultaneously by Friedrich Strohmeyer and Karl Samuel Leberecht Hermann, German chemists.

1817. Lithium. Metal No. 43. Discovered by Johan August Arfvedson, Swedish chemist at Uppsala. Lithium belongs to the alkali metal group of chemical elements.

1817. Selenium. Metal No. 44. Discovered by Jöns Jakob Berzelius. It is classed as one of the metalloids.

1818. First Iron Steamboat in America. The 18.29 m (60 ft) long steamboat *Codorus* was built with a solid wrought iron hull resulting in a vessel lighter than if it had been constructed of wood. It was important that the vessel should stand high in the water because it was to cruise the Susquehanna, a very shallow river during much of the year. The pieces of iron were made and formed at Brandywine Iron Works and Nail Factory in Coatesville, Pennsylvania., and then assembled upside down with rivets at a shop near the river. The hull was moved to the river on two wagons lashed together. The vessel was the attraction of the season.

1819. First Alloy Steel. Michael Faraday, English scientist, noticed that meteorites always contained iron with small amounts of nickel. After cutting and polishing a meteorite, he found that the polished surface did not tarnish. This led Faraday to attempt to make an alloy of iron and nickel that would not rust. This was the first alloy steel known to have been made.

1820. Wrought Iron Rails Invented. John Birkinshaw at the Bedlington Ironworks in Blyth Dene, Northumberland,

England, invented wrought iron rails that triggered the railway age. By 1822, Bedlington had delivered 1219 metric tons (1200 long tons) of malleable wrought iron rails.

1824. Silicon. Metal No. 45. Discovered by Jöns Jakob Berzelius, Swedish chemist. It is classed as one of the metalloids.

1824. Wilkinson's Gun Shop. Henry Wilkinson in England takes over his father's gun shop and begins developing stronger bayonet blade production techniques.

1825. Aluminum. Metal No. 46. First produced by Hans Christian Øersted, Danish physicist, in Copenhagen.

1826. First Large Chain Metal Suspension Bridge. Thomas Telford (1757–1834), a self-taught Scottish engineer, designs and builds a revolutionary chain suspension bridge over the Menai Strait between the Welsh mainland and the island of Anglesey. The bridge was built between two limestone towers 41 m (153 ft) tall, with a span of 176 m (577 ft). The bridge hung from 16 huge wrought iron chains, from which 444 wrought iron rods were suspended to support the roadway. The deck of the bridge is 30 m (100 ft) above the water to permit the passage of ships. This construction represents the beginning of the use of wrought iron for bridges instead of cast iron, which limits spans to 45.72 m (150 ft) in length and is subject to brittle failure. Three hundred men completed the bridge in only four years using only manpower and horses. The bridge was repaired in 1892, and between 1938 and 1942 the bridge underwent substantial renovation and the old wrought iron chains were replaced with chains of steel. The bridge is still in service.

1829. Corrugated Iron is Patented. Henry Robinson Palmer, architect and engineer for the London Dock Company, patents corrugated iron to provide stiffening of sheets, allowing a greater span over a lighter framework, as well as reducing installation time and labor.

1829. Faraday Discovers Sacrificial Corrosion. Michael Faraday, the noted English scientist, discovers the sacrificial action of zinc during an experiment between zinc, salt water, and iron nails. When the two metals are joined, the zinc becomes a sacrificial anode and dissolves before the iron is attacked.

1830. "Tom Thumb" Locomotive. The first American-built steam locomotive used on a common carrier is designed and built by Peter Cooper. He wanted to convince the newly organized Baltimore & Ohio Railroad that they should use steam engines rather than relying on horse-drawn cars. The Tom Thumb was a success and settled the question.

1830. England's Last Cast Iron Bridge. The last of 66 English cast iron bridges is erected, by the Coalbrookdale Works, in Staffordshire. It is the 42.67 m (140 ft) high bridge in Mavesyn Ridware. The first, and the most famous, cast iron bridge was the 30.48 m (100 ft) Ironbridge over the River Severn at Coalbrookdale in 1779. That bridge is still standing. Cast iron for bridges gave way to the more ductile wrought iron and then to steel.

1831. Baldwin Locomotive Works. Mathias W. Baldwin, Philadelphia jeweler and silversmith, establishes the Locomotive Works after creating a miniature locomotive for an exhibit. His company becomes the world's largest builder of steam locomotives. At one time there were over 18,000 employees. The production during their 125 years in business was approximately 75,000 locomotives.

1831. Hot Dip Galvanizing. The first metallic corrosion-resisting zinc coating process is developed.

1832. Muntz Metal. George Frederick Muntz, an English businessman, invents a 60% Cu-40% Zn alloy that is a low-cost, hot workable brass. It is to be used for many products and will replace copper as a sheathing material for wooden ship hulls.

1832. Baldwin's *Old Ironsides*. Matthias Baldwin put his first steam engine into service on the Philadelphia, Germantown and Norristown Railroad, the line used for horse-drawn cars. It was a four-wheel engine, weighing a little over 4.54 metric tons (5 U.S. tons). The wheels had heavy cast iron hubs with wooden spokes and rims and wrought iron tires. The engine remained in service for 20 years.

1834. First American Brass Mill. The Wolcottville Brass Company is set up in what is now Torrington, Connecticut, hiring skilled workers from England. The principal products are kettles and buttons. The company becomes a part of the American Brass Company in 1893.

1834. Metal Spinning Patented. William Porter of Taunton, Massachusetts, receives a patent on a spinning method for thin sheets of Britannia metal. Spinning lowered costs and increased production rates.

1838. Anthracite Iron is Made. William Henry is the first American to succeed in making iron with anthracite coal instead of charcoal. This was done by starting the furnace with hot air, a process that had been developed in Wales.

1839. Lanthanum. Metal No. 47. Discovered by Carl Gustav Mossander, Swedish chemist and mineralogist.

1839. Babbitt Metal for Axle Boxes. Babbitt metal is used for all railroad car axle boxes.

1840s. John Brown Company is Founded. In Sheffield, England, the John Brown Company is founded to manufacture files, a product used extensively in the local cutlery businesses. In the 1860s, the main business becomes the manufacturing of steel railroad rails using the recently invented Bessemer steel process. The company also began making coach springs and eventually was engaged in ship cladding and shipbuilding. In the 1930s the business was general construction.

1840. Commercial Electroplating Begins. John Wright, a surgeon in Birmingham, England, discovered that a solution of silver cyanide and potassium cyanide was excellent for the electrical deposition of silver, or what became known as silver plate. George Richardson Elkington and his cousin, Henry Elkington, were partners in a silver plating business in Birmingham. They purchased and patented Wright's process. Electroplated wares became very popular, and by 1880 the business had grown to 1800 employees at the main factory.

1841. Scranton Iron Works. William Henry, with his son-in-law, Seldon Scranton, and Sanford Grant purchase 503 acres of land that will become the town of Scranton, Pennsylvania. A blast furnace that will make iron with anthracite coal is completed in the autumn of 1841. By the summer of 1844, the furnace is producing 4.54 to 6.35 metric tons (5 to 7 U.S. tons) of pig iron a day. The company takes on the making of T-rails. By 1847 the company listed 800 employees, including many Welsh, Irish, and German immigrants. In 1853, the company name becomes the Lackawanna Iron & Coal Company. It is said that one of every six rails produced in America is made at Scranton. By 1900, they were shipping 272,155 metric tons (300,000 U.S. tons) of steel rails a year from the Iron Works.

1841. A Pin-making Machine Is Invented. A completely automatic pin-making machine is invented by John Howe, an American physician and founder of Howe Manufacturing Company. Pins are especially needed in the wool industry for carding.

1841. Roebling's First Wire Rope. John Augustus Roebling (1806–1869), a German-born American civil engineer, experimented for several years making wrought iron rope in the pasture behind his home in Saxonburg, Pennsylvania, before making a sale. Roebling had worked for the Pennsylvania Railroad Corporation on a canal construction project where seven inclines were built so that canal boats could be hauled on rails up and over seven mountain ranges. The hauling was with 7.62 cm (3 in.) hemp ropes that broke frequently, causing much damage. Roebling conceived of the idea of replacing the hemp with 2.54 cm (1 in.) diameter wire rope. Roebling's rope was flexible, very strong, and weighed no more than the hemp rope, assuring its acceptance as a new engineering material.

1842. Thomas Firth & Sons. Firth sets up a business with his two sons, Mark and Thomas. Firth had been head melter at Sanderson Brothers in Sheffield. Within 10 years the company moved to a larger site in Sheffield next to the John Brown Steel Works. Firth then had crucible furnaces and a file-making shop to make files for the cutlers. In time they had the largest rolling mill in Sheffield. In the 1860s they went into the armaments business making large guns. Firth merged with John Brown in 1930, forming Firth Brown.

1842. Roebling Wire Rope Patent. John Roebling receives a patent on his iron wire rope.

1843. Erbium. Metal No. 48. Discovered by Carl Gustaf Mossander, Swedish chemist and mineralogist.

1843. Terbium. Metal No. 49. Discovered by Carl Gustaf Mossander, Swedish chemist and mineralogist.

1843. SS *Great Britain*. The first Atlantic liner to be built of wrought iron and having a screw propeller is launched. At 88 m (322 ft) it is the world's largest ship. Isambard Kingdom Brunel built the ship at Bristol, England. The propellers were supplemented by sails on six masts.

The ship made a record crossing from Bristol to New York in 14 days. After being damaged in 1884, the ship was scuttled, but it has since been raised and is on display in Bristol.

1844. Iron Rails in America. Beginning in 1844, heavy wrought iron rails were produced by the Mt. Savage Rolling Mill at Frostburg, Maryland. The rails are 5.49 m (18 ft) long and weigh 18.14 kg (40 lb) per lineal yard.

1846. Pennsylvania Railroad. The Pennsylvania Railroad was organized and chartered to build a rail line from Harrisburg to Pittsburgh, an undertaking that involved the crossing of seven mountain ridges. The company gradually grew, and by 1900 had become the biggest railroad in America, eventually controlling approximately 16,000 km (10,000 miles) of rail lines. It was one of the largest users of metals for applications including railroad rails; locomotives; and passenger, freight, and coal cars. There were eventually 4,345 km (2,700 miles) of electrified lines requiring copper wire and steel for supporting towers and signals. At the peak there were a quarter of a million workers. In the 1920s, the "Pennsy" was operating hourly passenger trains between New York and Washington. In the 1950s, 18-car stainless steel Metroliners were on the Congressional and Senator runs between Boston and Washington. The Pennsylvania Railroad merged with the New York Central Railroad in 1968 to form the Penn Central, an alliance that went bankrupt in 1970.

1848. Niagara Gorge Suspension Bridge. Charles Ellet, Jr., the first civil engineer hired for this project, builds a temporary suspension footbridge as the first part of his plan to span Niagara Falls with a 244 m (800 ft) wrought iron suspension bridge. Following a dispute with the bridge companies, Ellet left the project.

1848. Roebling Wire Company. John Roebling moves from Saxonburg, Pennsylvania, to Trenton, New Jersey, to set up the wire company closer to his iron supplier.

1848. Spiegeleisen. Discovered by Robert Forester Mushet (1811–1891), English metallurgist, in a lump of white crystalline metal brought to him from Rhenish Prussia. It is an iron-manganese carbonate crystalline material speckled with minute spots of uncombined carbon. From its brightness it was called *spiegel glanz* or *spiegeleisen*, meaning looking glass iron. Ten years later, in 1856, Mushet would realize the value of spiegeleisen in perfecting the Bessemer Process.

1849. California Gold Rush. The world's greatest mining event came in 1849 when gold was discovered at Sutter's Mill in California. During the summer of 1849, some 80,000 prospectors arrived in San Francisco.

They came by ship around Cape Horn or across the Isthmus of Panama, and from Peru and Chile. They arrived by covered wagon, and others came from Europe. Gold dust could be found by washing dirt.

1849. First I-Beam. Alphonse Halbou, a Belgian engineer at the company Forge de la Providence, a steel producer based in the Hainaut region of Belgium, patents a method of producing an I-beam from a single piece of steel.

1850s. Electroplating Arrives. With the advent of electroplating, the trend is to shiny silver-plated tableware, surpassing the demand for the dull gray pewter products. The metal base for electroplated ware is usually nickel silver, also called German silver. The alloy actually contains no silver, but resembles it so that places where the thin silver plating may wear away are not too noticeable.

1850. Roebling's Suspension Bridge across the Niagara Gorge. John Roebling, hired to complete work which Charles Ellet, Jr. began several years earlier, becomes involved in an engineering project that startles the world. He erects a 251 m (825 ft.) span that is 61 m (200 ft) above the river using his own wire rope for the cables. The bridge has two levels: an upper level for trains and a road below for pedestrian and vehicular (carriage) traffic. It is the world's first working railway suspension bridge. When the bridge is finally finished, Roebling dares to move a fully-loaded freight train across it. Roebling's name skyrocketed to fame. The bridge was in service for 42 years before it was replaced to withstand the weight of heavier locomotives.

1850. First American Zinc. The first production of zinc in the United States is started in 1850 using the process developed in Belgium in 1805. American production of zinc became the largest in the world.

1850. First Nickel Silver Coins. Switzerland is the first country to adopt German silver for coinage. The alloy, which contains no silver, contains copper, nickel, and zinc.

1851. Krupp's Cannon. At the Great Exposition of 1851 in Hyde Park in London, England, Krupp's first cannon was exhibited.

1852. New Jersey Zinc. The company is founded to mine zinc deposits of Franklin and Sterling Hill in northwestern New Jersey. The ore is particularly rich in zinc and also contains manganese and iron. These are the constituents of a newly discovered zinc mineral on the site, which is named Franklinite for the town, not for Benjamin Franklin. The zinc occurred in veins that went down 0.8 km (0.5 miles).

Fifty-six kilometers (thirty-five miles) of tunnels were excavated. By the time mining was stopped, 13.6 million kg (30 million lb) of zinc had been removed. At its peak, the company owned and operated smelters across the United States and Canada. In 1966 the zinc company merged with Gulf & Western. In 1991, N.J. Zinc became a subsidiary of Horsehead Industries, which filed for bankruptcy in 2002 because it was saddled with environmental cleanup liabilities.

1855. First Transatlantic Telegraph Cable. The first undersea telegraph cable was laid between Ireland and Newfoundland. The laying of this cable on the floor of the Atlantic decreased the time needed to communicate between North America and Europe from 10 days—the time it took a ship to deliver a message—to a few minutes.

1856. Bessemer Process. Henry Bessemer (1813–1898), English inventor, invents the Bessemer converter wherein oxygen is blown into molten pig iron, reducing carbon to make steel in quantity by an inexpensive process in approximately 20 minutes.

1856. Open Hearth Steel Furnace. William Siemens (1834–1883), a German-born Englishman, patents the open hearth reverberatory furnace, which makes large heats of steel of higher quality than the Bessemer steel.

1856. Spiegeleisen and the Bessemer Process. Robert Mushet, English metallurgist, discovers that spiegeleisen, an Fe-C-Mg alloy, when added to the Bessemer converter, deoxidizes steel, eliminates blow holes, and ties up sulfur. Mushet's discovery turned the Bessemer Process into a success.

1857. Bethlehem Steel. Bethlehem Steel had its beginnings when the Saucona Iron Company was founded by Augustus Wolle in South Bethlehem, Pennsylvania. Construction was delayed, however, until 1861, at which time the name was changed to the Bethlehem Iron Company. The first blast furnace was in operation in that same year. By 1863, a rolling mill was built and the first railroad rails were being made.

During World Wars I and II, Bethlehem was a major supplier of armor plate and ordnance to the U.S. armed forces, including large-caliber guns for the Navy. Bethlehem became one of the world's major shipbuilders. After producing the first wide-flange structural shapes, which were largely responsible for the construction of skyscrapers, Bethlehem became a leading supplier for the construction industry. In the latter part of the 20th century, Bethlehem could not compete with foreign steel and went bankrupt.

1857. Galvanized U.S. Mint Roof. One of the first corrugated, galvanized iron roofs in the southern United States is installed on the U.S. Mint in New Orleans, Louisiana.

1857. Steel Rails Invented in England. Robert Forester Mushet (1811–1891), British metallurgist, is the first to produce steel railroad rails to replace much less durable cast iron rails.

1858. First Bessemer Commercial Production. Goran Fredrik Goransson, the founder of Sandvik in Sandviken, Sweden, is credited with being the first to succeed in making steel by the Bessemer Process on an industrial scale.

1859. Spectroscope Invented. German chemist Robert Bunsen and German physicist Gustav Kirchhoff invent the spectro-scope, a new device for use in determining the chemical composition of metals.

1859. First Ironclad Battleship. *La Gloire*, a French battleship, is sheathed with 12 cm (4.7 in.) of iron over 43 cm (17 in.) of timber.

1859. Cold Rolling of Iron Invented. Bernard Lauth of Pittsburgh invents and patents a method for cold rolling iron (wrought iron).

1860. Caesium. Metal No. 50. Discovered in Heidelberg by German chemist Robert Bunsen and German physicist Gustav Kirchhoff.

1860. Wood's Metal. A patent is filed by a Dr. Barnabas Wood of Nashville, Tennessee, covering "Metallic Composition for Fusible Alloy and other Purposes." This seems likely to be for the low-melting alloy known as "Wood's metal," which is 50% Bi, 25% Pb, 12.5% Cd, and 12.5% Sn. The melting range is 70 to 72 °C (158 to 166 °F).

1860. The First Cupro-nickel Coins. Belgium introduces a coin with 75% Cu and 25% Ni.

1860. HMS *Warrior* is Launched. The British Royal Navy launched the *Warrior*, which, at 127 m (418 ft), was almost twice as large as France's *La Gloire*. It was a steam-powered vessel with a screw propeller in addition to being a full-rigged ship. Her armament consisted of 26 muzzle-loading 68-pounder guns and 10 breech-loading 17.78 cm (7 in.) guns. It was the first armor-plated, iron-hulled warship. Iron armor plates 11.43 cm (4.5 in.) thick, with tongue-and-groove joints and backed with 43.18 cm (17 in.) of teak, were bolted to the iron hull. The ship's complement consisted of 705 offices and men. The great ship never engaged in combat.

1861. Rubidium. Metal No. 51. Discovered in Heidelberg by German chemist Robert Bunsen and German physicist Gustav Kirchhoff.

1861. Thallium. Metal No. 52. Discovered by William Crookes, English chemist, in London.

1861. Mushet Invents the Dozzle. Robert Forester Mushet (1811–1891) was aware that when steel is poured into an ingot mold, uneven cooling caused a central cavity or "pipe," requiring a large portion to be cut off. Mushet invented the "dozzle," a clay cone that was heated white hot and inserted into the top of the ingot mold during the end of the pour, and then filled with molten steel to maintain a reservoir of molten steel to fill the pipe of the ingot as it cooled. Dozzles are now called hot tops or feeder heads.

1863. Metallography Pioneered. Dr. Henry Clifton Sorby (1826–1908) in England pioneers the field of "microscopic metallurgy" and is the first to prepare pictures of metals at high magnifications that show their crystalline structures. He named seven crystal structures that he observed in iron and steel. The development of metallography was most important to stainless steel technology.

1863. Indium. Metal No. 53. Discovered by Ferdinand Reich and Hieronymus Theodor Richter, German chemists.

1864. St. Joseph Lead. A large plot of land is purchased in southeastern Missouri to set up the St. Joseph Lead Company. By 1892 a lead smelter is in operation in the town of Herculaneum.

1865. Wrought Iron Hull Plating. The tug *Margaret* was plated with iron in 1865 and was found in good condition 71 years later in 1936. (*Wrought Iron* by James Aston, 1939.)

1865. First Bessemer Steel Rails Produced in America. The production of steel rails is started in America to replace iron rails. The rails were rolled at the North Chicago Rolling Mill.

1866. First American Steel Producer. The Pennsylvania Steel Company at Steelton, Pennsylvania, 5 km (3 miles) south of Harrisburg, is the first in the country to produce steel exclusively. It later became a division of Bethlehem Steel.

1867. Electric Dynamo. German inventor Dr. Werner Siemens and English physicist Sir Charles Wheatstone invent the first practical electric dynamo, independently and simultaneously, with both having thier papers to the Royal Society read on January 17, 1867. The machine will provide a future source of electricity for new metal refining methods, the electric arc furnace, and the production of aluminum.

1867. Midvale Steel Co. The company began as the William Butcher Steel Works at Wissahickon Avenue and Roberts Street in the Nicetown Section of Philadelphia. It would make cast iron locomotive wheels for the Baldwin Locomotive Works, which was just a mile away in the Spring Garden section of the city. In 1872 William Sellers took over, changing the name to the Midvale Steel Works. Although never a large company, Midvale became an expert in making heavy artillery, coastal, and field guns. It also produced steam turbines, naval armor plate, and pressure vessels. In 1956, Heppenstal Steel Company of Pittsburgh merged with Midvale to become the Midvale-Heppenstal Steel Company. The company closed in 1976.

1867. First American Steel Rails Rolled on Order. The Cambria Iron Works, rolling mill technology industry leader in Johnstown, Pennsylvnia, becomes the first in America to roll steel railroad rails on order, in the way of regular business.

1867. Barbed Wire. The first patent in the United States for barbed wire is obtained by Lucien Smith of Kent, Ohio.

1867. Handy & Harman. A company is founded in New York City by Peter Hayden to trade in precious metals. By the end of the 19th century the company, which had become Handy & Harman, was the largest U.S. silver trading firm and a supplier of silver bullion to silversmiths and jewelers. Today the company is a diversified manufacturer producing alloy and stainless steel wire and cable, small diameter stainless steel tubing, and carbon steel refrigeration tubing. A Precious Metals Division is engaged in precision plating and surface finishing; electronics applications; and sterling silver, silver alloyed wire, strip, and brazing alloy fabrications.

1868. R. Mushet's Special Steel (RMS). Robert Forester Mushet (1811–1891), English metallurgist, invents the first true tool steel, an iron-tungsten alloy. It was an "air-hardening steel," also known as a "self-hardening" steel. The alloy could cut at higher speeds and cut harder metals than previously possible. It was a forerunner of high-speed steels.

1869. British Iron & Steel Institute. The British Iron & Steel Institute is formed with 292 members.

1869. Periodic Table of the Elements. Proposed independently by Lothar Meyer and Dimitri Ivanovich Mendeleev.

1870. Peter Cooper Awarded Bessemer Gold Medal. Peter Cooper (1791–1883) at the Canton Iron Works in Baltimore, Maryland, received the Gold Medal for rolling the first iron for fireproof buildings.

1871. AIME. The American Institute of Mining Engineers is founded by 22 mining engineers in Wilkes-Barre, Pennsylvania. The organization is ultimately enlarged to become the American Institute of Mining, Metallurgical and Petroleum Engineers, maintaining the AIME logo.

1871. Gallium. Metal No. 54. Predicted by Dmitry Ivanovich Mendeleev, Russian chemist, in 1871; discovered by Paul Émile Lecoq de Boisbaudran, French chemist, in 1875.

1871. Scandium. Metal No. 55. Predicted by Dmitry Ivanovich Mendeleev, Russian chemist, in 1871; discovered by Lars Fredrik Nilson, Swedish chemist, in 1879.

1871. Germanium. Metal No. 56. Predicted by Dmitry Ivanovich Mendeleev, Russian chemist, in 1871; discovered by Clemens Winkler, German chemist, in 1886 at Frieburg University. It is a metalloid in the carbon group.

1871. Protactinium. Metal No. 57. Predicted by Dmitry Ivanovich Mendeleev, Russian chemist, in 1871. Discovered by Kasimir Fajans and Otto Göhring in Karlsruhe, Germany, in 1913; by Frederic Soddy, John Cranston, and Andrew Fleck in Glasgow, Scotland, in 1918; and by Otto Hahn and Lise Meitner at the Kaiser-Wilhelm Institute, Berlin, Germany, in 1918.

1871. Thyssen & Company. August Thyssen founds Thyssen & Company in Styrum (today Mülheim Styrum), Germany.

1874. Mississippi's First Bridge. Steamboats on the Mississippi had made St. Louis merchants prosperous by handling shipments to the West. By 1860, however, that trade had ended for the most part because of competition in Chicago that was being served by 11 railroads. St. Louis badly needed a bridge across the Mississippi. James Buchanan Eads (1820–1887) suddenly arrived on the scene with a plan for a bridge. Eads had never built a bridge, nor was he an engineer. He had been a clerk on a steamboat and then became wealthy salvaging dozens of sunken ships. There were no other proposals, and his plan was accepted. It would be the first major construction of steel: a bridge that featured a double-deck design and three tubular steel arch spans of 153 m (502 ft), 158.5 m (520 ft), and 153 m (502 ft)—464.5 m (1524 ft) in all for the spans, which set an engineering precedent. The bridge was completed in 1874. It was a fine bridge and one still standing, now in the shadow of Eero Saarinen's Gateway Arch.

1875. Carnegie Builds His First Steel Mill. Andrew Carnegie, the Scottish-born industrialist, builds the Edgar Thomson Steel Works—which he named for the president of the Pennsylvania Railroad, his former employer—in Braddock, Pennsylvania.

1875. Ludlum Steel. The Pompton Steel & Iron Co. is formed by James Ludlum in Pompton Lakes, New Jersey. Upon his death in 1892, his son renamed the company Ludlum Steel & Spring Company. Ludlum and Allegheny merged in 1938 to form Allegheny Ludlum, the largest company in the United States specializing in stainless, electrical, tool, and other alloy steels and carbides.

1875. Nickel Ore in New Caledonia. Wide-scale nickel mining begins in 1875, following the discovery of nickel in New Caledonia, Melanesia in 1864 by engineer Jules Garnier. Garnier discovered nickel-oxide ores, which became the most important source of nickel for 30 years, until nickel-sulfide ores were discovered in the Canadian Sudbury Mines.

1876. Basic Bessemer Process. Welshman Sidney Gilchrist Thomas adds limestone (a base) to the Bessemer converter, creating the basic Bessemer process, making it possible to produce low-phosphorus steel without the need for low-phosphorus iron ore.

1876. Mushet Receives Bessemer Medal. Robert Forester Mushet (1811–1891), British metallurgist, receives the Bessemer Gold Medal from the Iron & Steel Institute for his numerous achievements. He perfected Bessemer's process, turning it into a great success; he invented the first commercially-produced alloy steel; and he invented steel railroad rails and the hot-topping of ingot molds.

1876. Monster Steam Engine Exhibited. At the United States Centennial Exposition in Philadelphia, a two-story-high Corliss steam engine from Hartford, Connecticut, is on display while, at the same time, providing power to approximately 100 machines also on display in Machinery Hall. The single-cylinder Corliss is the largest stationery steam engine. It is the star of the exposition, which is opened with the starting of the engine by the president of the United States and the emperor of Brazil. The Exposition runs for six months and is attended by over 10 million visitors.

1876. First Sandvik Steel Sales in America. Sandvik Steel agents book their first sales in America at the United States Centennial Exposition in Philadelphia. Attendance totaled well over 10 million during the six-month exposition.

1877. Chromium Steels. J.B. Boussingalt and Almé Brustlein, working at Aciéries Holtzer in Unieux, France, develop chromium steels for the first time in Europe. They recognize that their mechanical properties are dependent on their chromium and carbon contents.

1878. Chromiferous Spiegeleisen. The Terre Noire Company in France produces a "chromiferous spiegeleisen," an iron alloy with a brilliant crystalline fracture (hence the German name for brilliant iron) with 25% Cr and 13% Mn.

1878. Holmium. Metal No. 58. Discovered by Per Teodor Cleve (1840–1905), Swedish chemist, in Uppsala. J.L. Soret and M. Delafontaine, Swiss chemists, previously observed holmium's absorption spectrum.

1878. Ytterbium. Metal No. 59. Discovered by Jean Charles Galissard de Marignac, Swiss chemist. It is one of the rare earth metals.

1879. Samarium. Metal No. 60. Discovered by Paul Émile Lecoq de Boisbaudran, French chemist. It is a rare earth metal.

1879. Thulium. Metal No. 61. Discovered by Per Teodor Cleve (1840–1905), Swedish chemist, in Uppsala. It is a rare earth metal.

1880. First American Steel Mill. The Pennsylvania Steel Company is founded in Baldwin, Pennsylvania, 5 km (3 miles) west of Harrisburg. Steelmaking is conducted with a Bessemer converter that can convert several tons of molten pig to steel in 20 minutes. The first jobs are given to English and Irish immigrants who have had experience with the Bessemer process. The first products are steel railroad rails that last five times as long as wrought iron rails, which had been the only rails available. When the town is incorporated its name is changed to Steelton. Bethlehem Steel eventually buys Pennsylvania Steel and it becomes their railroad products division.

1880. ASME. The American Society for Mechanical Engineers is established. The main function will be to write codes and standards for mechanical devices.

1880. Gadolinium. Metal No. 62. Discovered by Jean Charles Galissard de Marignac, Swiss chemist, in 1880; isolated by Paul Émile (François) Lecoq de Boisbaudran, French chemist, in 1886. It is one of the rare earth metals.

1882. Wrought Iron Bridge Company. This bridge company of Canton, Ohio, is established to produce wrought iron truss bridge components that can be shipped to and erected on a site. The bridges are made in spans of 6.1 to 91.4 m (20 to 300 ft). Over a period of 18 years, approximately 4300 spans are built.

1882. Hadfield's Manganese Steel. Sir Robert A. Hadfield in Sheffield, England, invents a ferrous alloy containing 11–13% Mn, which he called a manganese steel. It was not an alloy that could be hardened by heat treatment but was extremely hardened by cold working. It found uses in mining equipment, power shovels, rail crossings and switches, and other places subject to heavy wear. Unknown to Hadfield, he had created an austenitic alloy that when work hardened became very wear resistant. The alloy is also known as mangalloy.

1882. Cable-Motor Cars in Philadelphia. The Union Passenger Railway Company adopts the cable-motor system for the propulsion of cars on Columbia Avenue in Philadelphia.

1883. Brooklyn Bridge. One of the oldest suspension bridges in the United States is completed. John Augustus Roebling (1806–1869), the German-born civil engineer who designed the Brooklyn Bridge, had already built half a dozen bridges including one across the Niagara Gorge. Roebling conceived of a great bridge that would span the East River, dividing Manhattan and the Borough of Brooklyn. He had developed the process of making wire for suspension bridge cables and had set up a factory in Trenton, New Jersey, to manufacture the wire. Cold drawn wire, and later steel, would make it possible to make the longest suspension bridges because of the great strength of the wire. Roebling worked out every detail for the bridge and for the construction of the two colossal stone towers that would support the cables. In 1869, Roebling's plan for the bridge was accepted. While Roebling was inspecting the site where one of the towers would be erected, his foot was crushed as a ferry entered the slip. He developed tetanus and within three weeks was dead. Roebling's son Washington, also a civil engineer, at the age of 32 undertook the building of the bridge, assisted eventually by his wife, Emily, when he became ill with caisson disease. The bridge, with a span of 486 m (1595 ft), was the world's largest by far. The bridge's opening day in 1883 was the grandest day in the history of New York. Emily Roebling was honored by being the first to cross the bridge, with a band following. She rode in a carriage with a red rooster, a symbol of victory, in her lap. The Brooklyn Bridge has become one of the world's most famous bridges.

1884. Washington Monument. The 169 m (555 ft) tall monument is completed with a cap made of cast aluminum in the shape of a pyramid that weighs approximately 2.72 kg (6 lb). It was decided that the aluminum would resist the weather and serve as part of a lightning rod system. The cap is almost 23 cm (9 in.) tall and is 12.95 cm (5.1 in.) square at the base. At that time, aluminum was a rare metal and cost approximately $1 an ounce. William Frishmuth in Philadelphia made the casting in his foundry for $225.

1885. Praseodymium. Metal No. 63. Discovered by Carl Auer von Freiherr von Welsbach, Austrian scientist and inventor. It is a rare earth metal.

1885. Neodymium. Metal No. 64. Discovered by Carl Auer Freiherr von Welsbach, Austrian scientist and inventor. It is one of the rare earth metals.

1885. First Skyscraper. The 10-story Home Insurance Building in Chicago is credited with being the world's first skyscraper. The 55 m (180 ft) building, which in 1890 had two floors added for a total of 12, was designed by architect William LeBaron Jenney. Unlike traditional buildings of the day, with self-supporting walls of masonry, the skyscraper was supported by an iron or steel skeleton of columns that permitted the weight of the floors to be distributed evenly to the columns, not the walls. Skyscrapers also had to have elevators.

1885. First Nickel Steel Armor. It is believed that the first nickel steel armor plate was first commercially produced at the Montataire Works in France, where Henri Marbeau, French metallurgist, and Le Chèsne, chemist and inventor, were collaborating on steelmaking.

1886. Hall-Héroult Process. A process that has come to be known as the Hall-Héroult process is discovered independently and nearly simultaneously by Charles Martin Hall (1863–1914), American chemist, in Ohio, and Paul Louis-Toussaint Héroult, French scientist, in Paris. The Hall-Héroult process is an electrolytic method of producing aluminum inexpensively, bringing the metal into wide commercial use.

1886. First Open-Hearth Installed by Carnegie. Andrew Carnegie installs what may have been the first open-hearth furnace in America at the Homestead Mill, opening the way to the production of structural steel beams as well as armor plate for the U.S. Navy.

1886. Jones & Laughlin Installs Bessemer Converters. The Jones & Laughlin Steel Company, located 6.5 km (4 miles) from Pittsburgh, installs two Bessemer converters in order to enter the steel business.

1886. Dysprosium. Metal No. 65. Discovered by Paul Émile Lecoq de Boisbaudran, French chemist, in 1886; isolated in 1906; and obtained in pure form by Frank H. Spedding and colleagues in 1950.

1888. Alcoa. Charles Martin Hall, who invented the electrolytic process for making aluminum only two years earlier, sets up the Pittsburgh Reduction Company to start producing aluminum commercially. In 1907, the name was changed to the Aluminum Company of America (Alcoa). The company, after relocation to New Kensington, Pennsylvania, became the largest producer of aluminum in the United States, and today is the world's third largest producer. Alcoa has plants in nine countries and, in 2010, had 59,000 employees worldwide. Applications include aircraft, autos, commercial transportation, packaging, building and construction, oil and gas, defense, and other industrial products.

1889. Eiffel Tower. The Eiffel Tower is completed on the Champ de Mars in Paris to serve as an entrance arch to the 1889 World's Fair. The tower was designed and built by Gustav Eiffel, who had won the design competition. It is the world's tallest tower at 324 m (1062 ft) and is built of over 18,000 pieces of wrought iron riveted together. The tower has become the greatest paid tourist attraction and a widely recognized symbol of Paris and France.

1889. Carpenter Steel Company. James Carpenter starts a company in Reading, Pennsylvania., to produce an air-hardening tool steel that he has patented. Carpenter went on to make special steels for the burgeoning automotive industry and also entered the stainless steel business in 1918. Carpenter remains one of the prominent companies making specialty steels.

1889. Nickel-Steel Tests. Tests at various steelworks in Great Britain revealed that steels containing 3% Ni developed remarkable projectile-breaking qualities. James Riley, manager of the steelworks of C. Tennant Sons and Co. Ltd., presented a paper on nickel steels to the Iron & Steel Institute of Great Britain. Although the actual nickel content does not seem to have been given, he reported that by hardening and tempering, breaking strengths up to nearly 96 tons per sq. in. (215,040 psi) were obtained.

1889. Wilkinson Sword Is Incorporated. The company began as a gun shop in 1772 and was run by James Wilkinson and his son, Henry, until 1856.

1890. Wilkinson Sword Makes Razors. The production of straight-edge razors is started.

1890. First Large-Span Steel Bridge. The railway bridge over the Firth of Forth in Scotland was designed by Sir Benjamin Barker (1840–1907) and Sir John Fowler (1817–1898), English civil engineers. It was the first large-span bridge to be completely built of the just-legalized steel, and the longest cantilever bridge ever built, with a total span of 2529 m (8296 ft). Uppermost in both men's minds, of course, was the collapse of a bridge over the Firth of Tay built 11 years earlier, killing an estimated 70 people a year after it was constructed. The new bridge would have open-truss construction and be supported 104 m (342 ft) in the air on giant tubes 36.6 m (12 ft) in diameter.

1892. Roberts-Austen. Sir William Chandler Roberts-Austen, English metallurgist, publishes the first known textbook on metallurgy, which ran to six editions. He did not invent, but greatly developed, the use of photomicrography in the study of metal crystal structures. An associate named the high-temperature phase of steel (austenite) in his honor.

1892. Carnegie Steel Is Formed. The Carnegie Steel Company is organized followed by the building of a 13-story Carnegie Building in downtown Pittsburgh.

1892. Hadfield Studies Chromium Alloys. Sir Robert A. Hadfield, English metallurgist, studies 1–9% Cr alloys with 1–2% C in 50% sulfuric acid solutions and concludes that chromium is deleterious. Hadfield is probably best known as the inventor of Hadfield's Manganese Steel, a steel with 11% Mn that is fully austenitic but transforms to martensite when cold worked and becomes highly abrasion resistant. (Note: Robert A. Hadfield is not to be confused with Dr. William H. Hatfield, the English metallurgist who, in 1924, became the inventor of 18-8 stainless steels.)

1892. Case-hardened Armor Plate. American steelmaker Hayward Augustus Harvey invents a process for nickel steel that hardens the face by packing plates in carbon and heating for up to 14 days. Similar work was carried out in Great Britain, with firing tests at Portsmouth showing that "Harveyized" plates were much superior. The hard surface broke the projectiles.

1892. First Trolley Cars in Philadelphia. The first electric trolley cars are put into operation in the city, replacing the cable motor cars.

1893. American Brass Company (1893–1969). The first large brass company in America is organized at Wolcottville (now Torrington), Connecticut, with a combination of Wolcottville Brass and the Ansonia Brass and Battery Company. It was the largest brass company in America for most of its existence and a holding company for six brass manufacturers: Plume & Atwood, Waterbury Brass, Scoville Manufacturing, Holms, Booths & Haydens, and Coe Brass.

1893. John Fritz Receives Bessemer Gold Medal. John Fritz (1822–1913) started his career as a blacksmith in Bethlehem, Pennsylvania. He became an innovator in the iron industry, developing a new technique for rolling iron railroad rails that was used throughout the United States and England. In 1860, he was hired as superintendent and chief engineer for the Bethlehem Iron Company, Bethlehem, Pennsylvania. He served as president of the American Institute of Mining Engineers in (AIME) in 1893. He built the well-equipped Fritz Engineering Laboratory at Lehigh University.

1894. Haynes Apperson Automobiles. Elwood Haynes and brothers Edgar and Elmer Apperson form a company in Kokomo, Ind., to build one of the first gasoline-powered cars in America. In 1902 the Appersons form their own company, the Apperson Automobile Company, and in 1905 the Haynes Apperson Company name was changed to the Haynes Automobile Company, which Haynes operated until the company went bankrupt in 1924.

1895. Discovery of Thermite Reaction. Hans Goldschmidt (1861–1923), German scientist, develops a low-carbon ferrochromium alloy, permitting the possibility of producing a high-chromium, low-carbon stainless steel alloy at a substantially reduced cost.

1895. Austenite Discovered. Floris Osmond (1849–1912), French metallurgist, discovers austenite and names it for William Chandler Roberts-Austen, the famous English metallographer.

1896. First Iron-Carbon Diagram. William Chandler Roberts-Austen (1843–1902), the English metallographer, produces the first iron-carbon constitutional diagram. The diagram was constructed by plotting temperatures as the ordinate (y axis) and the percent carbon in the steel as the abscissa (x axis). The diagram showed the range in composition and temperature within which the phase changes are stable and also the boundaries of the phases, identified by the Greek lowercase letters α through δ. The diagram became the key to the heat treatment of steel.

1896. First Safety Razor. Wilkinson Sword introduces the Pall Mall, the first safety razor.

1898. ASTM. The American Society for Testing and Materials is organized in Philadelphia to set up voluntary standard specifications and methods of testing. The first office is in a building at the University of Pennsylvania in Philadelphia. The first committee is Committee A-1 on Steel, which initially concentrates on writing specifications for steel railroad products and methods of testing.

1898. Polonium. Metal No. 66. Discovered by Marie Skłodowska-Curie, Polish-born French physicist and chemist, and her husband, Pierre Curie, French physicist.

1898. Radium. Metal No. 67. Discovered by Marie Skłodowska-Curie, Polish-born French physicist and chemist, and her husband, Pierre Curie, French physicist.

1898. International Silver Company. The company is an amalgamation, in Meriden, Connecticut, of 14 silver companies, primarily in Connecticut, but also the Standard Silver Company of Toronto. The parent company produces sterling silver hollowware and tableware, as well as electroplated hollowware and tableware, while the individual companies continue to maintain their separate identities. International Silver is still in business.

1899. Armco. The American Rolling Mill Company is established at Middletown, Ohio. The name was later changed to the Armco Company and then to A-K Steel Company in 1994.

1899. Actinium. Metal No. 68. Discovered by André-Louis Debierne, French chemist.

1899. First Commercial Electric Arc Furnace. Paul Héroult (1863–1914), one of the developers of the electrolytic process for producing aluminum, places into production the first successful commercial direct arc furnace. A year later, the first carload of metal was shipped from Héroult's plant in La Praz to the firm of Schneider and Company at Creusot, France.

1900. Open-hearth Furnaces. By 1900, most Bessemer converters are replaced by open-hearth furnaces.

1900. Brass Industry Workers. By 1900, there are approximately 10,000 workers in the brass industry in the United States, about half of whom work for the American Brass Company.

1900. Brinell Hardness Test. The first widely accepted indentation hardness test for metals is invented by Johan August Brinell (1849–1925), Swedish mechanical engineer.

1900. Sherardizing. Discovered by Sherard Cowper-Coles, this is a diffusion process in which steel components are heated in the presence of zinc dust at temperatures between 320 and 500 °C (608 and 932 °F), followed by phosphating.

1901. United States Steel Corporation. U.S. Steel was organized in one of the largest business enterprises ever created. A group headed by Elbert Gary and J.P. Morgan bought Andrew Carnegie's Carnegie Steel Company and combined it with the holdings of the Federal Steel Company. These companies became the nucleus of U.S. Steel, which also included American Steel & Wire, National Tube, American Tin Plate, American Steel Hoop, and American Sheet Steel. In its first full year of operation, U.S. Steel made 67% of all steel produced in the United States. In 2007, U.S. Steel purchased the Texas-based welded tubular products maker, Lone Star Technologies, making U.S. Steel the largest tubular goods producer in North America. Also in 2007, U.S. Steel increased its flat rolled products capacity by acquiring Canada's Stelco, Inc., which was renamed U.S. Steel Canada. U.S. Steel is now the fifth largest steelmaker, with a total capacity of 28.8 million net metric tons (31.7 million net U.S. tons).

1901. Europium. Metal No. 69. Discovered by Eugène Anatole Demarçay, French chemist. It is a member of the lanthanide series.

1902. ASTM Committee B-2 on Nonferrous Metals and Alloys. Committee B-2 writes standard specifications and test methods, concentrating in the beginning on materials for the railroad industry, including lead, tin, and zinc. Starting in the 1950s, standards for nickel alloys become the major work of the committee.

1902. INCO. The International Nickel Company, based in New York City, came into being as a result of a merger of the Canadian Copper Company and the Orford Copper Company of Bayonne, New Jersey. Canadian Copper had been mining nickel at the Sudbury Mines in Ontario and shipping it to Orford for refining. At the time, the largest use of nickel was for nickel steel armor plate for ships. The company was able to control a majority of the market and to eliminate competition. After World War I, INCO concentrated on finding new peace time uses for nickel and became, in effect, the research department for the entire nickel industry. Stainless steel, the industry that ultimately would use vast quantities of nickel, was not yet in the picture. INCO had high hopes for MONEL metal and made their first major investment in a mill for producing that metal at a new plant in Huntington, West Virginia. In 1928, INCO merged with Mond Nickel Company, a British firm that owned half of the best nickel deposits in Sudbury, Ontario. INCO became one of the top producers of nickel, operating mines in Sudbury and elsewhere in Canada as well as in the United Kingdom and Indonesia.

When stainless steel entered the picture, INCO began setting up regional offices staffed with metallurgists who would provide free advice to anyone with questions about stainless steel, and a corrosion testing site was set up at Kure Beach to collect data on the resistance of stainless steel to a marine environment.

Today the company, now Vale Inco, produces approximately 25% of the world's nickel and also produces precious metals, cobalt, copper, and specialty nickel products.

1902. Firth Brown Steels. In Sheffield, England, neighboring companies John Brown & Company and Thomas Firth & Sons exchange shares. In 1908, they build the Brown Firth Research Laboratories, replacing their individual laboratories. Harry Brearley is appointed the first director, and in 1912 the Laboratories discover high-chromium stainless steel.

1902. Engelhard Precious Metals. In Newark, New Jersey, Charles W. Engelhard, Sr. founds his company, which becomes the world's largest refiner and fabricator of the precious metals gold, silver, and platinum. The company is credited with producing the first commercial catalytic converter. In 2006, German chemical manufacturer BASF purchases Engelhard for $5 billion.

1902. Flatiron Building. New York's first skyscraper becomes the most recognizable building in the city: the Flatiron Building. The intended name of the building on a triangular lot at the intersection of Fifth Avenue and Broadway was the Fuller Building, but everyone began calling it the Flatiron because its shape resembles that of a cast-iron clothes iron. Daniel H. Burnham (1846–1912), American architect and urban planner, designed the building, which was 30.48 m (100 ft) taller than the Home Insurance Building in Chicago and had 22 stories. The Flatiron is New York's oldest skyscraper.

1903. ASTM Committee A-2 on Wrought Iron. Committee A-2 is established to develop specifications for wrought iron, at first particularly for railroad applications.

1903. Izod Impact Test. Edwin Gilbert Izod (1876–1946), English engineer, invents the Izod test for determining the impact strength of metals using a notched specimen.

1903. Duralumin. Duralumin is a trade name for one of the earliest types of age-hardenable aluminum alloys. The alloy, which contains 4% Cu, was developed by the German metallurgist Alfred Wilm at Dürener Metallwerke Aktien Gesellschaft. Wilm discovered that, after quenching, the alloy gradually hardened when left at room temperature for several days. The modern equivalent is AA2024, which contains 4.4% Cu, 1.5% Mg, 0.6% Mn, and 93.5% Al.

1904. Hadfield Awarded Bessemer Gold Medal. Robert Abbott Hadfield (1859–1940) is awarded the Bessemer Gold Medal for his outstanding metallurgical achievements, including the invention of an austenitic manganese steel and the invention of Era steel for armor plating.

1905. ASTM Committee A-3 on Cast Iron. Committee A-3 is organized to develop specifications for all types of cast iron.

1905. Nichrome. Albert Leroy Marsh (1877–1944), American metallurgist, discovers an alloy consisting of approximately 80% Ni and 20% Cr, which gets the name "Nichrome." It proves to be ideal as a heating element for toasters.

1905. The Charpy Impact Test. The Charpy impact test, also known as the Charpy V-notch test, was developed by French scientist Georges Charpy. The standard specimen size is 10 × 25 × 55 mm (0.39 × 0.98 × 2.17 in.). A V-notch is carefully machined on one edge. The energy absorbed when struck by a swinging pendulum is inferred by comparing the difference in the height of the pendulum before and after breaking the specimen.

1906. Léon Guillet Analyzes Iron-Nickel-Chrome Alloys. Guillet (1873–1946), professor of metallurgy and metal processing at the Conservatoire National des Arts et Métiers in Paris, undertakes a study of Fe-Cr and Fe-Cr-Ni alloys. He is the first to discover the mechanical and physical properties of the ferritic, martensitic, and austenitic stainless steels. He wrote papers on his findings but never applied for patents or caused any of the alloys to be commercially produced.

1906. MONEL Metal. David H. Brown, metallurgist at the International Nickel Company (INCO), creates an alloy of approximately 65% Ni and 35%Cu, which are the same proportions that occur in the ore at the Sudbury Mines in Ontario. Brown names the alloy MONEL for his company president, Ambrose Monell, with one L omitted because at that time the patent office did not accept family names. There are actually five grades of the alloy. All of them are highly corrosion resistant and strong at elevated temperatures.

1906. ASTM Committee A-5 on Metallic-Coated Iron and Steel Products. Committee A-5 is organized to develop specifications, initially for terne-coated and galvanized products.

1906. Tungsten Filament Patented. The General Electric Company patents a method of making filaments, and in 1911 uses ductile tungsten wire for making the first practical incandescent bulb.

1906. Baldwin's Best Year. Baldwin Locomotive Works, the company that had built "Old Ironsides" in 1832, reached its best year, turning out 2,666 locomotives with a workforce of 18,499. The company, which occupied eight city blocks, was Philadelphia's largest business.

1907. "Heat-resisting" Stainless Steel. Frederick Mark Becket (1875–1942), a metallurgist at Electro Metallurgical Company, Niagara Falls, New York, discovered a ferrous alloy with 25–27% Cr that was extremely resistant to oxidation in high-temperature furnaces. This was the first of the high-chromium alloys that became known as heat-resisting stainless steels.

1907. Lutetium. Metal No. 70. Discovered independently by Carl Auer Freiherr von Welsbach, Austrian chemist, and Georges Urbain, French chemist, at the Sorbonne. It is the last element in the lanthanide series.

1907. Stellite Patent. The first patent for Stellite is applied for by Elwood Haynes of Kokomo, Indiana, and granted the same year. The chromium-cobalt alloy is said to be ". . .a novel metal designed for use in the manufacture of cutlery knives."

1907. World Zinc Output. The world production of zinc reaches 668,595 metric tons (737,000 U.S. tons). The United States produced 31%, Germany 28%, Belgium 21%, United Kingdom 8%, and all others 12%.

1908. AISI Is Organized. The American Iron & Steel Institute (AISI) is organized in New York City to expand opportunities for collaborative research into manufacturing technologies and product development.

1908. Model T Fords. In Dearborn, Mich., Ford Motor Company's Model T Fords start rolling off the first automobile assembly lines. The cars cost $275 and were generally thought of as the first affordable automobile. Fifteen million were built before the model was discontinued in 1928.

1908. Corrosion Resistance is Discovered. Philipp Monnartz, German metallurgist, who studied the role of carbon content in the corrosion resistance of high-chromium iron alloys, discovered a precipitous drop in corrosion rate when the chromium content reached approximately 12%. He also discovered the dependence of corrosion resistance on oxidizing rather than reducing conditions.

1909. Sandvik's First Foreign Operation. Sandvik Steel of Sanviken, Sweden, begins the production of springs for watches in Switzerland.

1910. Haynes Licenses Stellite. Elwood Haynes (1857–1925), American metallurgist, licenses production of Stellite, his trademarked high cobalt-chromium alloy, to Deloro Smelting and Refining Company of Deloro Ontario, Canada.

1910. First Electric Arc Furnace Installed in United Kingdom. Edgar Allen & Company of Sheffield installs a Hérault arc furnace that replaces and vastly improves upon the crucible steel process.

1911. Dantsizen Discovers Stainless Iron. Christian Dantsizen, working at the General Electric Research Laboratories in Schenectady, New York, develops what appears to be the first stainless iron. It is a ferritic, non hardenable alloy containing 14–15% Cr and 0.07–0.15% C.

The alloy is used for lead-in wires for incandescent bulbs.

1911. Cathedral of St. John the Divine Opens. The grand cathedral in New York City is constructed with wrought iron piping services throughout. In the center of the Cathedral is the high altar, behind which is a wrought iron enclosure containing the tomb of the New York bishop who originally conceived of the cathedral.

1912. Austenitic Stainless Steel Is Patented. Eduard Maurer (1886–1969) and Benno Strauss (1873–1944), German metallurgists at the Krupp Steel Works, Essen, Germany, discovered and produced commercially the first austenitic Fe-Cr-Ni stainless steel. The alloy, which consisted of 20% Cr and 7% Ni, was the forerunner of the most popular stainless alloy, the modern 18% Cr and 8% Ni (type 304) stainless steel.

1912. A Stainless Steel Age Begins. The invention of stainless steel at Krupp in 1912 and other stainless classes to follow introduce a series of alloys where iron is changed into metals that do not rust as if by some alchemical magic. In 1949 Carl Zapffe, an American metallurgical consultant, called the chromium-nickel alloy "the miracle metal." The material has become the third most widely used of all metals, after steel and aluminum, with a production of 27 million metric tons (30 million U.S. tons) in 2010.

1912. Steel Automobile Body Invented. Edward Gowan Budd (1870–1949), owner of the E.G. Budd Manufacturing Company in Philadelphia, invented the steel automobile body. Budd became a major supplier of automobile bodies to companies in Detroit, England, France, and Germany. Budd became the largest manufacturer of stainless steel for transportation. Beginning in 1931, during the depths of the Great Depression when auto sales had plummeted, he built a small, three-seat seaplane of stainless steel as a promotional stunt and flew it across the English Channel and the Alps. This was followed a year later by a lightweight stainless steel self-propelled railcar equipped with rubber-tired wheels for Michelin et Cie in France. That experiment, in turn, led to an order for a stainless steel three-car train from the Chicago, Burlington & Quincy Railroad. That train was equipped with steel wheels and a 600-horsepower diesel engine in the leading car. It was streamlined and weighed just one-third that of a regular steel railcar. This weight reduction was possible largely because Budd had discovered how to fabricate the car body using cold rolled stainless steel that was three times as strong as structural steel using an electric resistance spot welding method that did not reduce the strength of the steel or impair its corrosion resistance adjacent to the weld.

The three-car stainless steel train, called the *Burlington Zephyr*, broke all records on a nonstop run from Denver to Chicago. The fuel cost was only $17.00, approximately 11% that of the fuel cost for the steam train it replaced. Orders for stainless steel trains began to pour in, and the Budd Company produced almost 11,000 stainless steel railcars and subway cars during the next 50 years.

During World War II, the Budd Company received orders for 600 stainless steel cargo planes from the U.S. Army and Navy. The company was also engaged in the manufacture of stainless steel trailer truck bodies for many years.

1912. *Metallography and Heat Treatment of Iron and Steel.* Albert Sauveur (1863–1939), a Belgian-born metallurgist, develops the science of metallography and physical metallurgy, founding the first metallographic laboratory in a university. In 1912 he publishes *Metallography and Heat Treatment of Iron and Steel*, a volume containing dozens of photomicrographs of metal crystalline structures. He was a professor of mining and metallurgy at Harvard.

1912. Dofasco. Clifton and Frank Sherman start the Dominion Foundries and Steel Company (Dofasco) in the city of Hamilton, Ontario, near Toronto. Dofasco became one of the major steel companies in North America. It also owned the Adams and Sherman Iron Mines in Northeastern Ontario until 1990, when the mines were closed. In 2006, Dofasco was purchased by Arcelor, the world's second-largest steel company by volume. Arcelor then was taken over by Mittal, the world's largest steel producer.

1913. Harry Brearley Discovers Stainless Steel in England. Brearley (1871–1948), a self-taught metallurgist at the Brown Firth Laboratories, Sheffield, England, discovers a ferrous alloy with approximately 12% Cr and 0.35% C that was rust resistant and hardenable by heat treatment. It was the same martensitic cutlery alloy that Elwood Haynes discovered at the same time in America. The alloy eventually created a large business in the cutlery trade and aircraft engine exhaust valves. Brearley became one of the greatest promoters of stainless steel. He formed the Firth-Brearley Stainless Steel Syndicate for the purpose of promoting the use of stainless steel and arranged that his name should be imprinted on every stainless steel knife blade made in Sheffield. He organized the American Stainless Steel Company, a patent-holding firm in Pittsburgh. For his work with stainless steel in 1920, he was awarded the Bessemer Gold Medal, the highest metallurgical honor.

1913. Elwood Haynes Discovers Stainless Steel in America. Elwood Haynes (1857–1925), founder of the Haynes Stellite Company, Kokomo, Indiana, experimented with five iron-chromium alloys that he found to be resistant to corrosion. He applied for a patent that at first was denied because the U.S. Patent Office already had patents for chromium steels. Also, because the patent claims were similar to those of Harry Brearley, a patent was not granted to Haynes until April 1, 1919.

1913. Woolworth Building. The second skyscraper in New York, a building at 233 Broadway, is completed. It is 57 stories tall and, at 241 m (792 ft), is well over twice as tall as New York's first skyscraper, the Flatiron, built in 1902. Frank Winfield Woolworth was the merchandising magnate of F.W. Woolworth Company, the five- and ten-cent store company. The building was designed by Cass Gilbert and remained the world's tallest until 1930, when it was eclipsed by the Chrysler Building a few streets away.

1913. First ASTM Wrought Iron Standard. "ASTM A42 on Wrought Iron Plates" is published. It covers plates up to 5.08 cm (2 in.) thick and 60.96 cm (24 in.) wide. The specification requires a minimum tensile strength of 330.95 MPa (48,000 psi) and a minimum yield point of 186.16 MPa (27,000 psi).

1913. Nitriding Is Patented. Adolph Machlet of American Gas Company in Elizabeth, New Jersey, receives a patent on the nitriding of steel.

1914. Stainless Steel Requisitioned by British Munitions Board. The first demonstration of the outstanding properties of chromium stainless steel occurred when the British Munitions Board requisitioned every pound of the new metal that could be made from Thomas Firth & Sons of Sheffield during World War I. The metal, which was marketed as Firth's Aeroplane Steel, was vital for Royal Air Force aircraft engine exhaust valves. (Ironically, Firth at first thought Harry Brearley's steel was of little use and refused to patent it.)

1914. Rockwell Hardness Tester. Co-invented by Hugh M. Rockwell (1890–1957) and Stanley P. Rockwell (1886–1940), Connecticut natives. The tester determines hardness by measuring the depth of penetration of an indenter under a large load compared to the penetration made by a preload. Different hardness scales are available, depending on the relative hardness of a metal, which use different loads and indenters. A dial shows the hardness number. It is the most commonly used tester because of its speed, reliability, robustness, resolution, and small area of indentation.

1915. First Stainless Steel in American Produced by Firth Sterling. The first commercial heat of chromium stainless steel is produced in McKeesport, Pennsylvania, by Firth Sterling, the American subsidiary of Thomas Firth & Sons of Sheffield, England.

1915. MONEL Yacht Scrapped Because of Galvanic Corrosion. A 65.5 m (215 ft) yacht with the first all-MONEL hull disintegrates after six weeks in the water. The hull had been fastened directly to the steel skeleton without any insulation to prevent galvanic corrosion. The boat had to be scrapped.

1916. ASTM Committee on Metallography. ASTM Committee E-4 on Metallography is established.

1916. Railroad Trackage in the United States. Trackage reached 370,902 km (230,468 miles).

1916, 1917. Metals' Atomic Structure Revealed by X-ray Diffraction. The atomic structure of metals is determined by Dutch physicist Peter Debye (1884–1966) and Swiss physicist Paul Scherrerin (1890–1969) in Europe in 1916, and by Albert Hull (1880–1966) in America in 1917.

1917. DIN (Deutsches Institut fur Normung e.V). The German standards organization is established in Berlin.

1918. Big Bertha. One of the four guns that the Krupp Steel Works built in germany to shell Paris in World War I. They were the largest guns ever built, consisting of a 38.1 cm (15 in.) naval gun to which a tube was added to make a barrel that was approximately 33.5 m (110 ft) long and required a support to keep it straight. The guns shelled Paris at a range of approximately 122 km (76 miles), a distance never before achieved. The guns were named for Frau Bertha von Bohlen of the Krupp family. The 367 shells that landed in or near Paris are reported to have killed 256 people.

1919. Cast Iron Molasses Tank Explodes. A tank explodes, spilling 9.5 million liters (2.5 million gallons) of molasses into Commercial Street in Boston's East End. Twenty-one people were killed and almost 150 people were injured in the 4.6 m (15 ft) high torrent. The tank, which was 15.24 m (50 feet high), was owned by the U.S. Industrial Alcohol Company. There was no obvious reason for the break, but the court ruled that there had been insufficient safety inspection, the construction was shoddy, and the tank was over-filled. The company paid nearly $1 million in claims.

1919. Sandvik Subsidiary in the United States. Sandvik Steel of Sandviken, Sweden, sets up its first subsidiary company in the United States.

1920. Brearley Awarded Bessemer Medal. Harry Brearley receives one of the highest awards for metallurgical achievement: the Bessemer Gold Medal is bestowed upon him by the British Iron & Steel Institute for his work on the discovery and commercialization of chromium stainless steel. The award was established and endowed by Sir Henry Bessemer of Sheffield. Harry Brearley received the fourth of the medals awarded beginning in 1874.

1920. Union Carbide Company Acquires Haynes Stellite Company.

1920. Garden Shears Introduced by Wilkinson. Pruning shears and other garden equipment make the company a leading manufacturer of this line of equipment.

1921. Victorinox Cutlery. Karl Elsener in Ibach, Switzerland, changes the name of his company from Victoria, his mother's name, to Victorinox, keeping part of her name and adding "inox" from the French *inoxidable,* meaning stainless. Elsener does this to attract attention to the fact that he is using stainless steel in all of the knife blades produced by his company, including the Swiss Army Knives.

1923. Hafnium. Metal No. 71. Co-discovered by Dirk Coster (1889–1950), Dutch physicist, and György Hevesy (1885–1966), Hungarian radiochemist, at the University of Copenhagen in Denmark.

1923. America's First Continuous Hot Rolling Mill. A continuous hot rolling mill is installed by the American Rolling Mill Company in Ashland, Kentucky. The mill can produce steel strip up to 91.44 cm (36 in.) wide and as thin as 0.17 cm (0.065 in.). This event signaled the demise of hand sheet rolling, except for some alloy sheet, and ushered in the age of modern hot mills that can roll sheet up to 2.44 m (96 in.) wide while operating at speeds of up to 1219 m (4000 ft) per minute.

1924. 18-8 Stainless Steel Is Invented. Dr. William H. Hatfield (1882–1943), chief metallurgist at the Brown Firth Laboratories, invented the 18% Cr-8% Ni austenitic alloy that was a modification of Krupp's original 20% Cr-7% Ni alloy. The Hatfield alloy is the modern type 304 stainless steel, the most commonly used of all stainless alloys. It is often called 18-8. (Hatfield's company called the metal Staybrite.)

1924. Vickers Hardness Test. Robert L. Smith and George E. Sandland at Vickers Ltd. in England develop an indentation hardness test with a diamond pyramid indenter. The test method is in ASTM E384, "Test Method for Knoop and Vickers Hardness Testing."

1925. Rhenium. Metal No. 72. Discovered by Walter Noddack, Ida Tacke-Noddack, and Otto Carl Berg in Berlin. They named the metal after the River Rhine (*Rhein* in German).

1925. Technetium (Masurium). Metal No. 73, and the last of the naturally occurring metals. Discovered by Walter Noddack, Ida Tacke-Noddack, and Otto Carl Berg in Berlin.

1925. Avesta Produces Stainless. Avesta Jernwerks produces the first austenitic stainless steel in Sweden under a Krupp patent.

1925. ASTM Committee B-5 on Copper Is Established.

1925. ASTM Committee B-7 on Light Metals Is Established.

1925. Spelling Change for Aluminium. Although "aluminium" was the accepted spelling in the United States, in 1925 the American Chemical Society decided to change it to "aluminum" in their future publications. It is not known why.

1926. Rustless Steel Process Invented. A.L. Feild of the Rustless Iron and Steel Company of Baltimore, Md., discovers a method of making stainless steel, called the Rustless Process. The process permits stainless steel to be made without the use of the expensive low-carbon ferrochromium. Feild produces the stainless steel either from chromite ore or stainless steel scrap. Rustless Iron and Steel becomes one of the three largest American stainless steel producers. They are bought in 1940 by the American Rolling Mills Company of Middletown, Ohio.

1926. Allegheny Metallurgists Visit Sheffield. The chromium-nickel stainless steel invented at Krupp in 1912 has not yet been produced in America, probably because American metallurgists were preoccupied learning the tricks of Harry Brearley's chromium stainless steel from Sheffield. Now an invitation comes from William Hatfield at Thomas Firth & Sons to learn about the 18-8 modification of the Krupp steel that he has discovered. Within a few years, the 18-8 steel is being made in the United States.

1926. *Stainless Iron and Steel.* John Henry Gill Monypenny, metallurgist at the Brown Firth Laboratories, Sheffield, England, writes the first definitive book on stainless steel in the English language, *Stainless Iron and Steel.* The book is republished in four editions.

1926. "ASTM A109 for Cold Rolled Strip." The first ASTM specification for cold rolled strip steel is published.

1926. Benjamin Franklin Bridge. The first bridge to cross the Delaware River at Philadelphia is constructed. The largest span, at 533 m (1750 ft), was the world's longest. The bridge has seven traffic lanes and two tracks for the high-speed line to Lindenwold, New Jersey.

1927. First 18-8 Steel Produced in America. The first chromium-nickel (18-8) stainless steel is produced in America under license from the Krupp Steel Works in Germany.

1927. World's First High-Frequency Induction Melting Furnace. Edgar Allen & Company installs the first high-frequency induction melting furnace at Sheffield.

1927. Leipzig Fair Exhibits Stainless Steel. The Leipzig Spring Fair in Germany exhibits stainless steel products made of Krupp Nirosta, including turbine blades, beer barrels, tableware, and kitchenware.

1928. Albert Sauveur Achievement Award. ASM International begins bestowing the Albert Sauveur Achievement Award for outstanding achievement in materials science and engineering, to honor Albert Sauveur, the Belgian-born metallurgist who developed the science of metallography.

1929. ASTM Committee on Stainless Steel. Committee B-10 is established to write specifications and test methods for wrought stainless steel products.

1929. "Precipitation-Hardening Stainless Steels" Discovered. William J. Kroll (1889–1973), a metallurgist in Luxembourg, discovers a new class of stainless steels that become known as "precipitation-hardening stainless steels." He later developed the "Kroll Process" for the refining of titanium and zirconium. Precipitation-hardening stainless steel was first produced commercially as Stainless W in 1945 by the Carnegie-Illinois Steel Company.

1930. Chrysler Building. Walter Percy Chrysler (1875–1940) and William Van Alen (1883–1954), owner and architect, respectively, of the Chrysler Building in New York City, were the first to use extensive amounts of stainless steel in architecture. The metal was used as a cladding for the 56.39 m (185 ft) tower, for heroic sculpture on the outside of the building, and for trim and banisters in the lobby. At a height of 318.82 m (1046 ft), the building was briefly the tallest skyscraper in the world. The building has become a metallurgical icon of stainless steel and an architectural icon of skyscrapers.

1930. Aston's Wrought Iron Process. Because of increasing demand for wrought iron, the A.M. Byers Company built a mill in Ambridge, Pa., that would use the new highly-mechanized Aston Process. They were able to produce starting sponge balls of 2722 to 3629 kg (6000 to 8000 lb) in a process that involved making pig iron in a blast furnace and refining it in a Bessemer converter before combining it with molten slag to make the final product. The plant, which was the last to produce wrought iron in North America, closed in 1969 because of competition with other materials.

1930. Duplex Stainless Steel. Avesta, in Sweden, developed two duplex stainless alloys, one of which became known as AISI type 329 alloy. General production of duplex stainless alloys did not begin, however, until the 1970s when the argon-oxygen-decarburization (AOD) process could be used.

1930. First Metal Airplane. The Boeing Company builds "Monomail," the first all-metal single-wing aircraft. It was a mail plane. Welded steel tubing was used for construction of the fuselage.

1931. Electric Resistance Spot Welding Process. Colonel Earl J.W. Ragsdale (1885–1946), Edward G. Budd Manufacturing Company's welding engineer, developed and applied for a patent for the first welding system capable of producing a spot weld in cold rolled 18-8 stainless steel without annealing or impairing the corrosion resistance of the metal adjacent to the weld. It was an electric resistance spot welding process for which Ragsdale applied for a patent in 1931, and was granted one in 1934.

Budd Manufacturing showcased its patented ribbed stainless steel design in its *Zephyr 9900* (also known as the *Pioneer Zephyr*) train. The train featured extensive use of stainless steel and included innovations such as shot welding to join the stainless steel. It was this patent that kept potential competitors from building stainless steel trains.

1931. Empire State Building. The 102-story Empire State Building in New York City is completed, topping the Chrysler Building, completed a year earlier, by 61.87 m (203 ft).

The 381 m (1250 ft) building remained the world's tallest building for 41 years until topped by the ill-fated World Trade Center Buildings.

1931. George Washington Bridge. In 1931, the gifted Swiss-born Othmar Hermann Ammann (1879–1965) completes what Le Corbusier called "the most beautiful bridge in the world." It was the first bridge to span the Hudson River at New York City. When built, it was the world's longest bridge, with a span of 1450 m (4760 ft) and towers rising to a height of 184 m (600 ft). Ammann revolutionized long-span design by eliminating the need for heavy and costly stiffening trusses. With great foresight he planned for future traffic needs by leaving space to add two more lanes to the original six. That need was fulfilled in 15 years with the addition of two lanes, and in 1962, after another 16 years, the bridge had become congested. The time had come to add the second deck below, which Ammann had anticipated. The six new lanes gave the bridge a total of 14 lanes.

1932. A Rubber-Tired Train. In 1931, Chauvette Michelin of the Michelin Tire Company met with Edward G. Budd of the Budd Manufacturing Company, the automobile body manufacturer. Michelin had the idea that passenger trains riding on pneumatic tires would become a wave of the future if only the trains were light enough to be supported by rubber tires. Budd was interested because the Depression had left his company with practically no work, and Michelin was hoping to find a new market for tires. Budd built some lightweight railway cars using thin high-strength stainless steel sheet to minimize the weight. He built self-propelled cars with gasoline engines and sold one to Michelin, and others to three American railroads for testing. The rubber-tired cars never caught on, but the design led the Budd Company into the production of streamlined stainless steel trains with steel wheels.

1934. *Burlington Zephyr's* Record Run. The three-car *Burlington Zephyr*, a streamlined, lightweight, stainless steel train built by the Budd Manufacturing Company in Philadelphia, makes its first scheduled run from Denver to Chicago. The train was purchased by the Chicago, Burlington and Quincy Railroad (CB&Q) in an effort to stimulate passenger travel, which had fallen greatly during the midst of the Great Depression. The body of the car is made of type 301 stainless steel, cold rolled to 1034 MPa (150,000 psi) minimum yield strength. The structure is assembled by the patented Budd shot weld process. This high-strength steel had never been used before the development of this welding method.

The three-car train, which weighs 94.35 metric tons (104 U.S. tons), including 20.87 metric tons (23 U.S. tons) of stainless steel, is about the weight of a single Pullman coach. The train is sleek and shiny and is powered by one of the newly developed diesel-electric Winton engines. The interior features comfortable and stylish modern furnishings as well as air conditioning. The train, with 72 passengers and mail on board, makes the nonstop, dawn-to-dusk, 1632 km (1014 mile) trip in a little over 13 hours, cutting in half the time for the normal steam train run. Fuel consumption for the trip was 1965 liters (519 gallons), and cost just $16.72, with the cost of diesel fuel at four cents per gallon.

In other words, thanks to the streamlining, the light weight, and a remarkably efficient diesel engine, the hundred-ton train got two miles per gallon and cost 1.6 cents per mile.

The *Burlington Zephyr* represented just the first of what would be hundreds of Budd-built stainless steel trains to be built in the coming years for 80 American and some foreign railroads.

Following its record-breaking run to Chicago, the train was exhibited for six months at the Chicago's Century of Progress Fair. Next, the train was taken on a nationwide tour and displayed in 222 cities. The train was in regular service for 26 years, having covered 5.1 million km (3.2 million miles). The train, which has been refurbished, is on permanent display at the Chicago Museum of Industry and Science.

1934. World Metal Index. The foundation of the World Metal Index is laid at the Sheffield Central Library when a copy of every British metal standard was bought, including standards dating back to 1915. Throughout the years, the collection has been expanded to include information on alloy grades worldwide and the establishment of the World Metal Index. An experienced library reference staff is available to provide information on both ferrous and nonferrous materials.

1935. First ASTM Standards for Stainless Steel. Specifications are published for stainless steel flat rolled products including ASTM A167 for Fe-Cr-Ni alloys and A176 for Fe-Cr alloys. The specifications are written by members of ASTM Committee A-10 on Stainless Steel.

1935. ASTM Committee E-3 on Chemical Analysis of Metals Is Established.

1935. First Aluminum Airplane. Douglas Aircraft builds the Douglas DC-3, the first commercially successful airplane, using high-strength aluminum alloy as the major construction material.

1936. RMS *Queen Mary*. The *Queen Mary* enters service, sailing from Southampton to New York. She was one of the largest liners afloat, with a length of 311 m (1019 ft), a beam of 36 m (118 ft), a height of 55 m (181 ft), and a draft of 11.9 m (39 ft). Gross tonnage was 81,237. The capacity was 2,139 passengers and 1,103 crew. The service speed was 53 km/hr (33 mph). The ship was outfitted elegantly, and stainless steel was used widely for the first time on any ship. It was seen in the kitchens, the swimming pools, and stairways.

The *Queen Mary* and her sister ship, the *Queen Elizabeth*, provided a two-ship weekly service between Southampton and New York, with the running time averaging approximately four and a half days.

During World War II, the *Queen Mary* was converted to a troop ship that carried 15,000 soldiers. The vessel did not travel in a convoy because her speed was too fast to be concerned about German submarines. Her orders were to go full speed ahead and never stop for any reason. On one trip the ship almost capsized when struck by a rogue wave. On another occasion the *Queen Mary* struck a light cruiser, slicing it in half and never stopping.

The *Queen Mary* was restored after the war and stayed in service until 1967 when she steamed to Long Beach, Calif., and was permanently moored to serve as a tourist attraction: the ship was turned into a hotel, restaurant, and museum.

1936. Old Wrought Iron Sheathing Inspected. Wrought iron sheathing on the tug *Margaret* is found in good condition after 71 years in service, according to a report published by Aston.

1937. Alloy Casting Research Institute Established. The organization is started in New York City by a major group of producers of corrosion-resisting and heat-resisting castings.

1937. Sir Frank Whittle's Jet Engine. Frank Whittle (1907–1996) invents the first gas turbine engine for jet propulsion.

1937. Golden Gate Bridge. Joseph B. Strauss designs and is chief engineer of what has become a magnet for bridge lovers the world over. With a total length of 2737.4 m (8981 ft), it was the longest suspension bridge until 1964. Steel for the bridge, which spans the opening of

the San Francisco Bay into the Pacific Ocean, came from the Bethlehem Steel Company in Pennsylvania through the Panama Canal. The steel wire for the cables came from John A. Roebling & Sons in New Jersey. On May 27, 1937, the opening day, 200,000 people streamed across the bridge.

1938. Promethium. Claimed to be discovered by H.B. Law, J.D. Pool, Andrei V. Kurbatov, and L.L. Quill at The Ohio State University in 1938, and proved in 1945 by Jacob A. Marinsky, Lawrence E. Glendenin, and Charles D. Coryell at Oak Ridge Laboratories, Tennessee.

1938. ASTM Committee E-7 on Nondestructive Testing Is Established.

1939. Francium. Discovered by Marguerite Perey at Curie Institute, Paris. Its existence was predicted by Dimitri Ivanovich Mendeleev during the 1870s.

1939. Revere Ware. Revere Ware copper-bottomed stainless steel pots and pans are introduced. In 1801, Paul Revere founded the Revere Copper Company, which became the Revere Copper and Brass Company, Inc., the makers of Revere Ware.

1939. Steam Locomotive Is a Wrought Iron Showcase. The ultra-modern S-1 locomotive built for the Pennsylvania Railroad is exhibited at the New York World's Fair of 1939. Wrought iron is used extensively to provide long-lasting trouble-free service, including pipe in all steam, air, water, and sand-delivery lines. The Pennsylvania Railroad collaborated with Baldwin Locomotive Works, the Lima Locomotive Works, and the American Locomotive Company on the experimental design of the S1.

1940. Astatine. Discovered by Dale R. Corson, Kenneth Ross Mackenzie, and Emilio Segré at University of California.

1940. Tacoma Narrows Bridge Failure. The bridge in the state of Washington opens on July 1. The slender suspension bridge, at 853 m (2800 ft), was the third-longest after the George Washington and the Golden Gate. No suspension bridge had been so long and slender. (The width was just 11.89 m, or 39 ft.) From the time the deck was installed, the bridge moved up and down to such an extent that the workers began calling her "Gallopin' Gertie." Even on the opening day the bridge was in motion. Attempted remedial measures failed. It is said that some drivers drove miles just to experience the crossing of the bridge, while others drove miles out of their way to avoid the crossing. Four months and seven days after the bridge opening, a 40 mph wind came up that caused such extreme vibration in the bridge that it literally tore itself apart in a matter of minutes. The few occupants of cars managed to run to safety. The failure of the bridge was attributed to "a lack of understanding of bridge aerodynamics."

A new bridge built 10 years later had a bridge deck width of 19.51 m (64 ft), approximately 50% wider than Gerti's.

1940. Neptunium is discovered by Edward M. McMillan and Philip H. Abelson in Berkeley, California, by bombarding ^{238}U with neutrons.

1940. Plutonium is discovered by Glenn T. Seaborg, Arthur C. Wahl, Joseph W. Kennedy, Michael Cefola, and Edwin M. McMillan at Lawrence Radiation Labs, University of California Berkeley by bombarding ^{238}U with deuterons.

1940. Budd Receives Largest Stainless Order. About 1935, the Edward G. Budd Manufacturing Company in Philadelphia set up a Truck Trailer Division in addition to its Auto Body and Railcar Divisions. An announcement in the April 24, 1940 edition of *Time Magazine* reveals that the Fruehauf Trucking Company of Detroit has become a national distributor for the stainless steel trailer sets, and has entered an initial order of 10,000 Budd trailers. This represents Budd's largest single order for stainless steel equipment: a $9 million order. The trailer sets, made of approximately 1.8 metric tons (2 U.S. tons) of type 301 stainless steel, are shipped disassembled.

1941. ACI Numbers. The Alloy Castings Institute of New York develops a numbering system, now widely used for numbering stainless steel and nickel alloy castings. The system divides the casting alloys into two groups, consisting of a C series for corrosion-resisting alloys that are used at temperatures below 650 °C (1200 °F), and an H series for heat-resistant alloys that are used at temperatures above 650 °C (1200 °F). A second letter, from A to Z, is used to indicate the approximate combined amounts of chromium and nickel. The next two digits indicate the maximum or average carbon content in hundredths. For example, CA15 is for a 12% Cr alloy with 0.15% C. HX50 is the designation for a nickel alloy with total chromium and nickel content of 85% and a carbon content of 0.40–0.60%.

1941. ASTM Committee B-8 on Metallic and Inorganic Coatings Is Established.

1943. Edward Budd Receives ASME Medal. Edward G. Budd, who has been called "the father of the stainless steel streamlined train," receives the highest award of the American Society for Mechanical Engineers, a medal for "outstanding engineering achievements." Budd received the medal just three years before his death at the age of 76.

1944. Curium is discovered by Glenn T. Seaborg, Ralph A. James, and Albert Ghiorso at University of California Berkeley, by bombarding ^{239}Pu with α particles.

1944. Americium is discovered by Glenn T. Seaborg, Leon O. Morgan, Ralph A. James, and Albert Ghiorso at Argonne National Laboratory, Chicago, by bombarding ^{239}Pu with neutrons.

1944. ASTM Committee B-9 on Metal Powders and Metal Powder Products Is Established.

1944. First Jet Aircraft. A Gloster E.28/29 airplane powered by Sir Frank Whittle's jet engine goes into service for the Royal Air Force. Whittle eventually was awarded a £100,000 tax-free gift by the British government for inventing the jet engine and received knighthood.

1946. A 36 Structural Steel Adopted. ASTM publishes A 36, a structural steel specification that has a minimum yield strength of 248.21 MPa (36,000 psi), in comparison with A 7 steel, which had been one of the most widely used structural steels for many years with a minimum yield strength of only 227.53 MPa (33,000 psi).

1946. ASTM Committee E-8 on Fatigue and Fracture Testing Is Established.

1948. AIME Expands. The organization expands to include metallurgical and petroleum engineers, creating the American Institution of Mining, Metallurgical and Petroleum Engineers—while retaining the AIME logo. The headquarters is in New York City.

1949. Grace Iron Mine. A new, rich iron ore deposit near Lebanon, Pennsylvania, is discovered by Bethlehem Steel geologists near the old Cornwall Mines, which are still operating. The discovery was made using airborne magnetomer searching and test drilling. The magnetite deposits are expected to yield over 90 million metric tons (100 million U.S. tons) of ore. The mine closed in 1977, however, because of flooding.

1949. Berkelium is discovered by Albert Ghiorso, Glenn T. Seaborg, Stanley G. Thompson, and Kenneth Street at Livermore National Labs, Berkeley, California, by bombarding ^{241}Am with α particles.

1949. *California Zephyr* Launched. In March the Chicago, Burlington & Quincy, Denver and Rio Grande Western, and Western Pacific Railroads jointly launch the *California Zephyr* between Chicago and San Francisco. The Budd-built streamlined stainless steel train is the first passenger train to include Vista Dome cars in regular service.

1950. Californium is discovered by Albert Ghiorso, Stanley G. Thompson, Kenneth Street, Jr., and Glenn T. Seaborg at Livermore Research Labs, Berkeley, California, by bombarding ^{242}Cm with α particles.

1951. ASTM Committee E-10 on Nuclear Technology and Applications Is Established.

1951. ASTM Committee F-1 on Electronics Is Established.

1952. Einsteinium is discovered by Albert Ghiorso et al. at Livermore Research Labs, Berkeley, California, and by G.R. Choppin et al. at Los Alamos National Lab., New Mexico, as a product of nuclear explosion.

1952. Fermium is discovered by Albert Ghiorso et al. at Livermore Research Labs, Berkeley, California, jointly with Argonne National and Los Alamos National Labs., as a product of nuclear explosion.

1953. ASTM Commercial Quality Carbon Steel Sheet. ASTM A366 on "Commercial Quality Cold Rolled Carbon Steel Sheet" is published.

1953. Alcoa Building. The 120 m (410 ft) steel and aluminum Alcoa Building in Pittsburgh is completed. Designed by Harrison and Abramovitch, it is a showcase of aluminum as well as a testing laboratory for the use of aluminum in architecture.

1954. Argon-Oxygen-Decarburization (AOD) Process Discovered. William A. Krivitz, a young metallurgist at the Metals Research Laboratory of the Union Carbide and Carbon Company, Niagara Falls, New York, who had just graduated from the Massachusetts Institute of Technology, discovers a better way to refine stainless steel than by blowing oxygen into the molten metal. He diluted the oxygen with argon gas to reduce the high temperature caused by the oxygen, which was undesirable for the steel and damaging to the vessel lining. Krivitz applied for a patent that was granted 10 years later, long after Krivitz had left to head the Beryllium Corporation. Work on his idea continued for 12 years, ultimately resulting in a highly successful AOD Process.

1954. Roll-Formed Stainless Jet Engine Compressor Blades. R. Wallace & Sons, Silversmiths, Wallingford, Conn., receives a contract from the U.S. Navy to produce stainless steel jet engine compressor blades and vanes using a roll-forming method long used in the silverware industry for making silverware (knives, forks, and spoons). There was a concern on the part of the U.S. Government that there were not enough drop hammers in the nation to drop forge enough blades for aircraft engines in the event of a national emergency.

1955. Mendelevium was first discovered by Albert Ghiorso, Glenn T. Seaborg, Gregory R. Choppin, Bernard G. Harvey, and Stanley G. Thompson by bombarding ^{253}Es with α particles. It was named after Dmitri Ivanovich Mendeleev, who created the periodic table.

1956. World's First Atomic Power Plant. The British Magnox Reactor at Calder Hall is connected to the grid. It is also used for producing plutonium for military applications.

1956. First Stainless Steel Razor Blades. Wilkinson Sword Ltd. in England is the first to produce stainless steel razor blades.

1956. Socony-Mobil Is First Stainless Steel Skyscraper. The world's first stainless steel skyscraper, the Socony-Mobil Building, a 42-story building, is erected in New York City. Sheathed with 7000 type 304 stainless steel panels, it is the headquarters for the Mobil Oil Corporation from 1956 to 1987.

1957. Mackinac Bridge. The bridge, with a total length of almost 8 km (5 miles), links Mackinaw City and St. Ignace, Michigan, and links Lakes Michigan and Huron. The suspension bridge has a total of 2625.55 m (8614 ft) in suspension, with the longest span being 1159 m (3800 ft), a record from 1957 until 1998. The bridge deck at the center can move as much as 10.67 m (35 ft) from side to side during a high wind. The bridge was designed by David B. Steinman of New York.

1957. First Commercial Atomic Power Plant in United States. The Shippingport Atomic Power Station on the Ohio River about 40 km (25 miles) from Pittsburgh goes on line on December 2. It is the world's first full-scale atomic electric power plant. The reactor was to serve as a prototype design for aircraft carriers and for commercial electric power generation. The plant was decommissioned in 1982 after 25 years of operation.

1959. ASM's Geodesic Dome. The new American Society for Metals headquarters building in Metals Park, Ohio, is enclosed by R. Buckminster Fuller's largest geodesic dome. The 76 m (250 ft) diameter semicircular structure stands 31 m (102 ft) tall and is made with 21 km (13 miles) of 7075 aluminum tubing.

1959. Explosive Cladding. The "explosive-cladding" process for metals is discovered by the DuPont Company in Wilmington, Delaware. In 1963, DuPont will establish the Detaclad Division to apply this process, bonding stainless steel and other metals.

1959. Verrazano—America's Greatest Bridge. In 1959, construction begins on the Verrazano Narrows Bridge to connect Staten Island with Manhattan. The bridge was named for Giovanni di Verrazano, the Italian explorer who discovered the New York harbor and the east coast of North America in 1524. The bridge was designed by Othmar Ammann, the Swiss-born, most famous bridge designer. With a main span of 1298 m (4260 ft), it was the world's longest bridge. Because of the great length, the steel towers supporting the cables were 211 m (693 ft) tall. The cables were 0.91 m (3 ft) in diameter and each contained 26,108 steel wires that were 0.5 cm (0.196 in.) in diameter. The total wire length was 230,000 km (143,000 miles). The bridge deck, which had six vehicle lanes, was 69.5 m (228 ft) above the water.

The bridge opened November 21, 1964. The cost was $320 million. During the first full year of operation, the bridge carried approximately 45,000 vehicles per day. According to traffic projections, the lower deck would be opened in 1975, but that opening was moved up to 1969, adding another six lanes. As of 2011, the bridge traffic reached approximately 190,000 vehicles per day.

1960s. Basic Oxygen Process. The basic oxygen process, developed in Europe, is gradually being used in North America, replacing the open hearth furnace.

1960. HY-80 Steel Submarine. The first submarine hull said to be made entirely of HY-80 steel is the USS *Permit*. HY-80 is a structural steel with a minimum yield strength of 551.6 MPa (80,000 psi).

1960. Metal Prices

Metal prices in U.S. dollars per pound (1960)

Metal	Price	Metal	Price	Metal	Price
Iron	0.062	Thallium	7.50	Germanium	136.00
Lead	0.12	Titanium	7.65	Yttrium	200.00
Zinc	0.13	Molybdenum	8.00	Palladium	235.00
Sodium	0.16	Strontium	8.00	Dysprosium	275.00
Aluminum	0.26	Lithium	10.00	Erbium	275.00
Antimony	0.29	Zirconium	11.00	Gadolinium	275.00
Copper	0.33	Silver	13.30	Holmium	275.00
Magnesium	0.35	Uranium	18.18	Samarium	275.00
Arsenic	0.60	Indium	18.25	Ytterbium	275.00
Nickel	0.74	Thorium	23.00	Gold	510.00
Tin	1.00	Hafnium	30.00	Rhenium	780.00
Chromium	1.15	Tantalum	35.00	Ruthenium	800.00
Cadmium	1.40	Vanadium	35.00	Iridium	1,090.00
Cobalt	1.50	Columbium	36.00	Osmium	1,160.00
Potassium	2.00	Beryllium	47.00	Platinum	1,210.00
Calcium	2.05	Cerium	100.00	Europium	1,300.00
Bismuth	2.25	Cesium	100.00	Lutetium	1,300.00
Mercury	2.84	Lanthanum	100.00	Terbium	1,300.00
Tellurium	3.50	Neodymium	100.00	Thulium	1,300.00
Tungsten	3.50	Praseodymium	100.00	Gallium	1,362.00
Barium	6.00	Rubidium	100.00	Rhodium	2,000.00
Selenium	6.50	Silicon	130.00	Plutonium	12,700.00

1962. ASTM Committee E-20 on Temperature Measurement Is Established.

1962. Disposable Stainless Needles Introduced in England. Gillette Surgical, Reading, England, begins manufacturing disposable stainless steel hypodermic needles.

1962. Oneida Silver Introduces Ornate Stainless Flatware. In the early 1960s, the quality and stature of stainless steel flatware improved and permitted the Oneida Silver Company, Oneida, New York, to introduce the first ornate traditional pierced pattern in stainless steel. The rise of stainless steel popularity sparked Oneida's recovery and led the company into a new era of growth.

1962. ASTM Committee F-4 on Medical and Surgical Materials and Devices Is Established.

1963. Phoenix Iron Delivers Last Wrought Iron Bridge. A final trestle is furnished by Phoenix Iron & Steel, Phoenixville, Pennsylvania., to span the Pennsylvania Turnpike at Gulph Mills, Pennsylvania.

1963. ASTM Committee G-1 on Corrosion Is Organized.

1963. Stainless Beer Kegs. The first stainless steel beer kegs appear on the U.S. market.

1965. ASTM Committee B-10 on Reactive and Refractory Metals Is Organized.

1965. Stainless Exhaust System Replacements. Stainless steel replacements for exhaust systems are launched into the U.S. car market.

1966. John Deere Office Building. The handsome John Deere Building, designed by Finnish-American architect Eero Saarinen (1910–1961), is the first major architectural use of U.S. Steel's COR-TEN steel siding, which is a *weathering steel*. It is a steel that, in a few years, produces its own cinnamon-brown protective coat that will never need painting. (See ASTM A242 and A588 for information on weathering steels.)

1966. St. Louis Arch. Eero Saarinen (1910–1961), the famed Finnish-American architect, won a contest in 1946 to design a monument for the Jefferson National Memorial in St. Louis. He designed an arch in the form of an inverted catenary, a design that had never before been executed. His arch consisted of hollow triangular legs that gradually decreased in size as it rose. The monument would be 192 m (630 ft) tall and 192 m (630 ft) wide at the base. Saarinen chose stainless steel as the cladding material for the arch because he said that he "wanted it to last for a thousand years."

1966. Nobelium. The synthetic element was first correctly identified by scientists at the Flerov Laboratory of Nuclear Reactions in Dubna, Soviet Union.

1968. AOD Process in Production. The Joslyn Manufacturing & Supply Co. of Ft. Wayne, Indiana, developed a new and greatly improved refining method for stainless steel. It was the argon-oxygen-decarburization (AOD) process that had been discovered by William Krivitz in 1954. It took 12 years for Joslyn to build a successful full-scale refining vessel, which the company immediately started to use for their own stainless steel production while also licensing the process for use worldwide. The process reduced carbon levels to 0.02% and lower as well as reduced phosphorus and sulfur contents. It produced better quality steels at reduced costs.

1968. ASTM Committee E-28 on Mechanical Testing Is Organized.

1969. Wrought Iron Production Ended. The closing of the A.M. Byers Company of Ambridge, Pennsylvania, marks the end of wrought iron production in North America. The ASTM Committee A-2 on Wrought Iron, organized in 1905, disbanded in 1970.

1970. U.S. Steel Tower. The 64-story building in Pittsburgh was to be the headquarters of the giant corporation as well as a showpiece of one of their products, COR-TEN steel. The weather-resisting steel has the unique property of developing a tightly-adhering brown oxide coating that never needs painting. The metal, however, tuned into something of an embarrassment when the sidewalks and buildings nearby became badly stained with rust that was not easily removed.

1970. John Hancock Center. This tallest of multi-purpose buildings, located in Chicago, is made of steel, concrete, and glass. It has an even 100 stories that rise 344 m (1127 ft). It is the home of 700 condominium dwellers plus offices, stores, and a hotel. The building is a truncated pyramid with floors diminishing in size as it rises.

1970. Final Meeting of Wrought Iron Committee. In 1970, the members of ASTM Committee A-2 on Wrought Iron meet for the last time. L. Stanley Crane chaired the meeting. There were just four other members in attendance. The last producer of wrought iron in North America had closed the previous year. The committee had been organized about 1905 and became very active writing the specifications needed, which amounted to 12 in all. The metal, which could be made only in small quantities compared to steel, had become too costly to be commercially viable.

1972. Nucor Mini-Mill. F. Kenneth Iverson takes over at Nuclear Corporation to create the first steel mini-mill, a company that will make bars starting with steel scrap melted in an electric arc furnace. His plant in rural Darlington, South Carolina, will hire farmers, salesmen, and sharecroppers, all of whom will work for non-union wages. It will be a stripped down, no-frills operation that can sell bars at the lowest prices. Iverson branches into the making of sheet steel, coming into direct competition with Big Steel and one of their most lucrative products. Nucor has become the second-largest steel manufacturer in the United States, while launching an entire mini-mill industry.

1973. ASTM Steel and Stainless Steel Committees Merge. ASTM Committee A-10 on Stainless Steel and Related Alloys merges with ASTM Committee A-1 on Steel to form ASTM Committee A-1 on Steel, Stainless Steel and Related Alloys, creating a committee with approximately 900 members. Committee A-10 had been organized in 1929. The merger was needed principally because specifications, including both alloy and stainless steels, were required by both committees.

1974. ASTM Committee F-16 on Fasteners Is Organized.

1975. Unified Numbering System (UNS) Handbook Is Published. The first edition of *Metals and Alloys in the Unified Numbering System (UNS)*, ASTM DS56, is published jointly by the American Society for Testing and Materials (ASTM) and the Society of Automotive Engineers (SAE). The UNS was established in 1974 for assigning designations to metals and alloys according to a standard system that divides all metals and alloys into 18 series of designations consisting of a letter and five digits. The system includes an "S" series for wrought stainless steels, and a "J" series for cast steel and stainless steel.

The book contains a listing of approximately 1000 metals and alloys that have been assigned UNS designations. For example, type 304L stainless steel is given the UNS designation S30403, alloy XM-1 is S20300, and 17-4-PH is S17400.

The Department of Defense (DoD) uses DS56 and discontinues publication of *MIL-HDBK-H1: Metals*.

1975. Metal Properties Council. A Council is organized by Adolf O. Schaefer, recently retired Philadelphia metallurgist, to obtain data on metal properties most urgently required by the members of the Council. Schaefer, who also is Chairman of the ASTM/ASME Joint Committee on the Effect of Temperature on Metals, makes contracts with testing organizations, and prepares and distributes reports to the members. The Council continued for 10 years, until Schaefer's death in 1985.

1975. Leeb Rebound Hardness Tester. Dietmar Leeb and Antonio Brandestini develop a portable tester for metals at Proceq SA, South Africa. The test becomes covered by ASTM A956, "Standard Test Method for Leeb Hardness Testing of Steel Products."

1976. Concorde. France and the United Kingdom collaborate on the building of the supersonic turbojet Concorde. Twenty planes were built at a cost of £23 million each. The planes were built primarily of aluminum. Their flight time was about half that of normal jets, and they could reach speeds of 2170 km/hr (1359 mph). The planes remained in service for 27 years, retiring in 2003.

1977. 3CR12. The first 25-ton trial heat of 3CR12, a ferritic-martensitic chromium stainless steel, is cast at Middleburg Steel and Alloys, Middleburg, South Africa. This was the "utilitarian," low-cost stainless steel that had the lowest possible chromium content (10.5% min) to be called a stainless steel. The alloy has been widely used for applications not requiring a shiny finish, such as coal hopper cars, grain hopper cars, and truck salt spreaders. It is a dual-phase ferrite-tempered martensitic steel. The chemical composition, in weight percent, is 0.03% C, 1.50% max Mn, 0.040% max P, 0.030% max S, 1.00% max Si, 10.5–12.5% Cr, 1.50% max Ni, and 0.030% max N.

The alloy was marketed by Lukens Steel in Coatesville, Pennsylvania, beginning in 1990 and sold under the trade name Duracorr. The Lukens plant is now an Arcelor-Mittal plant that continues to market the product as Duracorr (ASTM A 240 and A 1010, EN 1088).

1980. Stainless Buses in Italy. Italian buses start using type 304 stainless steel in construction. The buses are 10% lighter, have a 10% improvement in crash worthiness of the passenger compartment, have reduced maintenance, and are more fuel efficient. In 2008, 80% of the buses are stainless.

1981. St. Joe and Homestake Lead Merge to form the Doe Run Company. Doe Run becomes the largest integrated lead producer in the Western Hemisphere and the third-largest lead producer in the world. Over 136,000 metric tons (150,000 U.S. tons) of lead are recycled annually from car batteries and telephone cables.

1981. DeLorean Automobile. The only automobile to go into production with a stainless steel skin is built by the DeLorean Motor Company (Detroit, Michigan and Dunmurry, Northern Ireland). John Z. DeLorean, founder, had been one of General Motor's most respected engineers. Over a three-year period, DeLorean built 8563 cars clad with type 304 stainless steel. The company closed because of bankruptcy. Today the cars are collectors' items.

1984. Mass-produced Partial Stainless Exhaust Systems. The Ford Motor Company mass produces partial stainless steel exhaust systems. Before the turn of the century, all cars produced in North America have exhaust systems completely made of stainless steel.

1988. Nickel Development Institute Organized. The Nickel Development Institute (NiDI) is organized. See also 2004, Nickel Institute.

1990. Chromium Association. The International Chromium Development Association (ICDA) was organized with headquarters in Paris, France. The organization has 103 members from 26 countries.

1991. Tallest Building in United Kingdom Stainless Clad. The 243.84 m (800 ft) Canary Wharf Tower, the tallest building in the United Kingdom, is completed. It is completely clad with stainless steel and the first skyscraper to be clad in the metal. Cesar Pelli is the architect.

1995. High-Silicon Stainless. About 1995 a new alloy, 700 Si, containing 7% Si, the highest silicon content in an iron-base alloy, is produced especially for handling hot sulfuric acid. The Fe-Ni-Cr-Si alloy's UNS designation is S70003, and it is the only alloy to have an S7xxxx designation.

1995. Chinese Numbering System for Steel Developed. The Chinese introduce ISC, Iron and Steel Code, Unified Numbering System for Iron, Steel and Alloys. The system, which employs a letter followed by five digits, imitates, but does not duplicate, the U.S. Unified Numbering System for Metals and Alloys (UNS). The letter "S" is used for stainless steels.

1997. Types 304, 304L, 316, and 316L Stainless Steels Approved for Drinking Water Systems. The American National Standards Institute (ANSI)/NSF International Standard 61 lists requirements for drinking water system components and deals with contaminants that migrate or get extracted into the water, and their maximum allowable limits. Stainless steel types 304, 304L, 316, and 316L are approved for such systems.

1998. Last European Tin Mine Closes. South Crofty Mine, near Cambourne in Cornwall, England, closes in March.

1998. Akashi Kaikyō Suspension Bridge. The bridge between Kobe and Iwaya is the longest, with a central span of 1990 m (6527 ft).

2003. Petronas Towers Reach 452 Meters (1483 Feet). The Petronas Twin Towers in Kuala Lampur, Malaysia, are the tallest ever built. The buildings are encased with 32,000 windows and 700,000 square feet of stainless steel.

2004. Nickel Institute Organized. The Nickel Institute is established through a merger of NiDI (Nickel Development Institute) and the Nickel Producers Environmental Research Association. The nonprofit organization represents the interests of 24 companies that produce more than 90% of the world's annual nickel output.

2007. U.S. Steel Buys Stelco. In a move to greatly increase its flat rolled products capacity, U.S. Steel purchases Stelco in Hamilton, Ontario, changing its name to U.S. Steel Canada.

2008. Alloy Cross-Reference Database. The "International Metallic Materials Cross-Reference Database" is available from Genium Publishing on diskette. This database was originally prepared by metallurgists John G. Gensure and Daniel L. Potts of the General Electric Research Laboratory in Schenectady, New York. The authors compared the chemical compositions of 45,000 alloy designations and grouped them according to the Unified Numbering System designations using their best engineering judgment when the chemistries were not so close as to be obvious. The book includes cast iron, steel, stainless steel, aluminum, copper, and nickel alloy designations.

2008. *Metals and Alloys in the Unified Numbering System*. The 11th edition of the SAE publication includes the chemical compositions and pertinent specification numbers for over 6,000 registered alloys. There is a cross reference of the UNS numbers to 17,000 trade names and alloys.

2008. Extra-low Carbon Steels. For the first time in 90 years, new grades of carbon steel are added to the original AISI list. They are the extra-low carbon grades with a maximum carbon content of 0.01% for Grade 1000. The other new grades are 1002, 1003, 1004 and 1007 (SAE J403).

2009. Colossal Stainless Sculpture Rises on Mongolian Steppes. *The New York Times* reported in an article on August 2, 2009, the erection in Mongolia of a colossal statue of Genghis Khan, "the legendary horseman who conquered the known world in the thirteenth century." The 40 m (131 ft) tall giant on horseback is wrapped with 227 metric tons (250 U.S. tons) of stainless steel. It rests on top of a two-story circular building that serves as a visitors' center. The statue is said to be the largest in the world.

2009. First Free-Cutting Steel with Vanadium. SAE 11V41 (UNS G11411) is the number for a new free-cutting carbon steel added to the former list of AISI free-cutting carbon steels in 2009. The steel appears in SAE J403 with the following chemical composition: 0.37–0.45% C, 1.35–1.65% Mn, 0.030% max P, 0.08–0.13% S, and 0.04–0.08% V. It is the first free-cutting steel to contain vanadium.

APPENDIX 2

Bibliography

- H.M. Cobb, *Standard Wrought Steels*, AIST, Warrendale, PA, 2001
- H.M. Cobb, *Stainless Steels*, AIST, Warrendale, PA, 2008
- Steel Sheet Specifications, *Annual Book of ASTM Standards*, Vol. 01.03, ASTM, 2009
- Metals & Alloys in the Unified Numbering System (UNS), 11th ed., SAE/ASTM, 2008
- H.M. Cobb, Steel Sheet (A Steel Products Manual), AIST, Warrendale, PA, 2004
- D. Birchon, *Dictionary of Metallurgy*, George Newnes Limited, London, 1965
- A.H. Harrison, *Understanding Stainless Steel*, British Stainless Steel Association, Sheffield, U.K., 2009
- J.R. Davis, Ed., *ASM Materials Engineering Dictionary*, ASM International, 1992
- *Woldman's Engineering Alloys*, 9th ed., ASM International, 2000
- *Kirk-Othmer Encyclopedia of Chemical Technology*, John Wiley & Sons, Inc., 2010
- Lamb, *Stainless Steels and Nickel Alloys*, ASM International,1989
- R.B. Ross, *Metallic Materials Specifications*, Chapman & Hall, London, 1992
- J. Hubble, *The Original Steelmakers*, AIST, Warrendale, PA, 1984
- C.A. Hampel, *Rare Earth Metals*, Reinhold Publishing, London, 1961
- Glossary of Terms Relating to Metals and Metallurgy, *Metals Handbook Desk Edition*, ASM International, 1985
- "Standard Terminology Relating to Steel, Stainless Steel, Related Alloys and Ferroalloys," A 941, *Annual Book of ASTM Standards*, Vol. 01.03, ASTM, 2010
- "Standard Test Methods and Definitions for Mechanical Testing of Steel Products," A 370, *Annual Book of ASTM Standards*, Vol. 01.03, ASTM, 2010
- The Elements: Names and Origins, www.bbc.co.UK/dna/h2g2/A3768672 (accessed December 1, 2009)
- H.M. Cobb, *The History of Stainless Steel*, ASM International, 2010
- G. Agricola, *De Re Metallica*, 1556. Translated by H.C. Hoover and L.H. Hoover (London: *The Mining Magazine*, 1912; Mineola, NY: Dover Publications, 1986)
- J. Aston, *Wrought Iron*, A.M. Byers Co., Pittsburgh, PA, 1939
- *Encyclopædia Britannica*, Encyclopædia Britannica, Inc., Chicago, IL, 1985

- *The Pyrotechnia of Vannoccio Biringuccio*, 1540. Translated by C.S. Smith and M.T. Gnudi (Mineola, NY: Dover Publications, 1959)
- C.W. Wegst, *Stahlschlüssel* (*Key to Steel*), Germany, 2004
- J.G. Gensure and D.L. Potts, *International Metallic Materials Cross Reference*, Genium Publications Corp., Amsterdam, N.Y.
- *New Columbia Encyclopedia*, Columbia University Press, New York, 1975
- N. Cossons and B. Tinder, *The Iron Bridge*, Moonraker Press, Wiltshire, England, 1960
- H. Brearley, *Steel-Makers*, Longmans, Green & Co., London, 1933
- C.D. Brown, *Dictionary of Metallurgy*, John Wiley & Sons, Inc., Sussex, England, 1998
- A.D. Merriman, *Dictionary of Metallurgy*, London, 1958
- *ASTM: 1898–1998—A Century of Progress*, ASTM International, 1998

APPENDIX 3

Properties and Conversion Tables

Table 1   Periodic table of the elements

Key to chart:

50	+2 +4
Sn	← Symbol
118.69	← Atomic Weight
-18-18-4	← Electron Configuration

50 = Atomic Number, +2 +4 = Oxidation States

Labels: Orbit, Nonmetals, Metals, Transition Elements, Lanthanides*, Actinides**

Periodic table element data (Atomic number, Oxidation states, Symbol, Atomic weight, Electron configuration):

No.	Ox.	Sym	At. wt	Config
1	+1, -1	H	1.0079	1
2	0	He	4.00260	2
3	+1	Li	6.939	2-1
4	+2	Be	9.0122	2-2
5	+3	B	10.81	2-3
6	+2, +4, -4	C	12.011	2-4
7	+1,+2,+3,+4,+5,-1,-2,-3	N	14.0067	2-5
8	-2	O	15.9994	2-6
9	-1	F	18.998403	2-7
10	0	Ne	20.179	2-8
11	+1	Na	22.9898	2-8-1
12	+2	Mg	24.312	2-8-2
13	+3	Al	26.98154	2-8-3
14	+4, -4	Si	28.08	2-8-4
15	+3,+5,-3	P	30.97376	2-8-5
16	+4,+6,-2	S	32.06	2-8-6
17	+1,+5,+7,-1	Cl	35.453	2-8-7
18	0	Ar	39.948	2-8-8
19	+1	K	39.09	-8-8-1
20	+2	Ca	40.08	-8-8-2
21	+3	Sc	44.9559	-8-9-2
22	+2,+3,+4	Ti	47.9	-8-10-2
23	+2,+3,+4,+5	V	50.941	-8-11-2
24	+2,+3,+6	Cr	51.996	-8-13-1
25	+2,+3,+4,+6,+7	Mn	54.9380	-8-13-2
26	+2,+3	Fe	55.847	-8-14-2
27	+2,+3	Co	58.9332	-8-15-2
28	+2,+3	Ni	58.71	-8-16-2
29	+1,+2	Cu	63.54	-8-18-1
30	+2	Zn	65.38	-8-18-2
31	+3	Ga	39.72	-8-18-3
32	+2,+4	Ge	72.59	-8-18-4
33	+3,+5,-3	As	74.9216	-8-18-5
34	+4,+6,-2	Se	78.96	-8-18-6
35	+1,+5,-1	Br	79.904	-8-18-7
36	0	Kr	83.80	-8-18-8
37	+1	Rb	85.467	-18-8-1
38	+2	Sr	87.62	-18-8-2
39	+3	Y	88.9059	-18-9-2
40	+4	Zr	91.22	-18-10-2
41	+3,+5	Nb	92.9064	-18-12-1
42	+6	Mo	95.94	-18-13-1
43	+4,+6,+7	Tc	98.9062	-18-13-2
44	+3	Ru	101.07	-18-15-1
45	+3	Rh	102.905	-18-16-1
46	+2,+4	Pd	106.4	-18-18-0
47	+1	Ag	107.868	-18-18-1
48	+2	Cd	112.40	-18-18-2
49	+3	In	114.82	-18-18-3
50	+2,+4	Sn	118.69	-18-18-4
51	+3,+5,-3	Sb	121.75	-18-18-5
52	+4,+6,-2	Te	127.60	-18-18-6
53	+1,+5,+7,-3	I	126.9045	-18-18-7
54	0	Xe	131.30	-18-18-8
55	+1	Cs	132.9054	-18-8-1
56	+2	Ba	137.3	-18-8-2
57*	+3	La	138.9055	-18-9-2
72	+4	Hf	178.49	-32-10-2
73	+5	Ta	180.948	-32-11-2
74	+6	W	183.85	-32-12-2
75	+4,+6,+7	Re	186.207	-32-13-2
76	+3,+4	Os	190.2	-32-14-2
77	+3,+4	Ir	192.9	-32-15-2
78	+2,+4	Pt	195.09	-32-16-2
79	+1,+3	Au	196.9665	-32-18-1
80	+1,+2	Hg	200.59	-32-18-2
81	+1,+3	Tl	204.37	-32-18-3
82	+2,+4	Pb	207.19	-32-18-4
83	+3,+5	Bi	208.980	-32-18-5
84	+2,+4	Po	(209)	-32-18-6
85	-1	At	(210)	-32-18-7
86	0	Rn	(222)	-32-18-8
87	+1	Fr	(223)	-18-8-1
88	+2	Ra	226.0254	-18-8-2
89**,+3	+3	Ac	(227)	-18-9-2
104	+4	Rf	(261)	-32-10-2
105		Ha	(262)	-32-11-2
106			(263)	-32-12-2

Lanthanides:

No.	Ox.	Sym	At. wt	Config
58	+3,+4	Ce	140.12	-20-8-2
59	+3	Pr	140.9077	-21-8-2
60	+3	Nd	144.24	-22-8-2
61	+3	Pm	147	-23-8-2
62	+2,+3	Sm	150.4	-24-8-2
63	+2,+3	Eu	151.96	-25-8-2
64	+3	Gd	157.25	-25-9-2
65	+3	Tb	158.925	-27-8-2
66	+3	Dy	162.50	-28-8-2
67	+3	Ho	164.9304	-29-8-2
68	+3	Er	167.26	-30-8-2
69	+3	Tm	168.9342	-31-8-2
70	+2,+3	Yb	173.04	-32-8-2
71	+3	Lu	174.967	-32-9-2

Actinides:

No.	Ox.	Sym	At. wt	Config
90	+4	Th	232.038	-18-10-2
91	+5,+4	Pa	231.0359	-20-9-2
92	+3,+4,+5,+6	U	238.029	-21-9-2
93	+3,+4,+5,+6	Np	237.0482	-22-9-2
94	+3,+4,+5,+6	Pu	239.052	-24-8-2
95	+3,+4,+5,+6	Am	(243)	-25-8-2
96	+3	Cm	(247)	-25-9-2
97	+3,+4	Bk	(247)	-27-8-2
98	+3	Cf	(251)	-28-8-2
99	+3	Es	(254)	-29-8-2
100	+3	Fm	(257)	-30-8-2
101	+2,+3	Md	(258)	-31-8-2
102	+2,+3	No	(259)	-32-8-2
103	+3	Lr	(260)	-32-9-2

Orbit columns (right side): K; K-L; K-L-M; -L-M-N; -M-N-O; -N-O-P; -O-P-Q; -N-O-P; -O-P-Q

Numbers in parentheses are mass numbers of most stable isotope of that element.

Table 2 Physical properties of the elements

Element	Density(a), g/cm³ (lb/in.³)	Boiling point, °C (°F)	Specific heat(b), cal/g · °C (J/kg · K)
Actinium (Ac)	... (...)	... (...)	... (...)
Aluminum (Al)	2.70 (0.0974)	2450 (4442)	0.215 (900)
Americium (Am)	11.87 (0.4285)	... (...)	... (...)
Antimony (Sb)	6.65 (0.240)	1380 (2516)	0.049 (205)
Argon (A)	1.784(g) (0.6440)(g)	−185.8 (−302.4)	0.125 (523)
Arsenic (As)	5.72 (0.206)	613(j) (1135)(j)	0.082 (343)
Astatine (At)	... (...)	... (...)	... (...)
Barium (Ba)	3.6 (0.13)	1640 (2980)	0.068 (285)
Berkelium (Bk)	... (...)	... (...)	... (...)
Beryllium (Be)	1.85 (0.0668)	2770 (5020)	0.45 (190)
Bismuth (Bi)	9.80 (0.354)	1560 (2840)	0.0294 (123)
Boron (B)	2.45 (0.0884)	... (...)	0.309 (1290)
Bromine (Br)	3.12 (0.113)	58 (136)	0.070 (290)
Cadmium (Cd)	8.65 (0.312)	765 (1409)	0.055 (230)
Calcium (Ca)	1.55 (0.0560)	1440 (2625)	0.149(s) (624)(s)
Californium (Cf)	... (...)	... (...)	... (...)
Carbon, graphite (C)	2.25 (0.0812)	4830 (8730)	0.165 (691)
Cerium (Ce)	6.77 (0.244)	3470 (6280)	0.045 (190)
Cesium (Cs)	1.87 (0.0675)	690 (1273)	0.04817 (201.7)
Chlorine (Cl)	3.214(g) (0.1160)(g)	−34.7 (−30.5)	0.116 (486)
Chromium (Cr)	7.19 (0.260)	2665 (4829)	0.11 (460)
Cobalt (Co)	8.85 (0.319)	2900 (5250)	0.099(410)
Copper (Cu)	9.86 (0.323)	2595 (4703)	0.092 (380)
Curium (Cm)	7 (0.3)	... (...)	... (...)
Dysprosium (Dy)	8.55 (0.309)	2330 (4230)	0.041 (170)
Einsteinium (E)	... (...)	... (...)	... (...)
Erbium (Er)	9.15 (0.330)	2630 (4770)	0.040 (170)
Europium (Eu)	5.24 (0.189)	1490 (2710)	0.039 (160)
Fermium (Fm)	... (...)	... (...)	... (...)
Fluorine (F)	1.696(g) (0.06123)(g)	−188.2 (−306.8)	0.18 (750)
Francium (Fr)	... (...)	... (...)	... (...)
Gadolinium (Gd)	7.86 (0.284)	2730 (4950)	0.071 (300)
Gallium (Ga)	5.91 (0.213)	2237 (4059)	0.079 (330)
Germanium (Ge)	5.32 (0.192)	2830 (5125)	0.073 (310)
Gold (Au)	19.3 (0.0697)	2970 (5380)	0.0312(ee) (131)(ee)
Hafnium (Hf)	13.1 (0.473)	5400 (9750)	0.0351 (147)
Helium (He)	0.1785(g) (0.006444)(g)	−268.9 (−452.0)	1.25 (5230)
Holmium (Ho)	6.79 (0.245)	2330 (4230)	0.039 (160)
Hydrogen (H)	0.0899(g) (0.00325)(g)	−252.7 (−422.9)	3.45 (14 400)
Indium (In)	7.31 (0.264)	2000 (3632)	0.057 (240)
Iodine (I)	4.94 (0.178)	183 (361)	0.052 (220)
Iridium (Ir)	22.65 (0.8177)	5300 (9570)	0.0307 (129)

(continued)

(a) Density may depend considerably on previous treatment. (b) At 20 °C (68 °F). (c) Near 20 °C (68 °F). (d) From 20 to 100 °C (68 to 212 °F). (e) From 20 to 60 °C (68 to 140 °F). (f) At 0 °C (32 °F). (g) Gas, grams per litre at 20 °C (68 °F) and 760 mm (30 in.). (h) 28 atm. (j) Sublimes. (k) Estimated. (m) From 25 to 100 °C (77 to 212 °F). (n) Annealed, commercial purity. (p) Approximate. (q) From 20 to 750 °C (68 to 1380 °F). (r) Sand cast. (s) From 0 to 100 °C (32 to 212 °F). (t) For α at 0 to 400 °C (32 to 750 °F). (u) Annealed. (v) At 28 °C (82 °F). (w) At 25 °C (77 °F). (x) Measured from stress-strain relationship on as-cast metal. (y) From 0 to 26 °C (32 to 70°F). (z) Near 40° C (105 °F); the coefficient of expansion of gadolinium changes rapidly between −100 and +100 °C(−150 and +212 °F). (aa) From 0 to 30 °C(32 to 86 °F). (bb) At melting point. (cc)For *a*-axis; 8.1 for *b*-axis and 54.3 for *c*-axis. (dd) Ohm · cm of intrinsic germanium at 300 K. (ee) At 18 °C (64 °F). (ff) From 20 to 200 °C (68 to 390 °F). (gg) W/cm/°C at 50 °C (120 °F). (jj) At 25 °C (77 °F) for high-purity k iron. (kk) For ingot iron at 0 °C (32 °F). (mm) From 17 to 100 °C(63 to 212 °F). (nn) Along *a*-axis; 24.3 along *c*-axis. (pp) Dynamic; static, 5.77; both for 99.98% magnesium. (qq) α; γ0.120; both at 25.2 °C (77.3 °F).

Heat of fusion, cal/g (Btu/lb)	Coefficient of linear thermal expansion(c), μin./in. °C (μin./in. °F)	Thermal conductivity(c), cal/cm²/cm/s/°C	Electrical resistivity, μΩ · cm	Modulus of elasticity in tension, 10⁶ psi
... (...)	... (...)	...	...	...
94.5 (170)	23.6(d) (13.1)(d)	0.53	2.6548(b)	9
... (...)	... (...)	...	...	...
38.3 (68.9)	8.5 to 10.8(e) (4.7 to 6)(e)	0.045	39.0(f)	11.3
6.7 (12)	... (...)	0.406×10^{-4}	...	...
88.5 (159.3)	4.7 (2.6)	...	33.3(b)	...
... (...)	... (...)	...	...	...
... (...)	... (...)	...	...	...
... (...)	... (...)	...	...	...
260(470)	11.6(m) (6.4)(m)	0.35	4(b)(n)	40 to 44
12.5 (22.5)	13.3 (7.4)	0.020	106.8(f)	4.6
... (...)	8.3(q) 4.6(q)	...	1.8×10^{12}(f)	...
16.2 (29.2)	... (...)	...	...	...
13.2 (23.8)	29.8 (16.55)	0.22	6.83(f)	8(r)
52 (93.6)	22.3(t) (12.4)(t)	0.3	3.91(f)	3.2 to 3.8(u)
... (...)	... (...)	...	...	...
... (...)	0.6 to 4.3(d) (0.3 to 2.4)(d)	0.057	1375(f)	0.7
8.5 (15.9)	8 (4.44)	0.026(v)	75(w)	6(z)
3.8 (6.8)	97(y) (54)(y)	...	20(b)	...
21.6 (38.9)	... (...)	0.172×10^{-4}	...	...
96 (173)	6.2 (3.4)	0.16	12.9(f)	36
58.4 (105)	13.8 (7.66)	0.165	6.24(b)	30
50.6 (91.1)	16.5 (9.2)	0.941 ± 0.005	1.6730(b)	16
... (...)	... (...)	...	...	...
25.2 (45.4)	9 (5)	0.024(v)	57(w)	10 to 14(x)
... (...)	... (...)	...	...	...
24.5 (44.1)	9 (5)	0.023(v)	107(w)	16(x)
16.5 (29.6)	26 (14.44)	...	90(w)	...
... (...)	... (...)	...	...	...
10.1 (18.2)	... (...)	...	...	...
... (...)	... (...)	...	...	...
23.5 (42.4)	4(z) (2.22)(z)	0.021(v)	140.5(w)	8 to 14(x)
19.16 (34.49)	18(aa) (10)(aa)	0.07 to 0.09(bb)	17.4(cc)	...
... (...)	5.75 (3.19)	0.14	46(dd)	...
16.1 (29.0)	14.2 (7.9)	0.71	2.35(b)	11.6
... (...)	519(ff) (288)(ff)	0.223 (gg)	35.1(w)	...
... (...)	... (...)	3.32×10^{-4}	...	...
24.9 (44.7)	... (...)	...	87(w)	11(x)
15.0 (27.0)	... (...)	4.06×10^{-4}	...	...
6.8 (12.2)	33 (18)	0.057	8.37(b)	1.57
14.2 (25.6)	93 (52)	10.4×10^{-4}	1.3×10^{15}(b)	...
... (...)	6.8 (3.8)	0.14	5.3(b)	76

(continued)

(rr) α; γ, 14; both from 0 to 100 °C (32 to 212 °F). (ss) α at 20 °C (68 °F). (tt) At 50 °C (122 °F). (uu) At −2.22 °C (28°F). (vv) At 0 °C (32 °F), unmagnetized. (ww) At 50 °C (122 °F), parallel to *a*-axis, mean value; parallel to *c*-axis at 50 °C (122 °F, 5.8. (xx) At 11 °C (51.8 °F). (yy) At 17 °C (63 °F). (zz) For small cyclic strains. (aaa) For α at 25 °C (77 °F). (bbb) From 21 to 104 °C (70 to 219 °F). (ccc) At 107 °C (224.6 °F). (ddd) At 25 °C (77 °F), for cast metal. (eee) From 20 to 500 °C (68 to 930 °F). (fff) For hard wire. (ggg) At −173 °C (−279 °F). (hhh) Calculated. (jjj) Average value at 22 °C (72 °F), zone-refined bar. (kkk) Chill cast specimen 90.2 by 24.6 mm (3.55 by 0.97 by 0.97 in.). (mmm) At 23 °C (73 °F). (nnn) From 25 to 1000 °C (77 to 1830 °F), for iodide thorium. (ppp) At 100 °C (212 °F). (qqq) From 0 to 100 °C (32 to 212 °F), for polycrystalline metal. (rrr) At 0 °C (32 °F), for white tin. (sss) Cast tin. (ttt) Btu · ft/h · ft² · °F at −400 °F. (uuu) At 27 °C (80.6 °F). (vvv) At 27 °C (80 °F). (www) Rolled rods (xxx) At 70 °C (158 °F). (yyy) Crystallographic average. (zzz) From 23 to 100 °C (73 to 212 °F). (aaaa) From 20 to 250 °C (68 to 480 °F), for polycrystalline metal. (bbbb) Pure zinc has no clearly defined modulus of elasticity. (cccc) α, polycrystalline. (dddd) W/cm/°C at 27 °C (80.6 °F)

Table 2 Physical properties of the elements (*continued*)

Element	Density(a), g/cm³ (lb/in.³)	Boiling point, °C (°F)	Specific heat(b), cal/g · °C (J/kg · K)
Iron (Fe)	7.87 (0.284)	3000 ± 150 (5430 ± 270)	0.11 (460)
Krypton (Kr)	3.743(g) (0.1351)(g)	−152 (−242)	... (...)
Lanthanum (La)	6.15 (0.222)	3470 (6280)	0.048 (200)
Lawrencium (Lw)	... (...)	... (...)	... (...)
Lead (Pb)	11.34 (0.4094)	1725 (3137)	0.0309(f) (129)(f)
Lithium (Li)	0.534 (0.193)	1330 (2426)	0.79 (3300)
Lutetium (Lu)	9.85 (0.356)	1930 (3510)	0.037 (150)
Magnesium (Mg)	1.74 (0.0628)	1107 ± 10 (2025 ± 20)	0.245 (1030)
Manganese (Mn)	7.43 (0.268)	2150 (3900)	0.115(qq) (481)(qq)
Mendelevium (Mv)	... (...)	... (...)	... (...)
Mercury (Hg)	13.55 (0.4892)	357 (675)	0.033 (140)
Molybdenum (Mo)	10.2 (0.368)	5560 (10 040)	0.066 (280)
Neodymium (Nd)	7.00 (0.253)	3180 (5756)	0.045 (190)
Neon (Ne)	0.8999(g) (0.03249)(g)	−246.0 (−410.8)	... (...)
Neptunium (Np)	20.5 (0.740)	... (...)	... (...)
Nickel (Ni)	8.9 (0.32)	2730 (4950)	0.105 (440)
Niobium (Nb)	8.57 (0.309)	4927 (8901)	0.065(f) (270)(f)
Nitrogen (N)	1.250(g) (0.04513)(g)	−195.8 (−320.4)	0.247 (1030)
Nobelium (No)	... (...)	... (...)	... (...)
Osmium (Os)	22.61 (0.8162)	5500 (9950)	0.031 (130)
Oxygen (O)	1.429(g) (0.05159)(g)	−183.0 (−297.4)	0.218 (913)
Palladium (Pd)	12.02 (0.4339)	3980 (7200)	0.0584(f) (245)(f)
Phosphorus, white (P)	1.83 (0.0661)	280 (536)	0.177 (741)
Platinum (Pt)	21.45 (0.7743)	4530 (8185)	0.0314(f) (131)(f)
Plutonium (Pu)	19.4 (0.700)	3235 (6000)	0.033(aaa) (140)(aaa)
Polonium (Po)	9.40 (0.339)	... (...)	... (...)
Potassium (K)	0.86 (0.031)	760 (1400)	0.177 (741)
Praseodymium (Pr)	6.77 (0.244)	3020 (5468)	0.045 (188)
Promethium (Pm)	... (...)	... (...)	... (...)
Protactinium (Pa)	15.4 (0.556)	... (...)	... (...)
Radium (Ra)	5.0 (0.18)	... (...)	... (...)
Radon (Rn)	9.960(g) (0.3956)(g)	−61.8 (−79.2)	... (...)
Rhenium (Re)	21.0 (0.76)	5900 (10 650)	0.033 (140)
Rhodium (Rh)	12.41 (0.4480)	4500 (8130)	0.059(f) (250)(f)
Rubidium (Rb)	1.53 (0.0552)	688 (1270)	0.080 (330)
Ruthenium (Ru)	12.45 (0.4494)	4900 (8850)	0.057(f) 240(f)
Samarium (Sm)	7.49 (0.270)	1630 (2966)	0.042(hhh) (180)(hhh)
Scandium (Sc)	2.9 (0.10)	2730 (4946)	0.134 (561)
Selenium (Se)	4.8 (0.17)	685 ± 1 (1265 ± 2)	0.084(u) (350)(u)
Silicon (Si)	2.33 (0.0841)	2680 (4860)	0.162(f) (678)(f)
Silver (Ag)	10.49 (0.3787)	2210 (4010)	0.0559(f) (234)(f)
Sodium (Na)	0.9712 (0.03506)	892 (1638)	0.295 (1240)

(*continued*)

(a) Density may depend considerably on previous treatment. (b) At 20 °C (68 °F). (c) Near 20 °C (68 °F). (d) From 20 to 100 °C (68 to 212 °F). (e) From 20 to 60 °C (68 to 140 °F). (f) At 0 °C (32 °F). (g) Gas, grams per litre at 20 °C (68 °F) and 760 mm (30 in.). (h) 28 atm. (j) Sublimes. (k) Estimated. (m) From 25 to 100 °C (77 to 212 °F). (n) Annealed, commercial purity. (p) Approximate. (q) From 20 to 750 °C (68 to 1380 °F). (r) Sand cast. (s) From 0 to 100 °C (32 to 212 °F). (t) For α at 0 to 400 °C (32 to 750 °F). (u) Annealed. (v) At 28 °C (82 °F). (w) At 25 °C (77 °F). (x) Measured from stress-strain relationship on as-cast metal. (y) From 0 to 26 °C (32 to 70°F). (z) Near 40° C (105 °F); the coefficient of expansion of gadolinium changes rapidly between −100 and +100 °C(−150 and +212 °F). (aa) From 0 to 30 °C(32 to 86 °F). (bb) At melting point. (cc)For *a*-axis; 8.1 for *b*-axis and 54.3 for *c*-axis. (dd) Ohm · cm of intrinsic germanium at 300 K. (ee) At 18 °C (64 °F). (ff) From 20 to 200 °C (68 to 390 °F). (gg) W/cm/°C at 50 °C (120 °F). (jj) At 25 °C (77 °F) for high-purity k iron. (kk) For ingot iron at 0 °C (32 °F). (mm) From 17 to 100 °C(63 to 212 °F). (nn) Along *a*-axis; 24.3 along *c*-axis. (pp) Dynamic; static, 5.77; both for 99.98% magnesium. (qq) α; γ0.120; both at 25.2 °C (77.3 °F).

Heat of fusion, cal/g (Btu/lb)	Coefficient of linear thermal expansion(c), μin./in. °C (μin./in. °F)	Thermal conductivity(c), cal/cm²/cm/s/°C	Electrical resistivity, μΩ · cm	Modulus of elasticity in tension, 10⁶ psi
65.5 (117.9)	11.76(jj) (6.53)(jj)	0.18(kk)	9.71(b)	28.5 ± 0.5
... (...)	... (...)	0.21×10^{-4}	...	...
17.3 (31.1)	5 (2.77)	0.033(u)	57(y)	10 to 11(x)
... (...)	... (...)	...	...	...
6.26 (11.27)	29.3(mm) (16.3)(mm)	0.083(f)	20.648(b)	2
104.2 (187.6)	56 (31)	0.17	8.55(f)	...
26.29 (47.32)	... (...)	...	79(w)	...
88 ± 2 (158 ± 4)	27.1(nn) (15.05)(nn)	0.367	4.45(b)	6.35(pp)
63.7 (114.7)	22(rr) (12.22)(rr)	...	185(ss)	23
... (...)	... (...)	...	...	...
2.8 (5.0)	... (...)	0.0196(f)	98.4(tt)	...
69.8(k) (125.6)(k)	4.9(d) (2.7)(d)	0.34	5.3(f)	47
11.78 (21.20)	6 (3.33)	0.031(uu)	64(w)	...
... (...)	... (...)	0.00011	...	...
... (...)	... (...)	...	...	...
73.8 (132.8)	13.3(s) (7.39)(s)	0.22 (w)	6.84(b)	30(vv)
69 (124.2)	7.31 (4.06)	0.125(f)	12.5(f)	...
6.2 (11.2)	... (...)	0.000060	...	...
... (...)	... (...)	...	...	...
... (...)	4.6(ww) (2.6)(ww)	...	9.5(b)	81
3.3 (5.9)	... (...)	0.000059	...	...
34.2 (61.6)	11.76 (6.53)	1.68(ee)	10.8(b)	16.3
5.0 (9.0)	125 (70)	...	10^{17}(xx)	...
26.9 (48.4)	8.9 (4.9)	0.165(yy)	10.6(b)	21.3(zz)
... (...)	55(bbb) (30.55)(bbb)	0.020(w)	141.4(ccc)	14(ddd)
... (...)	... (...)	...	...	...
14.6 (26.3)	83 (46)	0.24	6.15(f)	...
11.71 (21.08)	4 (2.22)	0.28(uu)	68(w)	7 to 14(x)
... (...)	... (...)	...	...	...
... (...)	... (...)	...	...	...
... (...)	... (...)	...	...	...
... (...)	... (...)	...	...	...
... (...)	6.7(eee) (3.7)(eee)	0.17	19.3(b)	66.7(b)
... (...)	8.3 (4.6)	0.21(yy)	4.51(b)	42.5(fff)
6.5 (11.79)	90 (50)	...	12.5(b)	...
... (...)	9.1 (5.1)	...	7.6(f)	60(p)
17.29 (31.12)	... (...)	...	88(w)	8(x)
84.52 (152.14)	... (...)	...	61(jjj)	...
16.4 (29.5)	37(21)	7 to 18.3×10^{-4}	12(f)	8.4
432 (778)	2.8 to 7.3 (1.6 to 4.1)	0.20	10(f)	16.35(kkk)
25 (45)	19.68(s) (10.9)(s)	1.0(f)	1.59(b)	11
27.5 (49.5)	71 (39)	0.32	4.2(f)	...

(continued)

(rr) α; γ, 14; both from 0 to 100 °C (32 to 212 °F). (ss) α at 20 °C (68 °F). (tt) At 50 °C (122 °F). (uu) At −2.22 °C (28°F). (vv) At 0 °C (32 °F), unmagnetized. (ww) At 50 °C (122 °F), parallel to *a*-axis, mean value; parallel to *c*-axis at 50 °C (122 °F, 5.8. (xx) At 11 °C (51.8 °F). (yy) At 17 °C (63 °F). (zz) For small cyclic strains. (aaa) For α at 25 °C (77 °F). (bbb) From 21 to 104 °C (70 to 219 °F). (ccc) At 107 °C (224.6 °F). (ddd) At 25 °C (77 °F), for cast metal. (eee) From 20 to 500 °C (68 to 930 °F). (fff) For hard wire. (ggg) At −173 °C (−279 °F). (hhh) Calculated. (jjj) Average value at 22 °C (72 °F), zone-refined bar. (kkk) Chill cast specimen 90.2 by 24.6 mm (3.55 by 0.97 by 0.97 in.). (mmm) At 23 °C (73 °F). (nnn) From 25 to 1000 °C (77 to 1830 °F), for iodide thorium. (ppp) At 100 °C (212 °F). (qqq) From 0 to 100 °C (32 to 212 °F), for polycrystalline metal. (rrr) At 0 °C (32 °F), for white tin. (sss) Cast tin. (ttt) Btu · ft/h · ft² · °F at −400 °F. (uuu) At 27 °C (80.6 °F). (vvv) At 27 °C (80 °F). (www) Rolled rods (xxx) At 70 °C (158 °F). (yyy) Crystallographic average. (zzz) From 23 to 100 °C (73 to 212 °F). (aaaa) From 20 to 250 °C (68 to 480 °F), for polycrystalline metal. (bbbb) Pure zinc has no clearly defined modulus of elasticity. (cccc) α, polycrystalline. (dddd) W/cm/°C at 27 °C (80.6 °F)

Table 2 Physical properties of the elements (*continued*)

Element	Density(a), g/cm³ (lb/in.³)	Boiling point, °C (°F)	Specific heat(b), cal/g · °C (J/kg · K)
Strontium (Sr)	2.60 (0.0939)	1380 (2520)	0.176 (737)
Sulfur, yellow (S)	2.07 (0.0747)	444.6 (832.3)	0.175 (733)
Tantalum (Ta)	16.6 (0.599)	5425 ± 100 (9800 ± 200)	0.034(w) (140)
Technetium (Tc)	11.5 (0.415)	... (...)	... (...)
Tellurium (Te)	6.24 (0.225)	989.8 ± 3.8 (1813.6 ± 6.8)	0.047 (200)
Terbium (Tb)	8.25 (0.298)	2530 (4586)	0.044 (180)
Thallium (Ti)	11.85 (0.4278)	1457 (2655)	0.031 (130)
Thorium (Th)	11.5 (0.415)	3850 ± 350 (7000 ± 600)	0.034 (140)
Thulium (Tm)	9.31 (0.336)	1720(ggg) (3130)(ggg)	0.038 (160)
Tin (Sn)	7.30 (0.264)	2270 (4120)	0.054 (230)
Titanium (Ti)	4.51 (0.163)	3260 (5900)	0.124 (519)
Tungsten (W)	19.3 (0.697)	5930 (10 706)	0.033 (140)
Uranium (U)	19.07 (0.6884)	3818 (6904)	0.02709(vvv) (113.4)(vvv)
Vanadium (V)	6.11 (0.221)	3400 (6150)	0.119(t) (498)(t)
Xenon (Xe)	5.896(g) (0.2128)(g)	−108.0 (−162.4)	... (...)
Ytterbium (Yb)	6.96 (0.251)	1530 (2786)	0.035 (150)
Yttrium (Y)	4.47 (0.161)	3030 (5490)	0.071 (300)
Zinc (Zn)	7.13 (0.257)	906 (1663)	0.0915 (383)
Zirconium (Zr)	6.49 (0.234)	3580 (6470)	0.067 ± 0.001 (280 ± 4)

(a) Density may depend considerably on previous treatment. (b) At 20 °C (68 °F). (c) Near 20 °C (68 °F). (d) From 20 to 100 °C (68 to 212 °F). (e) From 20 to 60 °C (68 to 140 °F). (f) At 0 °C (32 °F). (g) Gas, grams per litre at 20 °C (68 °F) and 760 mm (30 in.). (h) 28 atm. (j) Sublimes. (k) Estimated. (m) From 25 to 100 °C (77 to 212 °F). (n) Annealed, commercial purity. (p) Approximate. (q) From 20 to 750 °C (68 to 1380 °F). (r) Sand cast. (s) From 0 to 100 °C (32 to 212 °F). (t) For α at 0 to 400 °C (32 to 750 °F). (u) Annealed. (v) At 28 °C (82 °F). (w) At 25 °C (77 °F). (x) Measured from stress-strain relationship on as-cast metal. (y) From 0 to 26 °C (32 to 70°F). (z) Near 40° C (105 °F); the coefficient of expansion of gadolinium changes rapidly between −100 and +100 °C(−150 and +212 °F). (aa) From 0 to 30 °C(32 to 86 °F). (bb) At melting point. (cc)For *a*-axis; 8.1 for *b*-axis and 54.3 for *c*-axis. (dd) Ohm · cm of intrinsic germanium at 300 K. (ee) At 18 °C (64 °F). (ff) From 20 to 200 °C (68 to 390 °F). (gg) W/cm/°C at 50 °C (120 °F). (jj) At 25 °C (77 °F) for high-purity k iron. (kk) For ingot iron at 0 °C (32 °F). (mm) From 17 to 100 °C(63 to 212 °F). (nn) Along *a*-axis; 24.3 along *c*-axis. (pp) Dynamic; static, 5.77; both for 99.98% magnesium. (qq) α; γ0.120; both at 25.2 °C (77.3 °F).

Heat of fusion, cal/g (Btu/lb)	Coefficient of linear thermal expansion(c), μin./in. °C (μin./in. °F)	Thermal conductivity(c), cal/cm²/cm/s/°C	Electrical resistivity, μΩ · cm	Modulus of elasticity in tension, 10^6 psi
25 (45)	... (...)	...	23(b)	...
9.3 (16.7)	64 (36)	6.31×10^{-4}	2×10^{23}(b)	...
38 (68)	6.5 (3.6)	0.130	12.45(w)	27(b)
... (...)	... (...)	...	...	...
32 (58)	16.75 (9.3)	0.014	436 000(mmm)	6
24.54 (44.17)	7 (3.88)	...	...	...
5.04 (9.07)	28 (16)	0.093	18(f)	...
<19.82 (<35.68)	12.5 (nnn) (6.9)(nnn)	0.090(ppp)	13(f)	...
26.04 (46.87)	... (...)	...	79(w)	...
14.5 (26.1)	23(qqq) (13)(qqq)	1.50(e)	11(rrr)	6 to 6.5(sss)
104(k) (188)(k)	8.41 (4.67)	6.6(ttt)	42(b)	16.8
44 (70)	4.6 (2.55)	0.397(e)	5.65(uuu)	50
... (...)	6.8 to 14.1(www) (3.8 to 7.8)(www)	0.07(xxx)	30(yyy)	24
... (...)	8.3(zzz) (4.6)(zzz)	0.074(ppp)	24.8 to 26.0(b)	18 to 20
... (...)	... (...)	1.24×10^{-4}	...	...
12.71 (22.88)	25 (13.9)	...	29(w)	...
46 (83)	... (...)	0.035(uu)	57(aaaa)	17(x)
24.09 (43.36)	39.7(aaaa) (22.0)(aaaa)	0.27(w)	5.916(b)	(bbb)
60(k) (110)(k)	5.85(cccc) (3.2)(cccc)	0.211(dddd)	40	13.7

(rr) α; γ, 14; both from 0 to 100 °C (32 to 212 °F). (ss) α at 20 °C (68 °F). (tt) At 50 °C (122 °F). (uu) At −2.22 °C (28°F). (vv) At 0 °C (32 °F), unmagnetized. (ww) At 50 °C (122 °F), parallel to *a*-axis, mean value; parallel to *c*-axis at 50 °C (122 °F, 5.8. (xx) At 11 °C (51.8 °F). (yy) At 17 °C (63 °F). (zz) For small cyclic strains. (aaa) For α at 25 °C (77 °F). (bbb) From 21 to 104 °C (70 to 219 °F). (ccc) At 107 °C (224.6 °F). (ddd) At 25 °C (77 °F), for cast metal. (eee) From 20 to 500 °C (68 to 930 °F). (fff) For hard wire. (ggg) At −173 °C (−279 °F). (hhh) Calculated. (jjj) Average value at 22 °C (72 °F), zone-refined bar. (kkk) Chill cast specimen 90.2 by 24.6 mm (3.55 by 0.97 by 0.97 in.). (mmm) At 23 °C (73 °F). (nnn) From 25 to 1000 °C (77 to 1830 °F), for iodide thorium. (ppp) At 100 °C (212 °F). (qqq) From 0 to 100 °C (32 to 212 °F), for polycrystalline metal. (rrr) At 0 °C (32 °F), for white tin. (sss) Cast tin. (ttt) Btu · ft/h · ft² · °F at −400 °F. (uuu) At 27 °C (80.6 °F). (vvv) At 27 °C (80 °F). (www) Rolled rods (xxx) At 70 °C (158 °F). (yyy) Crystallographic average. (zzz) From 23 to 100 °C (73 to 212 °F). (aaaa) From 20 to 250 °C (68 to 480 °F), for polycrystalline metal. (bbbb) Pure zinc has no clearly defined modulus of elasticity. (cccc) α, polycrystalline. (dddd) W/cm/°C at 27 °C (80.6 °F)

Table 3 Density of metals and alloys

Metal or alloy	Density g/cm^3	lb/in.^{3}
Aluminum and aluminum alloys		
Aluminum (99.996%)	2.6989	0.0975
Wrought alloys		
EC, 1060 alloys	2.70	0.098
1100	2.71	0.098
2011	2.82	0.102
2014	2.80	0.101
2024	2.77	0.100
2218	2.81	0.101
3003	2.73	0.099
4032	2.69	0.097
5005	2.70	0.098
5050	2.69	0.097
5052	2.68	0.097
5056	2.64	0.095
5083	2.66	0.096
5086	2.65	0.096
5154	2.66	0.096
5357	2.70	0.098
5456	2.66	0.096
6061, 6063	2.70	0.098
6101, 6151	2.70	0.098
7075	2.80	0.101
7079	2.74	0.099
7178	2.82	0.102
Casting alloys		
A13	2.66	0.096
43	2.69	0.097
108, A108	2.79	0.101
A132	2.72	0.098
D132	2.76	0.100
F132	2.74	0.099
138	2.95	0.107
142	2.81	0.101
195, B195	2.81	0.101
214	2.65	0.096
220	2.57	0.093
319	2.79	0.101
355	2.71	0.098
356	2.68	0.097
360	2.64	0.095
380	2.71	0.098
750	2.88	0.104
40E	2.81	0.101
Copper and copper alloys		
Wrought coppers		
Pure copper	8.96	0.324
Electrolytic tough pitch copper (ETP)	8.89	0.321
Deoxidized copper, high residual phosphorus (DHP)	8.94	0.323
Free-machining copper		
0.5% Te	8.94	0.323
1.0% Pb	8.94	0.323

Metal or alloy	Density g/cm^3	lb/in.^{3}
Wrought alloys		
Gilding, 95%	8.86	0.320
Commercial bronze, 90%	8.80	0.318
Jewelry bronze, 87.5%	8.78	0.317
Red brass, 85%	8.75	0.316
Low brass, 80%	8.67	0.313
Cartridge brass, 70%	8.53	0.308
Yellow brass	8.47	0.306
Muntz metal	8.39	0.303
Leaded commercial bronze	8.83	0.319
Low-leaded brass (tube)	8.50	0.307
Medium-leaded brass	8.47	0.306
High-leaded brass (tube)	8.53	0.308
High-leaded brass	8.50	0.307
Extra-high-leaded brass	8.50	0.307
Free-cutting brass	8.50	0.307
Leaded Muntz metal	8.41	0.304
Forging brass	8.44	0.305
Architectural bronze	8.47	0.306
Inhibited admiralty	8.53	0.308
Naval brass	8.41	0.304
Leaded naval brass	8.44	0.305
Manganese bronze (A)	8.36	0.302
Phosphor bronze, 5% (A)	8.86	0.320
Phosphor bronze, 8% (C)	8.80	0.318
Phosphor bronze, 10% (D)	8.78	0.317
Phosphor bronze, 1.25%	8.89	0.321
Free-cutting phosphor bronze	8.89	0.321
Cupro-nickel, 30%	8.94	0.323
Cupro-nickel, 10%	8.94	0.323
Nickel silver, 65–18	8.73	0.315
Nickel silver, 55–18	8.70	0.314
High-silicon bronze (A)	8.53	0.308
Low-silicon bronze (B)	8.75	0.316
Aluminum bronze, 5% Al	8.17	0.294
Aluminum bronze, (3)	7.78	0.281
Aluminum-silicon bronze	7.69	0.278
Aluminum bronze, (1)	7.58	0.274
Aluminum bronze, (2)	7.58	0.274
Beryllium copper	8.23	0.297
Casting alloys		
Chromium copper (1% Cr)	8.7	0.31

Metal or alloy	Density g/cm^3	lb/in.^{3}
88Cu-10Sn-2Zn	8.7	0.31
88Cu-8Sn-4Zn	8.8	0.32
89Cu-11Sn	8.78	0.317
88Cu-6Sn-1.5Pb-4.5Zn	8.7	0.31
87Cu-8Sn-1Pb-4Zn	8.8	0.32
87Cu-10Sn-1Pb-2Zn	8.8	0.32
80Cu-10Sn-10Pb	8.95	0.323
83Cu-7Sn-7Pb-3Zn	8.93	0.322
85Cu-5Sn-9Pb-1Zn	8.87	0.320
78Cu-7Sn-15Pb	9.25	0.334
70Cu-5Sn-25Pb	9.30	0.336
85Cu-5Sn-5Pb-5Zn	8.80	0.318
83Cu-4Sn-6Pb-7Zn	8.6	0.31
81Cu-3Sn-7Pb-9Zn	8.7	0.31
76Cu-2.5Sn-6.5Pb-15Zn	8.77	0.317
72Cu-1Sn-3Pb-24Zn	8.50	0.307
67Cu-1Sn-3Pb-29Zn	8.45	0.305
61Cu-1Sn-1Pb-37Zn	8.40	0.304
Manganese bronze		
60 ksi	8.2	0.30
65 ksi	8.3	0.30
90 ksi	7.9	0.29
110 ksi	7.7	0.28
Aluminum bronze		
Alloy 9A	7.8	0.28
Alloy 9B	7.55	0.272
Alloy 9C	7.5	0.27
Alloy 9D	7.7	0.28
Nickel silver		
12% Ni	8.95	0.323
16% Ni	8.95	0.323
20% Ni	8.85	0.319
25% Ni	8.8	0.32
Silicon bronze	8.30	0.300
Silicon brass	8.30	0.300
Iron and iron alloys		
Pure iron	7.874	0.2845
Ingot iron	7.866	0.2842
Wrought iron	7.7	0.28
Gray cast iron	7.15(a)	0.258(a)
Malleable iron	7.27(b)	0.262(b)
0.06% C steel	7.871	0.2844
0.23% C steel	7.859	0.2839
0.435% C steel	7.844	0.2834
1.22% C steel	7.830	0.2829
Low-carbon chromium-molybdenum steels		
0.5% Mo steel	7.86	0.283
1Cr-0.5Mo steel	7.86	0.283
1.25Cr-0.5Mo steel	7.86	0.283
2.25Cr-1.0Mo steel	7.86	0.283
5Cr-0.5Mo steel	7.78	0.278
7Cr-0.5Mo steel	7.78	0.278
9Cr-1Mo steel	7.67	0.276

(continued)

(a) 6.95 to 7.35 g/cm^3 (0.251 to 0.265 lb/in.^{3}). (b) 7.20 to 7.34 g/cm^3 (0.260 to 0.265 lb/in^3). (c) Annealed. (d) As cast. (e) Face-centered cubic. (f) Hexagonal. (g) Body-centered cubic. (h) Close-packed hexagonal. (j) Rhombohedral

Table 3 Density of metals and alloys (*continued*)

Metal or alloy	Density g/cm³	lb/in.³	Metal or alloy	Density g/cm³	lb/in.³	Metal or alloy	Density g/cm³	lb/in.³
Medium-carbon alloy steels			Type 303	7.9	0.29	**Molybdenum-base alloy**		
1Cr-035Mo-0.25V steel	7.86	0.283	Type 304	7.9	0.29	Mo-0.5Ti	10.2	0.368
H11 die steel (5Cr-	7.79	0.281	Type 305	8.0	0.29	**Lead and lead alloys**		
1.5Mo-0.4V)			Type 308	8.0	0.29	Chemical lead	11.34	0.4097
Other iron-base alloys			Type 309	7.9	0.29	(99.90 + % Pb)		
A-286	7.94	0.286	Type 310	7.9	0.29	Corroding lead	11.36	0.4104
16-25-6 alloy	8.08	0.292	Type 314	7.72	0.279	(99.73 + % Pb)		
RA-330	8.03	0.290	Type 316	8.0	0.29	Arsenical lead	11.34	0.4097
Incoloy	8.02	0.290	Type 317	8.0	0.29	Calcium lead	11.34	0.4097
Incoloy T	7.98	0.288	Type 321	7.9	0.29	5-95 solder	11.0	0.397
Incoloy 901	8.23	0.297	Type 347	8.0	0.29	20-80 solder	10.2	0.368
T1 tool steel	8.67	0.313	Type 403	7.7	0.28	50-50 solder	8.89	0.321
M2 tool steel	8.16	0.295	Type 405	7.7	0.28	**Antimonial lead alloys**		
H41 tool steel	7.88	0.285	Type 410	7.7	0.28	1% antimonial lead	11.27	0.407
20W-4Cr-2V-12Co steel	8.89	0.321	Type 416	7.7	0.28	Hard lead (96Pb-4Sb)	11.04	0.399
Invar (35% Ni)	8.00	0.289	Type 420	7.7	0.28	Hard lead (94Pb-6Sb)	10.88	0.393
Hipernik (50% Ni)	8.25	0.298	Type 430	7.7	0.28	8% antimonial lead	10.74	0.388
4% Si	7.6	0.27	Type 430F	7.7	0.28	9% antimonial lead	10.66	0.385
10.27% Si	6.97	0.252	Type 431	7.7	0.28	**Lead-base babbitt alloys**		
Stainless steels			Types 440A, 440B,	7.7	0.28	Lead-base babbitt		
and heat-resistant alloys			440C			SAE 13	10.24	0.370
			Type 446	7.6	0.27	SAE 14	9.73	0.352
Corrosion-resistant steel castings			Type 501	7.7	0.28	Alloy 8	10.04	0.363
CA-15	7.612	0.2750	Type 502	7.8	0.28	Arsenical lead		
CA-40	7.612	0.2750	19-9DL	7.97	0.29	Babbitt (SAE 15)	10.1	0.365
CB-30	7.53	0.272	**Precipitation-hardening stainless steels**			"G" Babbitt	10.1	0.365
CC-50	7.53	0.272	PH15-7 Mo	7.804	0.2819	**Magnesium and**		
CE-30	7.67	0.277	17-4 PH	7.8	0.28	**magnesium alloys**		
CF-8	7.75	0.280	17-7 PH	7.81	0.282			
CF-20	7.75	0.280	**Nickel-base alloys**			Magnesium (99.8%)	1.738	0.06279
CF-8M, CF-12M	7.75	0.280	D-979	8.27	0.299	**Casting alloys**		
CF-8C	7.75	0.280	Nimonic 80A	8.25	0.298	AM100A	1.81	0.065
CF-16F	7.75	0.280	Nimonic 90	8.27	0.299	AZ63A	1.84	0.066
CH-20	7.72	0.279	M-252	8.27	0.298	AZ81A	1.80	0.065
CK-20	7.75	0.280	Inconel	8.51	0.307	AZ91A,B,C	1.81	0.065
CN-7M	8.00	0.289	Inconel "X" 550	8.30	0.300	AZ92A	1.82	0.066
Heat-resistant alloy castings			Inconel 700	8.17	0.295	HK31A	1.79	0.065
HA	7.72	0.279	Inconel "713C"	7.913	0.2859	HZ32A	1.83	0.066
HC	7.53	0.272	Waspaloy	8.23	0.296	ZH42, ZH62A	1.86	0.067
HD	7.58	0.274	René 41	8.27	0.298	ZK51A	1.81	0.065
HE	7.67	0.277	Hastelloy alloy B	9.24	0.334	ZE41A	1.82	0.066
HF	7.75	0.280	Hastelloy alloy C	8.94	0.323	EZ33A	1.83	0.066
HH	7.72	0.279	Hastelloy alloy X	8.23	0.297	EK30A	1.79	0.065
HI	7.72	0.279	Udimet 500	8.07	0.291	EK41A	1.81	0.065
HK	7.75	0.280	GMR-235	8.03	0.290	**Wrought alloys**		
HL	7.72	0.279	**Cobalt-chromium-nickel-base alloys**			MIA	1.76	0.064
HN	7.83	0.283	N-155 (HS-95)	8.23	0.296	A3A	1.77	0.064
HT	7.92	0.286	S-590	8.36	0.301	AZ31B	1.77	0.064
HU	8.04	0.290	**Cobalt-base alloys**			PE	1.76	0.064
HW	8.14	0.294	S-816	8.68	0.314	AZ61A	1.80	0.065
HX	8.14	0.294	V-36	8.60	0.311	AZ80A	1.80	0.065
Wrought stainless and heat-resisting			HS-25	9.13	0.330	ZK60A, B	1.83	0.066
steels			HS-36	9.04	0.327	ZE10A	1.76	0.064
Type 301	7.9	0.29	HS-31	8.61	0.311	HM21A	1.78	0.064
Type 302	7.9	0.29	HS-21	8.30	0.300	HM31A	1.81	0.065
Type 302B	8.0	0.29						

(*continued*)

(a) 6.95 to 7.35 g/cm³ (0.251 to 0.265 lb/in.³). (b) 7.20 to 7.34 g/cm³ (0.260 to 0.265 lb/in³). (c) Annealed. (d) As cast. (e) Face-centered cubic. (f) Hexagonal. (g) Body-centered cubic. (h) Close-packed hexagonal. (j) Rhombohedral

Table 3   Density of metals and alloys (*continued*)

Metal or alloy	Density g/cm³	Density lb/in.³	Metal or alloy	Density g/cm³	Density lb/in.³	Metal or alloy	Density g/cm³	Density lb/in.³
Nickel and nickel alloys			Ti-4Al-3Mo-IV	4.507	0.1628	Bismuth	9.80	0.354
			Ti-2.5Al-16V	4.65	0.168	Cadmium	8.65	0.313
Nickel (99.95% Ni + Co)	8.902	0.322	**Zinc and zinc alloys**			Calcium	1.55	0.056
"A" Nickel	8.885	0.321	Pure zinc	7.133	0.2577	Cesium	1.903	0.069
"D" Nickel	8.78	0.317	AG40A alloy	6.6	0.24	Chromium	7.19	0.260
Duranickel	8.26	0.298	AC41A alloy	6.7	0.24	Cobalt	8.85	0.322
Cast nickel	8.34	0.301	Commercial rolled zinc			Gallium	5.907	0.213
Monel	8.84	0.319	0.08% Pb	7.14	0.258	Germanium	5.323	0.192
"K" Monel	8.47	0.306	0.06 Pb, 0.06 Cd	7.14	0.258	Hafnium	13.1	0.473
Monel (cast)	8.63	0.312	0.3 Pb, 0.3 Cd	7.14	0.258	Indium	7.31	0.264
"H" Monel (cast)	8.5	0.31	Copper-hardened, rolled zinc (1% Cu)	7.18	0.259	Iridium	22.5	0.813
"S" Monel (cast)	8.36	0.302				Lithium	0.534	0.019
Inconel	8.51	0.307	Rolled zinc alloy (1 Cu, 0.010 Mg)	7.18	0.259	Manganese	7.43	0.270
Inconel (cast)	8.3	0.30				Mercury	13.546	0.489
Ni-o-nel	7.86	0.294	Zn-Cu-Ti alloy (0.8 Cu, 0.15 Ti)	7.18	0.259	Molybdenum	10.22	0.369
Nickel-molybdenum-chromium-iron alloys						Niobium	8.57	0.310
			Precious metals			Osmium	22.583	0.816
Hastelloy B	9.24	0.334	Silver	10.49	0.379	Plutonium	19.84	0.717
Hastelloy C	8.94	0.323	Gold	19.32	0.698	Potassium	0.86	0.031
Hastelloy D	7.8	0.282	70Au-30Pt	19.92	...	Rhenium	21.04	0.756
Hastelloy F	8.17	0.295	Platinum	21.45	0.775	Rhodium	12.44	0.447
Hastelloy N	8.79	0.317	Pt-3.5Rh	20.9	...	Ruthenium	12.2	0.441
Hastelloy W	9.03	0.326	Pt-5Rh	20.65	...	Selenium	4.79	0.174
Hastelloy X	8.23	0.297	Pt-10Rh	19.97	...	Silicon	2.33	0.084
Nickel-chromium-molybdenum-copper alloys			Pt-20Rh	18.74	...	Silver	10.49	0.379
			Pt-30Rh	17.62	...	Sodium	0.97	0.035
Illium G	8.58	0.310	Pt-40Rh	16.63	...	Tantalum	16.6	0.600
Ilium R	8.58	0.310	Pt-5Ir	21.49	...	Thalium	11.85	0.428
Electrical resistance alloys			Pt-10Ir	12.53	...	Thorium	11.72	0.423
80Ni-20Cr	8.4	0.30	Pt-15Ir	21.57	...	Tungsten	19.3	0.697
60Ni-24Fe-16Cr	8.247	0.298	Pt-20Ir	21.61	...	Uranium	19.07	0.689
35Ni-45Fe-20Cr	7.95	0.287	Pt-25Ir	21.66	...	Vanadium	6.1	0.22
Constantan	8.9	0.32	Pt-30Ir	21.70	...	Zirconium	6.5	0.23
Tin and tin alloys			Pt-35Ir	21.79	...	**Rare earth metals**		
Pure tin	7.3	0.264	Pt-5Ru	20.67	...	Cerium	8.23(c)	...
Soft solder (30% Pb)	8.32	0.301	Pt-10Ru	19.94	...	Cerium	6.66(d)	...
Soft solder (37% Pb)	8.42	0.304	Palladium	12.02	0.4343	Cerium	6.77(e)	...
Tin babbitt			60Pd-40Cu	10.6	0.383	Dysprosium	8.55(f)	...
Alloy 1	7.34	0.265	95.5Pd-4.5Ru	12.07(a)	...	Erbium	9.15(f)	...
Alloy 2	7.39	0.267	95.5Pd-4.5Ru	11.62(b)	...	Europium	5.245(e)	...
Alloy 3	7.46	0.269	**Permanent magnet materials**			Gadolinium	7.86(f)	...
Alloy 4	7.53	0.272	Cunico	8.30	0.300	Holmium	6.79(f)	...
Alloy 5	7.75	0.280	Cunife	8.61	0.311	Lanthanum	6.19(d)	...
White metal	7.28	0.263	Comol	8.16	0.295	Lanthanum	6.18(c)	...
Pewter	7.28	0.263	Alnico I	6.89	0.249	Lanthanum	5.97(e)	...
Titanium and titanium alloys			Alnico II	7.09	0.256	Lutetium	9.85(f)	...
99.9% Ti	4.507	0.1628	Alnico III	6.89	0.249	Neodymium	7.00(d)	...
99.2% Ti	4.507	0.1628	Alnico IV	7.00	0.253	Neodymium	6.80(e)	...
99.0% Ti	4.52	0.163	Alnico V	7.31	0.264	Praseodymium	6.77(d)	...
Ti-6Al-4V	4.43	0.160	Alnico VI	7.42	0.268	Praseodymium	6.64(e)	...
Ti-5Al-2.5Sn	4.46	0.161	Barium ferrite	4.7	0.17	Samarium	7.49(g)	...
Ti-2Fe-2Cr-2Mo	4.65	0.168	Vectolite	3.13	0.113	Scandium	2.99(f)	...
Ti-8Mn	4.71	0.171	**Pure metals**			Terbium	8.25(f)	...
Ti-7Al-4Mo	4.48	0.162	Antimony	6.62	0.239	Thulium	9.31(f)	...
Ti-4Al-4Mn	4.52	0.163	Beryllium	1.848	0.067	Ytterbium	6.96(c)	...
						Yttrium	4.47(f)	...

(a) 6.95 to 7.35 g/cm³ (0.251 to 0.265 lb/in.³). (b) 7.20 to 7.34 g/cm³ (0.260 to 0.265 lb/in.³). (c) Face-centered cubic. (d) Hexagonal. (e) Body-centered cubic. (f) Close-packed hexagonal. (g) Rhombohedral

Table 4 Linear thermal expansion of metals and alloys

Metal or alloy	Temp., °C	Coefficient of expansion, μin./in. · °C	Metal or alloy	Temp., °C	Coefficient of expansion, μin./in. · °C
Aluminum and aluminum alloys			Red brass, 85%	20 to 300	18.7
			Low brass, 80%	20 to 300	19.1
Aluminum (99.996%)	20 to 100	23.6	Cartridge brass, 70%	20 to 300	19.9
Wrought alloys			Yellow brass	20 to 300	20.3
EC, 1060, 1100	20 to 100	23.6	Muntz metal	20 to 300	20.8
2011, 2014	20 to 100	23.0	Leaded commercial bronze	20 to 300	18.4
2024	20 to 100	22.8	Low-leaded brass	20 to 300	20.2
2218	20 to 100	22.3	Medium-leaded brass	20 to 300	20.3
3003	20 to 100	23.2	High-leaded brass	20 to 300	20.3
4032	20 to 100	19.4	Extra-high-leaded brass	20 to 300	20.5
5005, 5050, 5052	20 to 100	23.8	Free-cutting brass	20 to 300	20.5
5056	20 to 100	24.1	Leaded Muntz metal	20 to 300	20.8
5083	20 to 100	23.4	Forging brass	20 to 300	20.7
5086	60 to 300	23.9	Architectural bronze	20 to 300	20.9
5154	20 to 100	23.9	Inhibited admiralty	20 to 300	20.2
5357	20 to 100	23.7	Naval brass	20 to 300	21.2
5456	20 to 100	23.9	Leaded naval brass	20 to 300	21.2
6061, 6063	20 to 100	23.4	Manganese bronze (A)	20 to 300	21.2
6101, 6151	20 to 100	23.0	Phosphor bronze, 5% (A)	20 to 300	17.8
7075	20 to 100	23.2	Phosphor bronze, 8% (C)	20 to 300	18.2
7079, 7178	20 to 100	23.4	Phosphor bronze, 10% (D)	20 to 300	18.4
Casting alloys			Phosphor bronze, 1.25%	20 to 300	17.8
A13	20 to 100	20.4	Free-cutting phosphor bronze	20 to 300	17.3
43 and 108	20 to 100	22.0	Cupro-nickel, 30%	20 to 300	16.2
A108	20 to 100	21.5	Cupro-nickel, 10%	20 to 300	17.1
A132	20 to 100	19.0	Nickel silver, 65-18	20 to 300	16.2
D132	20 to 100	20.05	Nickel silver, 55-18	20 to 300	16.7
F132	20 to 100	20.7	Nickel silver, 65-12	20 to 300	16.2
138	20 to 100	21.4	High-silicon bronze (a)	20 to 300	18.0
142	20 to 100	22.5	Low-silicon bronze (b)	20 to 300	17.9
195	20 to 100	23.0	Aluminum bronze (3)	20 to 300	16.4
B195	20 to 100	22.0	Aluminum-silicon bronze	20 to 300	18.0
214	20 to 100	24.0	Aluminum bronze (1)	20 to 300	16.8
220	20 to 100	25.0	Beryllium copper	20 to 300	17.8
319	20 to 100	21.5	**Casting alloys**		
355	20 to 100	22.0	88Cu-8Sn-4Zn	21 to 177	18.0
356	20 to 100	21.5	89Cu-11Sn	20 to 300	18.4
360	20 to 100	21.0	88Cu-6Sn-1.5Pb-4.5Zn	21 to 260	18.5
750	20 to 100	23.1	87Cu-8Sn-1Pb-4Zn	21 to 177	18.0
40E	21 to 93	24.7	87Cu-10Sn-1pb-2Zn	21 to 177	18.0
Copper and copper alloys			80Cu-10Sn-10Pb	21 to 204	18.5
			78Cu-7Sn-15Pb	21 to 204	18.5
Wrought coppers			85Cu-5Sn-5Pb-5Zn	21 to 204	18.1
Pure copper	20	16.5	72Cu-1Sn-3Pb-24Zn	21 to 93	20.7
Electrolytic tough pitch copper (ETP)	20 to 100	16.8	67Cu-1Sn-3Pb-29Zn	21 to 93	20.2
Deoxidized copper, high residual phosphorus (DHP)	20 to 300	17.7	61Cu-1Sn-1Pb-37Zn	20 to 260	21.6
Oxygen-free copper	20 to 300	17.7	Manganese bronze		
Free machining copper, 0.5% Te or 1% Pb	20 to 300	17.7	60 ksi	21 to 204	20.5
Wrought alloys			65 ksi	21 to 93	21.6
Gilding, 95%	20 to 300	18.1	110 ksi	21 to 260	19.8
Commercial bronze, 90%	20 to 300	18.4	Aluminum bronze		
Jewelry bronze, 87.5%	20 to 300	18.6	Alloy 9A	...	17
			Alloy 9B	20 to 250	16.2
			Alloys 9C, 9D	...	16.2

(continued)

(a) Longitudinal; 23.4 transverse. (b) Longitudinal; 21.1 transverse. (c) Longitudinal; 19.4 transverse

Table 4   Linear thermal expansion of metals and alloys (*continued*)

Metal or alloy	Temp., °C	Coefficient of expansion, μin./in. · °C	Metal or alloy	Temp., °C	Coefficient of expansion, μin./in. · °C
Iron and iron alloys			Inconel	20 to 100	11.5
			Ni-o-nel	27 to 93	12.9
Pure iron	20	11.7	Hastelloy B	0 to 100	10.0
Fe-C alloys			Hastelloy C	0 to 100	11.3
0.06% C	20 to 100	11.7	Hastelloy D	0 to 100	11.0
0.22% C	20 to 100	11.7	Hastelloy F	20 to 100	14.2
0.40% C	20 to 100	11.3	Hastelloy N	21 to 204	10.4
0.56% C	20 to 100	11.0	Hastelloy W	23 to 100	11.3
1.08% C	20 to 100	10.8	Hastelloy X	26 to 100	13.8
1.45% C	20 to 100	10.1	Illium G	0 to 100	12.19
Invar (36% Ni)	20	0–2	Illium R	0 to 100	12.02
13Mn-1.2C	20	18.0	80Ni-20Cr	20 to 1000	17.3
13Cr-0.35C	20 to 100	10.0	60Ni-24Fe-16Cr	20 to 1000	17.0
12.3Cr-0.4Ni-0.09C	20 to 100	9.8	35Ni-45Fe-20Cr	20 to 500	15.8
17.7Cr-9.6Ni-0.06C	20 to 100	16.5	Constantan	20 to 1000	18.8
18W-4Cr-1V	0 to 100	11.2	**Tin and tin alloys**		
Gray cast iron	0 to 100	10.5	Pure tin	0 to 100	23
Malleable iron (pearlitic)	20 to 400	12	Solder (70Sn-30Pb)	15 to 110	21.6
Lead and lead alloys			Solder (63Sn-37Pb)	15 to 110	24.7
Corroding lead (99.73 + % Pb)	17 to 100	29.3	**Titanium and titanium alloys**		
5-95 solder	15 to 110	28.7	99.9% Ti	20	8.41
20-80 solder	15 to 110	26.5	99.0% Ti	93	8.55
50-50 solder	15 to 110	23.4	Ti-5Al-2.5Sn	93	9.36
1% antimonial lead	20 to 100	28.8	Ti-8Mn	93	8.64
Hard lead (96Pb-4Sb)	20 to 100	27.8	**Zinc and zinc alloys**		
Hard lead (94Pb-6Sb)	20 to 100	27.2	Pure zinc	20 to 250	39.7
8% antimonial lead	20 to 100	26.7	AG40A alloy	20 to 100	27.4
9% antimonial lead	20 to 100	26.4	AC41A alloy	20 to 100	27.4
Lead-base babbitt			**Commercial rolled zinc**		
SAE 14	20 to 100	19.6	0.08 Pb	20 to 40	32.5
Alloy 8	20 to 100	24.0	0.3 Pb, 0.3 Cd	20 to 98	33.9(a)
Magnesium and magnesium alloys			Rolled zinc alloy (1 Cu, 0.010 Mg)	20 to 100	34.8(b)
Magnesium (99.8%)	20	25.2	Zn-Cu-Ti alloy (0.8 Cu, 0.15 Ti)	20 to 100	24.9(c)
Casting alloys			**Pure metals**		
AM100A	18 to 100	25.2	Beryllium	25 to 100	11.6
AZ63A	20 to 100	26.1	Cadmium	20	29.8
AZ91A, B, C	20 to 100	26	Calcium	0 to 400	22.3
AZ92A	18 to 100	25.2	Chromium	20	6.2
HZ32A	20 to 200	26.7	Cobalt	20	13.8
ZH42	20 to 200	27	Gold	20	14.2
ZH62A	20 to 200	27.1	Iridium	20	6.8
ZK51A	20	26.1	Lithium	20	56
EZ33A	20 to 100	26.1	Manganese	0 to 100	22
EK30A, EK41A	20 to 100	26.1	Palladium	20	11.76
Wrought alloys			Platinum	20	8.9
M1A, A3A	20 to 100	26	Rhenium	20 to 500	6.7
AZ31B, PE	20 to 100	26	Rhodium	20 to 100	8.3
AZ61A, AZ80A	20 to 100	26	Ruthenium	20	9.1
ZK60A, B	20 to 100	26	Silicon	0 to 1400	5
HM31A	20 to 93	26.1	Silver	0 to 100	19.68
Nickel and nickel alloys			Tungsten	27	4.6
Nickel (99.95% Ni + Co)	0 to 100	13.3	Vanadium	23 to 100	8.3
Duranickel	0 to 100	13.0	Zirconium	...	5.85
Monel	0 to 100	14.0			
Monel (cast)	25 to 100	12.9			

(a) Longitudinal; 23.4 transverse. (b) Longitudinal; 21.1 transverse. (c) Longitudinal; 19.4 transverse

Table 5   Thermal conductivity of metals and alloys

Metal or alloy	Thermal conductivity near room temperature, cal/cm^2 · cm · s · °C	Metal or alloy	Thermal conductivity near room temperature, cal/cm^2 · cm · s · °C
Aluminum and aluminum alloys		**Wrought alloys**	
Wrought alloys		Gilding, 95%	0.56
EC(O)	0.57	Commercial bronze, 90%	0.45
1060 (O)	0.56	Jewelry bronze, 87.5%	0.41
1100	0.53	Red brass, 85%	0.38
2011 (T3)	0.34	Low brass, 80%	0.33
2014 (O)	0.46	Cartridge brass, 70%	0.29
2024 (O)	0.45	Yellow brass	0.28
2218 (T72)	0.37	Muntz metal	0.29
3003 (O)	0.46	Leaded-commercial bronze	0.43
4032 (O)	0.37	Low-leaded brass (tube)	0.28
5005	0.48	Medium-leaded brass	0.28
5050 (O)	0.46	High-leaded brass (tube)	0.28
5052 (O)	0.33	High-leaded brass	0.28
5056 (O)	0.28	Extra-high-leaded brass	0.28
5083	0.28	Leaded Muntz metal	0.29
5086	0.30	Forging brass	0.28
5154	0.30	Architectural bronze	0.29
5357	0.40	Inhibited admiralty	0.26
5456	0.28	Naval brass	0.28
6061 (O)	0.41	Leaded naval brass	0.28
6063 (O)	0.52	Manganese bronze (A)	0.26
6101 (T6)	0.52	Phosphor bronze, 5% (A)	0.17
6151 (O)	0.49	Phosphor bronze, 8% (C)	0.15
7075 (T6)	0.29	Phosphor bronze, 10% (D)	0.12
7079 (T6)	0.29	Phosphor bronze, 1.25%	0.49
7178	0.29	Free-cutting phosphor bronze	0.18
Casting alloys		Cupro-nickel, 30%	0.07
A13	0.29	Cupro-nickel, 10%	0.095
43 (F)	0.34	Nickel silver, 65-18	0.08
108 (F)	0.29	Nickel silver, 55-18	0.07
A108	0.34	Nickel silver, 65-12	0.10
A132 (T551)	0.28	High-silicon bronze (A)	0.09
D132 (T5)	0.25	Low-silicon bronze (B)	0.14
F132	0.25	Aluminum bronze, 5% Al	0.198
138	0.24	Aluminum bronze, (3)	0.18
142 (T21, sand)	0.40	Aluminum-silicon bronze	0.108
195 (T4, T62)	0.33	Aluminum bronze, (1)	0.144
B195 (T4, T6)	0.31	Aluminum bronze, (2)	0.091
214	0.33	Beryllium copper	0.20(a)
200 (T4)	0.21	**Casting alloys**	
319	0.26	Chromium copper (1% Cr)	0.4(a)
355 (T51, sand)	0.40	89Cu-11Sn	0.121
356 (T51, sand)	0.40	88Cu-6Sn-1.5Pb-4.5Zn	(b)
360	0.35	87Cu-8Sn-1Pb-4Zn	(c)
380	0.23	87Cu-10Sn-1Pb-2Zn	(c)
750	0.44	80Cu-10Sn-10Pb	(c)
40E	0.33	Manganese bronze, 110 ksi	(d)
Copper and copper alloys		Aluminum bronze	
Wrought coppers		Alloy 9A	(e)
Pure copper	0.941	Alloy 9B	(f)
Electrolytic tough pitch copper (ETP)	0.934	Alloy 9C	(b)
Deoxidized copper, high residual phosphorus (DHP)	0.81	Alloy 9D	(c)
		Propeller bronze	(g)
Free-machining copper (0.5% Te)	0.88	Nickel silver	
Free-machining copper (1% Pb)	0.92	12% Ni	(h)
		16% Ni	(h)

(continued)

(a) Depends on processing. (b) 18% of Cu. (c) 12% of Cu. (d) 9.05% of Cu. (e) 15 % of Cu. (f) 16% of Cu. (g) 11% of Cu. (h) 7% of Cu. (j) 6% of Cu. (k) 6.5% of Cu

Table 5   Thermal conductivity of metals and alloys (*continued*)

Metal or alloy	Thermal conductivity near room temperature, cal/cm^2 · cm · s · °C	Metal or alloy	Thermal conductivity near room temperature, cal/cm^2 · cm · s · °C
Nickel silver (*continued*)		Monel	0.062
20% Ni	(j)	"K" Monel	0.045
25% Ni	(k)	Inconel	0.036
Silicon bronze	(h)	Hastelloy B	0.027
Iron and iron alloys		Hastelloy C	0.03
		Hastelloy D	0.05
Pure iron	0.178	Illium G	0.029
Cast iron (3.16 C, 1.54 Si, 0.57 Mn)	0.112	Illium R	0.031
Carbon steel (0.23 C, 0.64 Mn)	0.124	60Ni-24Fe-16Cr	0.032
Carbon steel (1.22 C, 0.35 Mn)	0.108	35Ni-45Fe-20Cr	0.031
Alloy steel (0.34 C, 0.55 Mn, 0.78 Cr, 3.53 Ni, 0.39 Mo, 0.05 Cu)	0.079	Constantan	0.051
Type 410	0.057	**Tin and tin alloys**	
Type 304	0.036	Pure tin	0.15
T1 tool steel	0.058	Soft solder (63Sn-37Pb)	0.12
Lead and lead alloys		Tin foil (92Sn-8Zn)	0.14
Corroding lead (99.73% + % Pb)	0.083	**Titanium and titanium alloys**	
5-95 solder	0.085	Titanium (99.0%)	0.043
20-80 solder	0.089	Ti-5Al-2.5Sn	0.019
50-50 solder	0.111	Ti-2Fe-2Cr-2Mo	0.028
1% antimonial lead	0.080	Ti-8Mn	0.026
Hard lead (96Pb-4Sb)	0.073	**Zinc and zinc alloys**	
Hard lead (94Pb-6Sb)	0.069		
8% antimonial lead	0.065	Pure zinc	0.27
9% antimonial lead	0.064	AG40A alloy	0.27
Lead-base babbitt (SAE 14)	0.057	AC41A alloy	0.26
Lead-base babbitt (alloy 8)	0.058	Commercial rolled zinc	
Magnesium and magnesium alloys		0.08 Pb	0.257
		0.06 Pb, 0.06 Cd	0.257
Magnesium (99.8%)	0.367	Rolled zinc alloy (1 Cu, 0.010 Mg)	0.25
Casting alloys		Zn-Cu-Ti alloy (0.8 Cu, 0.15 Ti)	0.25
AM100A	0.17	**Pure metals**	
AZ63A	0.18	Beryllium	0.35
AZ81A (T4)	0.12	Cadmium	0.22
AZ91A, B, C	0.17	Chromium	0.16
AZ92A	0.17	Cobalt	0.165
HK31A (T6, sand cast)	0.22	Germanium	0.14
HZ32A	0.26	Gold	0.71
ZH42	0.27	Indium	0.057
ZH62A	0.26	Iridium	0.14
ZK51A	0.26	Lithium	0.17
ZE41A (T5)	0.27	Molybdenum	0.34
EZ33A	0.24	Niobium	0.13
EK30A	0.26	Palladium	0.168
EK41A (T5)	0.24	Platinum	0.165
Wrought alloys		Plutonium	0.020
M1A	0.33	Rhenium	0.17
AZ31B	0.23	Rhodium	0.21
AZ61A	0.19	Silicon	0.20
AZ80A	0.18	Silver	1.0
ZK60A, B (F)	0.28	Sodium	0.32
ZE10A (O)	0.33	Tantalum	0.130
HM21A (O)	0.33	Thallium	0.093
HM31A	0.25	Thorium	0.090
Nickel and nickel alloys		Tungsten	0.397
Nickel (99.95% Ni + Co)	0.22	Uranium	0.071
"A" nickel	0.145	Vanadium	0.074
"D" nickel	0.115	Yttrium	0.035

(a) Depends on processing. (b) 18% of Cu. (c) 12% of Cu. (d) 9.05% of Cu. (e) 15% of Cu. (f) 16% of Cu. (g) 11% of Cu. (h) 7% of Cu. (j) 6% of Cu. (k) 6.5% of Cu

Table 6 Electrical conductivity and resistivity of metals and alloys

Metal or alloy	Conductivity, % IACS	Resistivity, μΩ · cm	Metal or alloy	Conductivity, % IACS	Resistivity, μΩ · cm
Aluminum and aluminum alloys			80Pt-20Ir	5.6	31
			75Pt-25Ir	5.5	33
Aluminum (99.996%)	64.95	2.65	70Pt-30Ir	5	35
EC (O, H19)	62	2.8	65Pt-35Ir	5	36
5052 (O, H38)	35	4.93	95Pt-5Ru	5.5	31.5
5056 (H38)	27	6.4	90Pt-10Ru	4	43
6101 (T6)	56	3.1	89Pt-11Ru	4	43
Copper and copper alloys			86Pt-14Ru	3.5	46
Wrought copper			96Pt-4W	5	36
Pure copper	103.06	1.67	**Palladium and palladium alloys**		
Electrolytic (ETP)	101	1.71	Palladium	16	10.8
Oxygen-free copper (OF)	101	1.71	95.5Pd-4.5Ru	7	24.2
Free-machining copper 0.5% Te	95	1.82	90Pd-10Ru	6.5	27
1.0% Pb	98	1.76	70Pd-30Ag	4.3	40
Wrought alloys			60Pd-40Ag	4.0	43
Cartridge brass, 70%	28	6.2	50Pd-50Ag	5.5	31.5
Yellow brass	27	6.4	72Pd-26Ag-2Ni	4	43
Leaded commercial bronze	42	4.1	60Pd-40Cu	5	35(c)
Phosphor bronze, 1.25%	48	3.6	45Pd-30Ag-20Au-5Pt	4.5	39
Nickel silver, 55-18	5.5	31	35Pd-30Ag-14Cu-10Pt-10Au-1Zn	5	35
Low-silicon bronze (B)	12	14.3	**Gold and gold alloys**		
Beryllium copper	22 to 30(a)	5.7 to 7.8(a)	Gold	75	2.35
Casting alloys			90Au-10Cu	16	10.8
Chromium copper (1% Cr)	80 to 90(a)	2.10	75Au-25Ag	16	10.8
88Cu-8Sn-4Zn	11	15	72.5Au-14Cu-8.5Pt-4Ag-1Zn	10	17
87Cu-10Sn-1Pb-2Zn	11	15	69Au-25Ag-Pt	11	15
Electrical contact materials			41.7Au-32.5Cu-18.8Ni-7Zn	4.5	39
Copper alloys			**Electrical heating alloys**		
0.04 oxide	100	1.72	**Ni-Cr and Ni-Cr-Fe alloys**		
1.25 Sn + P	48	3.6	78.5Ni-20Cr-1.5Si (80-20)	1.6	108.05
5 Sn + P	18	11	73.5Ni-20Cr-5Al-1.5Si	1.2	137.97
8 Sn + P	13	13	68Ni-20Cr-8.5Fe-2Si	1.5	116.36
15 Zn	37	4.7	60Ni-16Cr-22.5Fe-1.5Si	1.5	112.20
20 Zn	32	5.4	35Ni-20Cr-43.5Fe-1.5Si	1.7	101.4
35 Zn	27	6.4	**Fe-Cr-Al alloys**		
2 Be + Ni or Co(b)	17 to 21	9.6 to 11.5	72Fe-23Cr-5Al	1.3	138.8
Silver and silver alloys			55Fe-37.5Cr-7.5Al	1.2	166.23
Fine silver	106	1.59	**Pure metals**		
92.5Ag-7.5Cu	85	2	Molybdenum	34	5.2
90Ag-10Cu	85	2	Platinum	16	10.64
72Ag-28Cu	87	2	Tantalum	13.9	12.45
72Ag-26Cu-2Ni	60	2.9	Tungsten	30	5.65
85Ag-15Cd	35	4.93	**Nonmetallic heating element materials**		
97Ag-3Pt	50	3.5	Silicon carbide, SiC	1 to 1.7	100 to 200
97Ag-3Pd	60	2.9	Molybdenum dislicide, $MoSi_2$	4.5	37.24
90Ag-10Pd	30	5.3	Graphite	...	910.1
90Ag-10Au	40	4.2	**Instrument and control alloys**		
60Ag-40Pd	8	23	**Cu-Ni alloys**		
70Ag-30Pd	12	14.3	98Cu-2Ni	35	4.99
Platinum and platinum alloys			94Cu-6Ni	17	9.93
Platinum	16	10.6	89Cu-11Ni	11	14.96
95Pt-5Ir	9	19	78Cu-22Ni	5.7	29.92
90Pt-10Ir	7	25	55Cu-45Ni (constantan)	3.5	49.87
85Pt-15Ir	6	28.5			

(*continued*)

(a) Precipitation hardened; depends on processing. (b) A heat-treated alloy. (c) Annealed and quenched. (d) At low field strength and high electrical resistance. (e) At higher field strength; annealed for optimum magnetic properties.

Table 6   Electrical conductivity and resistivity of metals and alloys (*continued*)

Metal or alloy	Conductivity, % IACS	Resistivity, μΩ · cm	Metal or alloy	Conductivity, % IACS	Resistivity, μΩ · cm
Instrument and control alloys (*continued*)			**Moderately high-permeability materials(d)**		
Cu-Mn-Ni alloys			Thermenol	0.5	162
87Cu-13Mn (manganin)	3.5	48.21	16 Alfenol	0.7	153
83Cu-13Mn-4Ni (manganin)	3.5	48.21	Sinimax	2	90
85Cu-10Mn-4Ni (manganin)	4.5	38.23	Monimax	2.5	80
70Cu-20Ni-10Mn	3.6	48.88	Supemalloy	3	65
67Cu-5Ni-27Mn	1.8	99.74	4-79 Moly Permalloy, Hymu 80	3	58
Ni-base alloys			Mumetal	3	60
99.8 Ni	23	7.98	1040 alloy	3	56
71Ni-29Fe	9	19.95	High Permalloy 49, A-L 4750, Armco 48	3.6	48
80Ni-20Cr	1.5	112.2	45 Permalloy	3.6	45
75Ni-20C-3Al+Cu or Fe	1.3	132.98	**High-permeability materials(e)**		
76Ni-17Cr-4Si-3Mn	1.3	132.98	Supermendur	4.5	40
60Ni-16Cr-24Fe	1.5	112.2	2V Permendur	4.5	40
35Ni-20Cr-45Fe	1.7	101.4	35% Co, 1% Cr	9	20
Fe-Cr-Al alloy			Ingot iron	17.5	10
72Fe-23Cr-5Al-0.5Co	1.3	135.48	0.5% Si steel	6	28
Pure metals			1.75% Si steel	4.6	37
Iron (99.99%)	17.75	9.71	3.0% Si steel	3.6	47
Thermostat metals			Grain-oriented 3.0% Si steel	3.5	50
75Fe-22Ni-3Cr	3	78.13	Grain-oriented 50% Ni iron	3.6	45
72Mn-18Cu-10Ni	1.5	112.2	50% Ni iron	3.5	50
67Ni-30Cu-1.4Fe-1Mn	3.5	56.52	**Relay steels and alloys after annealing**		
75Fe-22Ni-3Cr	12	15.79	**Low-carbon iron and steel**		
66.5Fe-22Ni-8.5Cr	3.3	58.18	Low-carbon iron	17.5	10
Permanent magnet materials			1010 steel	14.5	12
Carbon steel (0.65% C)	9.5	18	**Silicon steels**		
Carbon steel (1% C)	8	20	1% Si	7.5	23
Chromium steel (3.5% Cr)	6.1	29	2.5% Si	4	41
Tungsten steel (6% W)	6	30	3% Si	3.5	48
Cobalt steel (17% Co)	6.3	28	3% Si, grain-oriented	3.5	48
Cobalt steel (36% Co)	6.5	27	4% Si	3	59
Intermediate alloys			**Stainless steels**		
Cunico	7.5	24	Type 410	3	57
Cunife	9.5	18	Type 416	3	57
Comol	3.6	45	Type 430	3	60
Alnico alloys			Type 443	3	68
Alnico I	3.3	75	Type 446	3	61
Alnico II	3.3	65	**Nickel irons**		
Alnico III	3.3	60	50% Ni	3.5	48
Alnico IV	3.3	75	78% Ni	11	16
Alnico V	3.5	47	77% Ni (Cu, Cr)	3	60
Alnico VI	3.5	50	79% Ni (Mo)	3	58
Magnetically soft materials			**Stainless and heat-resisting alloys**		
Electrical steel sheet			Type 302	3	72
M-50	9.5	18	Type 309	2.5	78
M-43	6 to 9	20 to 28	Type 316	2.5	74
M-36	5.5 to 7.5	24 to 33	Type 317	2.5	74
M-27	3.5 to 5.5	32 to 47	Type 347	2.5	73
M-22	3.5 to 5	41 to 52	Type 403	3	57
M-19	3.5 to 5	41 to 56	Type 405	3	60
M-17	3 to 3.5	45 to 58	Type 501	4.5	40
M-15	3 to 3.5	45 to 69	HH	2.5	80
M-14	3 to 3.5	58 to 69	HK	2	90
M-7	3 to 3.5	45 to 52	HT	1.7	100
M-6	3 to 3.5	45 to 52			
M-5	3 to 3.5	45 to 52			

(a) Precipitation hardened; depends on processing. (b) A heat-treated alloy. (c) Annealed and quenched. (d) At low field strength and high electrical resistance. (e) At higher field strength; annealed for optimum magnetic properties.

Table 7   Vapor pressures of the metallic elements

| Element | Pressure, atm | | | | | | | | | | | |
| | 0.0001 | | 0.001 | | 0.01 | | 0.1 | | 0.50 | | 1.0 | |
	°C	°F	°C	°F	°C	°F	°C	°F	°C	°F	°C	°F
Aluminum	1110	2030	1263	2305	1461	2662	1713	3115	1940	3524	2056	3773
Antimony	759	1398	872	1602	1013	1855	1196	2185	1359	2478	1440	2624
Arsenic	308	586	363	685	428	802	499	930	578	1072	610	1130
Bismuth	914	1677	1008	1846	1121	2050	1254	2289	1367	2493	1420	2588
Cadmium	307(a)	585(a)	384(b)	723(b)	471	880	594	1101	708	1306	765	1409
Calcium	688	1270	802(c)	1476(c)	958(b)	1756(b)	1175	2149	1380	2516	1487	2709
Carbon	3257	5895	3547	6417	3897	7047	4317	7803	4667	8433	4827	8721
Chromium	1420(a)	2588(a)	1594(b)	2901(b)	1813	3295	2097	3807	2351	4264	2482	4500
Copper	1412	2574	1602	2916	1844	3351	2162	3924	2450	4442	2595	4703
Gallium	1178	2152	1329	2424	1515	2759	1751	3184	1965	3569	2071	3760
Gold	1623	2953	1839	3342	2115	3839	2469	4476	2796	5065	2966	5371
Iron	1564	2847	1760	3200	2004	3639	2316	4201	2595	4703	2735	4955
Lead	815	1499	953	1747	1135	2075	1384	2523	1622	2952	1744	3171
Lithium	592	1098	707	1305	858	1576	1064	1947	1266	2311	1372	2502
Magnesium	516	961	608(a)	1126(a)	725(b)	1337(b)	886	1627	1030	1886	1107	2025
Manganese	1115(d)	2039(d)	1269(b)	2316(b)	1476	2889	1750	3182	2019	3666	2151	3904
Mercury	77.9(b)	172.2(b)	120.8	249.4	176.1	349.0	251.3	484.3	321.5	610.7	357	675
Molybdenum	2727	4941	3057	5535	3477	6291	4027	7281	4537	8199	4804	8679
Nickel	1586	2887	1782	3240	2025	3677	2321	4210	2593	4699	2732	4950
Platinum	2367	4293	2687	4869	3087	5589	3637	6579	4147	7497	4407	7965
Potassium	261	502	332	630	429	804	565	1051	704	1299	774	1425
Rubidium	223	433	288	550	377	711	497	927	617	1143	679	1254
Selenium	282	540	347	657	430	806	540	1004	634	1173	680	1256
Silicon	1572	2862	1707	3105	1867	3393	2057	3735	2217	4023	2287	4149
Silver	1169	2136	1334	2433	1543	2809	1825	3317	2081	3778	2212	4014
Sodium	349	660	429	804	534	993	679	1254	819	1506	892	1638
Strontium	...	...	(a)	(a)	877(b)	1629(b)	1081	1978	1279	2334	1384	2523
Tellurium	(a)	(a)	509(b)	948(b)	632	1170	810	1490	991	1816	1087	1989
Thallium	692	1277	809	1488	962	1764	1166	2131	1359	2478	1457	2655
Tin	...	...	...	...	...	...	1932(b)	3510(b)	2163	3925	2270	4118
Tungsten	3547	6417	3937	7119	4437	8019	5077	9171	5647	10197	5927	10701
Zinc	399(a)	750(a)	477(b)	891(b)	579	1074	717	1323	842	1548	907	1665

(a) In the solid state. (b) In the liquid state. (c) β. (d) γ

Table 8 Standard reduction potentials of metals

Add the correction given at 18 °C (64 °F) if the activity (effective concentration of metallic ions) $C \times$ standard;
If activity is less than standard, log C will be negative and the potential will be reduced.

Metal	Ion considered	Standard reduction potential, V	Correction(a)
Gold	Au^{3+}	+1.50	0.019 log$C$
Platinum	Pt^{2+}	+1.2	0.029 log$C$
Silver	Ag^+	+0.799	0.058 log$C$
Mercury	$(Hg)_2 2+$	+0.789	0.029 log$C$
Copper	Cu^{2+}	+0.337	0.020 log$C$
Hydrogen (1 atm)	H^+	+0.000	0.058 log$C$
Lead	Pb^{2+}	−0.126	0.029 log$C$
Tin	Sn^{2+}	−0.136	0.029 log$C$
Nickel	Ni^{2+}	−0.250	0.029 log$C$
Cobalt	Co^{2+}	−0.28	0.029 log$C$
Cadmium	Cd^{2+}	−0.403	0.029 log$C$
Iron	Fe^{2+}	−0.440	0.029 log$C$
Chromium	Cr^{3+}	−0.74	0.019 log$C$
Zinc	Zn^{2+}	−0.763	(a)
Manganese	Mn^{2+}	−1.18	(a)
Titanium(b)	Ti^{2+}	−1.63	(a)
Aluminum(b)	Al^{3+}	−1.66	(a)
Magnesium(b)	Mg^{2+}	−2.37	(a)
Sodium(b)	Na^{2+}	−2.71	(a)
Calcium(b)	Ca^{2+}	−2.87	(a)
Potassium(b)	K^+	−2.92	(a)
Lithium(b)	Li^+	−3.02	(a)

(a) Potential almost independent of original concentration of metal ions. (b) Calculated values, of theoretical interest only. Aluminum, unless amalgamated, gives more positive values owing to the presence of an oxide film. The other metals evolve hydrogen freely, and potential measurements made directly would not represent equilibrium values.

Table 9 The 45 most abundant elements in the earth's crust

Relative abundance	Element	Abundance, ppm wt	Relative abundance	Element	Abundance, ppm wt
1	Oxygen	466 000	24	Zinc	70
2	Silicon	277 000	25	Cerium	60
3	Aluminum	81 300	26	Copper	55
4	Iron	50 000	27	Yttrium	33
5	Calcium	36 300	28	Lanthanum	30
6	Sodium	28 300	29	Neodymium	28
7	Potassium	25 900	30	Cobalt	25
8	Magnesium	20 900	31	Scandium	22
9	Titanium	4 400	32	Lithium	20
10	Hydrogen	1 400	33	Niobium	20
11	Phosphorus	1 050	34	Nitrogen	20
12	Manganese	950	35	Gallium	15
13	Fluorine	625	36	Lead	13
14	Barium	425	37	Radium	13
15	Strontium	375	38	Boron	10
16	Sulfur	260	39	Krypton	9.8
17	Carbon	200	40	Praseodymium	8.2
18	Zirconium	165	41	Protoactinium	8.0
19	Vanadium	135	42	Thorium	7.2
20	Chlorine	130	43	Neon	7.0
21	Chromium	100	44	Samarium	6.0
22	Rubidium	90	45	Gadolinium	5.4
23	Nickel	75			

Table 10 The electrochemical series

In this table, the elements are electropositive to the ones which follow them and will displace them from solutions of their salts

1	Cesium	18	Didymium	35	Mercury	52	Boron
2	Rubidium	19	Lanthanum	36	Palladium	53	Tungsten
3	Potassium	20	Manganese	37	Ruthenium	54	Molybdenum
4	Sodium	21	Zinc	38	Rhodium	55	Vanadium
5	Lithiuim	22	Iron	39	Platinum	56	Chromium
6	Barium	23	Nickel	40	Iridium	57	Arsenic
7	Strontium	24	Cobalt	41	Osmium	58	Phosphorus
8	Calcium	25	Thallium	42	Gold	59	Selenium
9	Magnesium	26	Cadmium	43	Hydrogen	60	Iodine
10	Beryllium	27	Lead	44	Tin	61	Bromine
11	Ytterbium	28	Germanium	45	Silicon	62	Chlorine
12	Erbium	29	Indium	46	Titanium	63	Fluorine
13	Scandium	30	Gallium	47	Niobium	64	Nitrogen
14	Aluminum	31	Bismuth	48	Tantalum	65	Sulfur
15	Zirconium	32	Uranium	49	Tellurium	66	Oxygen
16	Thorium	33	Copper	50	Antimony		
17	Cerium	34	Silver	51	Carbon		

Table 11 Metal melting range and color scale

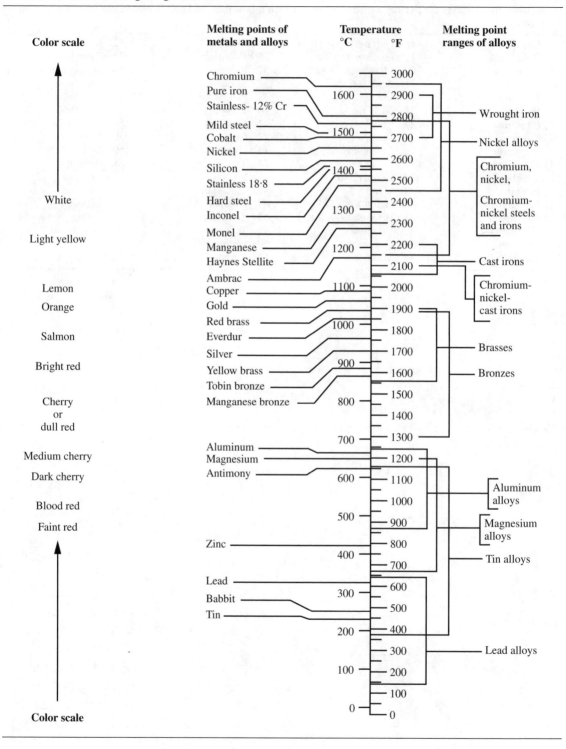

Table 12   Predominant flame colors of metallic elements

Element	Color	Element	Color	Element	Color
Lithium	Deep red	Boron	Green	Selenium	Blue
Strontium	Crimson	Tellurium	Deep green	Indium	Deep blue
Calcium	Yellow-red	Thallium	Greenish blue	Potassium	Purple-red
Sodium	Bright yellow	Antimony	Blue-green	Rubidium	Violet
Barium	Yellow-green	Copper	Green-blue	Cesium	Bluish purple
Molybdenum	Green-yellow	Arsenic	Light blue		
Zinc	Light green	Lead	Light blue		

Table 13   Average percentage of metals in igneous rocks

Metal	Percentage	Metal	Percentage	Metal	Percentage
Silicon	27.72	Chromium	0.037	Niobium, tantalum	0.003
Aluminum	8.13	Zirconium	0.026	Hafnium	0.003
Iron	5.01	Nickel	0.020	Thorium	0.002
Calcium	3.63	Vanadium	0.017	Lead	0.002–
Sodium	2.85	Rare earths	0.015	Cobalt	0.001
Potassium	2.60	Copper	0.010	Beryllium	0.001
Titanium	0.63	Tungsten	0.005	Strontium	0.001–
Manganese	0.10	Lithium	0.004	Uranium	0.001–
Barium	0.05	Zinc	0.004		

Table 14 Temperature conversions

The general arrangement of this table was devised by Sauveur and Boylston more than 60 years ago. The middle column of figures (in bold-faced type) contains the reading (°F or °C) to be converted. If converting from degrees Fahrenheit to degrees Centigrade, read the Centigrade equivalent in the column head "°C." If converting from Centigrade to Fahrenheit, read the Fahrenheit equivalent in the column headed "°F." $°C = \frac{5}{9}(°F - 32)$

°F		°C	°F		°C	°F		°C	°F		°C
...	**−458**	−272.22	...	**−350**	−212.22	−403.6	**−242**	−152.22	−209.2	**−134**	−92.22
...	**−456**	−271.11	...	**−348**	−211.11	−400.0	**−240**	−151.11	−205.6	**−132**	−91.11
...	**−454**	−270.00	...	**−346**	−210.00	−396.4	**−238**	−150.00	−202.0	**−130**	−90.00
...	**−452**	−268.89	...	**−344**	−208.89	−392.8	**−236**	−148.89	−198.4	**−128**	−88.89
...	**−450**	−267.78	...	**−342**	−207.78	−389.2	**−234**	−147.78	−194.8	**−126**	−87.78
...	**−448**	−266.67	...	**−340**	−206.67	−385.6	**−232**	−146.67	−191.2	**−124**	−86.67
...	**−446**	−265.56	...	**−338**	−205.56	−382.0	**−230**	−145.56	−187.6	**−122**	−85.56
...	**−444**	−264.44	...	**−336**	−204.44	−378.4	**−228**	−144.44	−184.0	**−120**	−84.44
...	**−442**	−263.33	...	**−334**	−203.33	−374.8	**−226**	−143.33	−180.4	**−118**	−83.33
...	**−440**	−262.22	...	**−332**	−202.22	−371.2	**−224**	−142.22	−176.8	**−116**	−82.22
...	**−438**	−261.11	...	**−330**	−201.11	−367.6	**−222**	−141.11	−173.2	**−114**	−81.11
...	**−436**	−260.00	...	**−328**	−200.00	−364.0	**−220**	−140.00	−169.6	**−112**	−80.00
...	**−434**	−258.89	...	**−326**	−198.89	−360.4	**−218**	−138.89	−166.0	**−110**	−78.89
...	**−432**	−257.78	...	**−324**	−197.78	−356.8	**−216**	−137.78	−162.4	**−108**	−77.78
...	**−430**	−256.67	...	**−322**	−196.67	−353.2	**−214**	−136.67	−158.8	**−106**	−76.67
...	**−428**	−255.56	...	**−320**	−195.56	−349.6	**−212**	−135.56	−155.2	**−104**	−75.56
...	**−426**	−254.44	...	**−318**	−194.44	−346.0	**−210**	−134.44	−151.6	**−102**	−74.44
...	**−424**	−253.33	...	**−316**	−193.33	−342.4	**−208**	−133.33	−148.0	**−100**	−73.33
...	**−422**	−252.22	...	**−314**	−192.22	−338.8	**−206**	−132.22	−144.4	**−98**	−72.22
...	**−420**	−251.11	...	**−312**	−191.11	−335.2	**−204**	−131.11	−140.8	**−96**	−71.11
...	**−418**	−250.00	...	**−310**	−190.00	−331.6	**−202**	−130.00	−137.2	**−94**	−70.00
...	**−416**	−248.89	...	**−308**	−188.89	−328.0	**−200**	−128.89	−133.6	**−92**	−68.89
...	**−414**	−247.78	...	**−306**	−187.78	−324.4	**−198**	−127.78	−130.0	**−90**	−67.78
...	**−412**	−246.67	...	**−304**	−186.67	−320.8	**−196**	−126.67	−126.4	**−88**	−66.67
...	**−410**	−245.56	...	**−302**	−185.56	−317.2	**−194**	−125.56	−122.8	**−86**	−65.56
...	**−408**	−244.44	...	**−300**	−184.44	−313.6	**−192**	−124.44	−119.2	**−84**	−64.44
...	**−406**	−243.33	...	**−298**	−183.33	−310.0	**−190**	−123.33	−115.6	**−82**	−63.33
...	**−404**	−242.22	...	**−296**	−182.22	−306.4	**−188**	−122.22	−112.0	**−80**	−62.22
...	**−402**	−241.11	...	**−294**	−181.11	−302.8	**−186**	−121.11	−108.4	**−78**	−61.11
...	**−400**	−240.00	...	**−292**	−180.00	−299.2	**−184**	−120.00	−104.8	**−76**	−60.00
...	**−398**	−238.89	...	**−290**	−178.89	−295.6	**−182**	−118.89	−101.2	**−74**	−58.89
...	**−396**	−237.78	...	**−288**	−177.78	−292.0	**−180**	−117.78	−97.6	**−72**	−57.78
...	**−394**	−236.67	...	**−286**	−176.67	−288.4	**−178**	−116.67	−94.0	**−70**	−56.67
...	**−392**	−235.56	...	**−284**	−175.56	−284.8	**−176**	−115.56	−90.4	**−68**	−55.56
...	**−390**	−234.44	...	**−282**	−174.44	−281.2	**−174**	−114.44	−86.8	**−66**	−54.44
...	**−388**	−233.33	...	**−280**	−173.33	−277.6	**−172**	−113.33	−83.2	**−64**	−53.33
...	**−386**	−232.22	...	**−278**	−172.22	−274.0	**−170**	−112.22	−79.6	**−62**	−52.22
...	**−384**	−231.11	...	**−276**	−171.11	−270.4	**−168**	−111.11	−76.0	**−60**	−51.11
...	**−382**	−230.00	...	**−274**	−170.00	−266.8	**−166**	−110.00	−72.4	**−58**	−50.00
...	**−380**	−228.89	−457.6	**−272**	−168.89	−263.2	**−164**	−108.89	−68.8	**−56**	−48.89
...	**−378**	−227.78	−454.0	**−270**	−167.78	−259.6	**−162**	−107.78	−65.2	**−54**	−47.78
...	**−376**	−226.67	−450.4	**−268**	−166.67	−256.0	**−160**	−106.67	−61.6	**−52**	−46.67
...	**−374**	−225.56	−446.8	**−266**	−165.56	−252.4	**−158**	−105.56	−58.0	**−50**	−45.56
...	**−372**	−224.44	−443.2	**−264**	−164.44	−248.8	**−156**	−104.44	−54.4	**−48**	−44.44
...	**−370**	−223.33	−439.6	**−262**	−163.33	−245.2	**−154**	−103.33	−50.8	**−46**	−43.33
...	**−368**	−222.22	−436.0	**−260**	−162.22	−241.6	**−152**	−102.22	−47.2	**−44**	−42.22
...	**−366**	−221.11	−432.4	**−258**	−161.11	−238.0	**−150**	−101.11	−43.6	**−42**	−41.11
...	**−364**	−220.00	−428.8	**−256**	−160.00	−234.4	**−148**	−100.00	−40.0	**−40**	−40.00
...	**−362**	−218.89	−425.2	**−254**	−158.89	−230.8	**−146**	−98.89	−36.4	**−38**	−38.89
...	**−360**	−217.78	−421.6	**−252**	−157.78	−227.2	**−144**	−97.78	−32.8	**−36**	−37.78
...	**−358**	−216.67	−418.0	**−250**	−156.67	−223.6	**−142**	−96.67	−29.2	**−34**	−36.67
...	**−356**	−215.56	−414.4	**−248**	−155.56	−220.0	**−140**	−95.56	−25.6	**−32**	−35.56
...	**−354**	−214.44	−410.8	**−246**	−154.44	−216.4	**−138**	−94.44	−22.0	**−30**	−34.44
...	**−352**	−213.33	−407.2	**−244**	−153.33	−212.8	**−136**	−93.33	−18.4	**−28**	−33.33

(continued)

Table 14 Temperature conversions (*continued*)

The general arrangement of this table was devised by Sauveur and Boylston more than 60 years ago. The middle column of figures (in bold-faced type) contains the reading (°F or °C) to be converted. If converting from degrees Fahrenheit to degrees Centigrade, read the Centigrade equivalent in the column head "°C." If converting from Centigrade to Fahrenheit, read the Fahrenheit equivalent in the column headed "°F." °C = $\frac{5}{9}$ (°F −32)

°F		°C	°F		°C	°F		°C	°F		°C
−14.8	**−26**	−32.22	179.6	**82**	27.78	374.0	**190**	87.78	568.4	**298**	147.78
−11.2	**−24**	−31.11	183.2	**84**	28.89	377.6	**192**	88.89	572.0	**300**	148.89
−7.6	**−22**	−30.00	186.8	**86**	30.00	381.2	**194**	90.00	575.6	**302**	150.00
−4.0	**−20**	−28.89	190.4	**88**	31.11	384.8	**196**	91.11	579.2	**304**	151.11
−0.4	**−18**	−27.78	194.0	**90**	32.22	388.4	**198**	92.22	582.8	**306**	152.22
3.2	**−16**	−26.67	197.6	**92**	33.33	392.0	**200**	93.33	586.4	**308**	153.33
6.8	**−14**	−25.56	201.2	**94**	34.44	395.6	**202**	94.44	590.0	**310**	154.44
10.4	**−12**	−24.44	204.8	**96**	35.56	399.2	**204**	95.56	593.6	**312**	155.56
14.0	**−10**	−23.33	208.4	**98**	36.67	402.8	**206**	96.67	597.2	**314**	156.67
17.6	**−8**	−22.22	212.0	**100**	37.78	406.4	**208**	97.78	600.8	**316**	157.78
21.2	**−6**	−21.11	215.6	**102**	38.89	410.0	**210**	98.89	604.4	**318**	158.89
24.8	**−4**	−20.00	219.2	**104**	40.00	413.6	**212**	100.00	608.0	**320**	160.00
28.4	**−2**	−18.89	222.8	**106**	41.11	417.2	**214**	101.11	611.6	**322**	161.11
32.0	**±0**	−17.78	226.4	**108**	42.22	420.8	**216**	102.22	615.2	**324**	162.22
35.6	**2**	−16.67	230.0	**110**	43.33	424.4	**218**	103.33	618.8	**326**	163.33
39.2	**4**	−15.56	233.6	**112**	44.44	428.0	**220**	104.44	622.4	**328**	164.44
42.8	**6**	−14.44	237.2	**114**	45.56	431.6	**222**	105.56	626.0	**330**	165.56
46.4	**8**	−13.33	240.8	**116**	46.67	435.2	**224**	106.67	629.6	**332**	166.67
50.0	**10**	−12.22	244.4	**118**	47.78	438.8	**226**	107.78	633.2	**334**	167.78
53.6	**12**	−11.11	248.0	**120**	48.89	442.4	**228**	108.89	636.8	**336**	168.89
57.2	**14**	−10.00	251.6	**122**	50.00	446.0	**230**	110.00	640.4	**338**	170.00
60.8	**16**	−8.89	255.2	**124**	51.11	449.6	**232**	111.11	644.0	**340**	171.11
64.4	**18**	−7.78	258.8	**126**	52.22	453.2	**234**	112.22	647.6	**342**	172.22
68.0	**20**	−6.67	262.4	**128**	53.33	456.8	**236**	113.33	651.2	**344**	173.33
71.6	**22**	−5.56	266.0	**130**	54.44	460.4	**238**	114.44	654.8	**346**	174.44
75.2	**24**	−4.44	269.6	**132**	55.56	464.0	**240**	115.56	658.4	**348**	175.56
78.8	**26**	−3.33	273.2	**134**	56.67	467.6	**242**	116.67	662.0	**350**	176.67
82.4	**28**	−2.22	276.8	**136**	57.78	471.2	**244**	117.78	665.6	**352**	177.78
86.0	**30**	−1.11	280.4	**138**	58.89	474.8	**246**	118.89	669.2	**354**	178.89
89.6	**32**	±0.00	284.0	**140**	60.00	478.4	**248**	120.00	672.8	**356**	180.00
93.2	**34**	1.11	287.6	**142**	61.11	482.0	**250**	121.11	676.4	**358**	181.11
96.8	**36**	2.22	291.2	**144**	62.22	485.6	**252**	122.22	680.0	**360**	182.22
100.4	**38**	3.33	294.8	**146**	63.33	489.2	**254**	123.33	683.6	**362**	183.33
104.0	**40**	4.44	298.4	**148**	64.44	492.8	**256**	124.44	687.2	**364**	184.44
107.6	**42**	5.56	302.0	**150**	65.56	496.4	**258**	125.56	690.8	**366**	185.56
111.2	**44**	6.67	305.6	**152**	66.67	500.0	**260**	126.67	694.4	**368**	186.67
114.8	**46**	7.78	309.2	**154**	67.78	503.6	**262**	127.78	698.0	**370**	187.78
118.4	**48**	8.89	312.8	**156**	68.89	507.2	**264**	128.89	701.6	**372**	188.89
122.0	**50**	10.00	316.4	**158**	70.00	510.8	**266**	130.00	705.2	**374**	190.00
125.6	**52**	11.11	320.0	**160**	71.11	514.4	**268**	131.11	708.8	**376**	191.11
129.2	**54**	12.12	323.6	**162**	72.22	518.0	**270**	132.22	712.4	**378**	192.22
132.8	**56**	13.33	327.2	**164**	73.33	521.6	**272**	133.33	716.0	**380**	193.33
136.4	**58**	14.44	330.8	**166**	74.44	525.2	**274**	134.44	719.6	**382**	194.44
140.0	**60**	15.56	334.4	**168**	75.56	528.8	**276**	135.56	723.2	**384**	195.56
143.6	**62**	16.67	338.0	**170**	76.67	532.4	**278**	136.67	726.8	**386**	196.67
147.2	**64**	17.78	341.6	**172**	77.78	536.0	**280**	137.78	730.4	**388**	197.78
150.8	**66**	18.89	345.2	**174**	78.89	539.6	**282**	138.89	734.0	**390**	198.89
154.4	**68**	20.00	348.8	**176**	80.00	543.2	**284**	140.00	737.6	**392**	200.00
158.0	**70**	21.11	352.4	**178**	81.11	546.8	**286**	141.11	741.2	**394**	201.11
161.6	**72**	22.22	356.0	**180**	82.22	550.4	**288**	142.22	744.8	**396**	202.22
165.2	**74**	23.33	359.6	**182**	83.33	554.0	**290**	143.33	748.4	**398**	203.33
168.8	**76**	24.44	363.2	**184**	84.44	557.6	**292**	144.44	752.0	**400**	204.44
172.4	**78**	25.56	366.8	**186**	85.56	561.2	**294**	145.56	755.6	**402**	205.56
176.0	**80**	26.67	370.4	**188**	86.67	564.8	**296**	146.67	759.2	**404**	206.67

(*continued*)

Table 14 Temperature conversions (*continued*)

The general arrangement of this table was devised by Sauveur and Boylston more than 60 years ago. The middle column of figures (in bold-faced type) contains the reading (°F or °C) to be converted. If converting from degrees Fahrenheit to degrees Centigrade, read the Centigrade equivalent in the column head "°C." If converting from Centigrade to Fahrenheit, read the Fahrenheit equivalent in the column headed "°F." $°C = \frac{5}{9}(°F - 32)$

°F		°C	°F		°C	°F		°C	°F		°C
762.8	406	207.78	957.2	514	267.78	1670.0	910	487.78	2642.0	1450	787.78
766.4	408	208.89	960.8	516	268.89	1688.0	920	493.33	2660.0	1460	793.33
770.0	410	210.00	964.4	518	270.00	1706.0	930	498.89	2678.0	1470	798.89
773.6	412	211.11	968.0	520	271.11	1724.0	940	504.44	2696.0	1480	804.44
777.2	414	212.22	971.6	522	272.22	1742.0	950	510.00	2714.0	1490	810.00
780.8	416	213.33	975.2	524	273.33	1760.0	960	515.56	2732.0	1500	815.56
784.4	418	214.44	978.8	526	274.44	1778.0	970	521.11	2750.0	1510	821.11
788.0	420	215.56	982.4	528	275.56	1796.0	980	526.67	2768.0	1520	826.67
791.6	422	216.67	986.0	530	276.67	1814.0	990	532.22	2786.0	1530	832.22
795.2	424	217.78	989.6	532	277.78	1832.0	1000	537.78	2804.0	1540	837.78
798.8	426	218.89	993.2	534	278.89	1850.0	1010	543.33	2822.0	1550	843.33
802.4	428	220.00	996.8	536	280.00	1868.0	1020	548.89	2840.0	1560	848.89
806.0	430	221.11	1000.4	538	281.11	1886.0	1030	554.44	2858.0	1570	854.44
809.6	432	222.22	1004.0	540	282.22	1904.0	1040	560.00	2876.0	1580	860.00
813.2	434	223.33	1007.6	542	283.33	1922.0	1050	565.56	2894.0	1590	865.56
816.8	436	224.44	1011.2	544	284.44	1940.0	1060	571.11	2912.0	1600	871.11
820.4	438	225.56	1014.8	546	285.56	1958.0	1070	576.67	2930.0	1610	876.67
824.0	440	226.67	1018.4	548	286.67	1976.0	1080	582.22	2948.0	1620	882.22
827.6	442	227.78	1022.0	550	287.78	1994.0	1090	587.78	2966.0	1630	887.78
831.2	444	228.89	1040.0	560	293.33	2012.0	1100	593.33	2984.0	1640	893.33
834.8	446	230.00	1058.0	570	298.89	2030.0	1110	598.89	3002.0	1650	898.89
838.4	448	231.11	1076.0	580	304.44	2048.0	1120	604.44	3020.0	1660	904.44
842.0	450	232.22	1094.0	590	310.00	2066.0	1130	610.00	3038.0	1670	910.00
845.6	452	233.33	1112.0	600	315.56	2084.0	1140	615.56	3056.0	1680	915.56
849.2	454	234.44	1130.0	610	321.11	2102.0	1150	621.11	3074.0	1690	921.11
852.8	456	235.56	1148.0	620	326.67	2120.0	1160	626.67	3092.0	1700	926.67
856.4	458	236.67	1166.0	630	332.22	2138.0	1170	632.22	3110.0	1710	932.22
860.0	460	237.78	1184.0	640	337.78	2156.0	1180	637.78	3128.0	1720	937.78
863.6	462	238.89	1202.0	650	343.33	2174.0	1190	643.33	3146.0	1730	943.33
867.2	464	240.00	1220.0	660	348.89	2192.0	1200	648.89	3164.0	1740	948.89
870.8	466	241.11	1238.0	670	354.44	2210.0	1210	654.44	3182.0	1750	954.44
874.4	468	242.22	1256.0	680	360.00	2228.0	1220	660.00	3200.0	1760	960.00
878.0	470	243.33	1274.0	690	365.56	2246.0	1230	665.56	3218.0	1770	965.56
881.6	472	244.44	1292.0	700	371.11	2264.0	1240	671.11	3236.0	1780	971.11
885.2	474	245.56	1310.0	710	376.67	2282.0	1250	676.67	3254.0	1790	976.67
888.8	476	246.67	1328.0	720	382.22	2300.0	1260	682.22	3272.0	1800	982.22
892.4	478	247.78	1346.0	730	387.78	2318.0	1270	687.78	3290.0	1810	987.78
896.0	480	248.89	1364.0	740	393.33	2336.0	1280	693.33	3308.0	1820	993.33
899.6	482	250.00	1382.0	750	398.89	2354.0	1290	698.89	3326.0	1830	998.89
903.2	484	251.11	1400.0	760	404.44	2372.0	1300	704.44	3344.0	1840	1004.4
906.8	486	252.22	1418.0	770	410.00	2390.0	1310	710.00	3362.0	1850	1010.0
910.4	488	253.33	1436.0	780	415.56	2408.0	1320	715.56	3380.0	1860	1015.6
914.0	490	254.44	1454.0	790	421.11	2426.0	1330	721.11	3398.0	1870	1021.1
917.6	492	255.56	1472.0	800	426.67	2444.0	1340	726.67	3416.0	1880	1026.7
921.2	494	256.67	1490.0	810	432.22	2462.0	1350	732.22	3434.0	1890	1032.2
924.8	496	257.78	1508.0	820	437.76	2480.0	1360	737.78	3452.0	1900	1037.8
928.4	498	258.89	1526.0	830	443.33	2498.0	1370	743.33	3470.0	1910	1043.3
932.0	500	260.00	1544.0	840	448.89	2516.0	1380	748.89	3488.0	1920	1048.9
935.6	502	261.11	1562.0	850	454.44	2534.0	1390	754.44	3506.0	1930	1054.4
939.2	504	262.22	1580.0	860	460.00	2552.0	1400	760.00	3524.0	1940	1060.0
942.8	506	263.33	1598.0	870	465.56	2570.0	1410	765.56	3542.0	1950	1065.6
946.4	508	264.44	1616.0	880	471.11	2588.0	1420	771.11	3560.0	1960	1071.1
950.0	510	265.56	1634.0	890	476.67	2606.0	1430	776.67	3578.0	1970	1076.7
953.6	512	266.67	1652.0	900	482.22	2624.0	1440	782.22	3596.0	1980	1082.2

(*continued*)

Table 14   Temperature conversions (*continued*)

The general arrangement of this table was devised by Sauveur and Boylston more than 60 years ago. The middle column of figures (in bold-faced type) contains the reading (°F or °C) to be converted. If converting from degrees Fahrenheit to degrees Centigrade, read the Centigrade equivalent in the column head "°C." If converting from Centigrade to Fahrenheit, read the Fahrenheit equivalent in the column headed "°F." °C $= \frac{5}{9}$(°F -32)

°F		°C	°F		°C	°F		°C	°F		°C
3614.0	**1990**	1087.8	4352.0	**2400**	1315.6	5090.0	**2810**	1543.3	6692.0	**3700**	2037.7
3632.0	**2000**	1093.3	4370.0	**2410**	1321.1	5108.0	**2820**	1548.9	6782.0	**3750**	2065.5
3650.0	**2010**	1098.9	4388.0	**2420**	1326.7	5126.0	**2830**	1554.4	6872.0	**3800**	2093.3
3668.0	**2020**	1104.4	4406.0	**2430**	1332.2	5144.0	**2840**	1560.0	6962.0	**3850**	2121.1
3686.0	**2030**	1110.0	4424.0	**2440**	1337.8	5162.0	**2850**	1565.6	7052.0	**3900**	2148.8
3704.0	**2040**	1115.6	4442.0	**2450**	1343.3	5180.0	**2860**	1571.1	7142.0	**3950**	2176.6
3722.0	**2050**	1121.1	4460.0	**2460**	1348.9	5198.0	**2870**	1576.7	7232.0	**4000**	2204.4
3740.0	**2060**	1126.7	4478.0	**2470**	1354.4	5216.0	**2880**	1582.2	7322.0	**4050**	2232.2
3758.0	**2070**	1132.2	4496.0	**2480**	1360.0	5234.0	**2890**	1587.8	7412.0	**4100**	2260.0
3776.0	**2080**	1137.8	4514.0	**2490**	1365.6	5252.0	**2900**	1593.3	7502.0	**4150**	2287.7
3794.0	**2090**	1143.3	4532.0	**2500**	1371.1	5270.0	**2910**	1598.9	7592.0	**4200**	2315.5
3812.0	**2100**	1148.9	4550.0	**2510**	1376.7	5288.0	**2920**	1604.4	7682.0	**4250**	2343.3
3830.0	**2110**	1154.4	4568.0	**2520**	1382.2	5306.0	**2930**	1610.0	7772.0	**4300**	2371.1
3848.0	**2120**	1160.0	4586.0	**2530**	1387.8	5324.0	**2940**	1615.6	7862.0	**4350**	2398.8
3866.0	**2130**	1165.6	4604.0	**2540**	1393.3	5342.0	**2950**	1621.1	7952.0	**4400**	2426.6
3884.0	**2140**	1171.1	4622.0	**2550**	1398.9	5360.0	**2960**	1626.7	8042.0	**4450**	2454.4
3902.0	**2150**	1176.7	4640.0	**2560**	1404.4	5378.0	**2970**	1632.2	8132.0	**4500**	2482.2
3920.0	**2160**	1182.2	4658.0	**2570**	1410.0	5396.0	**2980**	1637.8	8222.0	**4550**	2510.0
3938.0	**2170**	1187.8	4676.0	**2580**	1415.6	5414.0	**2990**	1643.3	8312.0	**4600**	2537.7
3956.0	**2180**	1193.3	4694.0	**2590**	1421.1	5432.0	**3000**	1648.9	8402.0	**4650**	2565.5
3974.0	**2190**	1198.9	4712.0	**2600**	1426.7	5450.0	**3010**	1654.4	8492.0	**4700**	2593.3
3992.0	**2200**	1204.4	4730.0	**2610**	1432.2	5468.0	**3020**	1660.0	8582.0	**4750**	2621.1
4010.0	**2210**	1210.0	4748.0	**2620**	1437.8	5486.0	**3030**	1665.6	8672.0	**4800**	2648.8
4028.0	**2220**	1215.6	4766.0	**2630**	1443.3	5504.0	**3040**	1671.1	8762.0	**4850**	2676.6
4046.0	**2230**	1221.1	4784.0	**2640**	1448.9	5522.0	**3050**	1676.7	8852.0	**4900**	2704.4
4064.0	**2240**	1226.7	4802.0	**2650**	1454.4	5540.0	**3060**	1682.2	8942.0	**4950**	2732.2
4082.0	**2250**	1232.2	4820.0	**2660**	1460.0	5558.0	**3070**	1687.8	9032.0	**5000**	2760.0
4100.0	**2260**	1237.8	4838.0	**2670**	1465.6	5576.0	**3080**	1693.3	9122.0	**5050**	2787.7
4118.0	**2270**	1243.3	4856.0	**2680**	1471.1	5594.0	**3090**	1698.9	9212.0	**5100**	2815.5
4136.0	**2280**	1248.9	4874.0	**2690**	1476.7	5612.0	**3100**	1704.4	9302.0	**5150**	2843.3
4154.0	**2290**	1254.4	4892.0	**2700**	1482.2	5702.0	**3150**	1732.2	9392.0	**5200**	2871.1
4172.0	**2300**	1260.0	4910.0	**2710**	1487.8	5792.0	**3200**	1760.0	9482.0	**5250**	2898.8
4190.0	**2310**	1265.6	4928.0	**2720**	1493.3	5882.0	**3250**	1787.7	9572.0	**5300**	2926.6
4208.0	**2320**	1271.1	4946.0	**2730**	1498.9	5972.0	**3300**	1815.5	9662.0	**5350**	2954.4
4226.0	**2330**	1276.7	4964.0	**2740**	1504.4	6062.0	**3350**	1843.3	9752.0	**5400**	2982.2
4244.0	**2340**	1282.2	4982.0	**2750**	1510.0	6152.0	**3400**	1871.1	9842.0	**5450**	3010.0
4262.0	**2350**	1287.8	5000.0	**2760**	1515.6	6242.0	**3450**	1898.8	9932.0	**5500**	3037.7
4280.0	**2360**	1293.3	5018.0	**2770**	1521.1	6332.0	**3500**	1926.6	10022.0	**5550**	3065.5
4298.0	**2370**	1298.9	5036.0	**2780**	1526.7	6422.0	**3550**	1954.4	10112.0	**5600**	3093.3
4316.0	**2380**	1304.4	5054.0	**2790**	1532.2	6512.0	**3600**	1982.2			
4334.0	**2390**	1310.0	5072.0	**2800**	1537.8	6602.0	**3650**	2010.0			

Table 15 Metric stress or pressure conversions

The middle column of figures (in bold-faced type) contains the reading (in MPa or ksi) to be converted. If converting from ksi to MPa, read the MPa equivalent in the column headed "MPa." If converting from MPa to ksi, read the ksi equivalent in the column headed "ksi." 1 ksi = 6.894757 MPa. 1 psi = 6.894757 kPa

ksi		MPa	ksi		MPa	ksi		MPa	ksi		MPa
0.14504	1	6.895	7.3969	51	351.63	15.954	110	758.42	88.473	610	...
0.29008	2	13.790	7.5420	52	358.53	17.405	120	827.37	89.923	620	...
0.43511	3	20.684	7.6870	53	365.42	18.855	130	896.32	91.374	630	...
0.58015	4	27.579	7.8320	54	372.32	20.305	140	965.27	92.824	640	...
0.72519	5	34.474	7.9771	55	379.21	21.756	150	1034.2	94.275	650	...
0.87023	6	41.369	8.1221	56	386.11	23.206	160	1103.2	95.725	660	...
1.0153	7	48.263	8.2672	57	393.00	24.656	170	1172.1	97.175	670	...
1.1603	8	55.158	8.4122	58	399.90	26.107	180	1241.1	98.626	680	...
1.3053	9	62.053	8.5572	59	406.79	27.557	190	1310.0	100.08	690	...
1.4504	10	68.948	8.7023	60	413.69	29.008	200	1379.0	101.53	700	...
1.5954	11	75.842	8.8473	61	420.58	30.458	210	1447.9	102.98	710	...
1.7405	12	82.737	8.9923	62	427.47	31.908	220	1516.8	104.43	720	...
1.8855	13	89.632	9.1374	63	434.37	33.359	230	1585.8	105.88	730	...
2.0305	14	96.527	9.2824	64	441.26	34.809	240	1654.7	107.33	740	...
2.1756	15	103.42	9.4275	65	448.16	36.259	250	1723.7	108.78	750	...
2.3206	16	110.32	9.5725	66	455.05	37.710	260	1792.6	110.23	760	...
2.4656	17	117.21	9.7175	67	461.95	39.160	270	1861.6	111.68	770	...
2.6107	18	124.11	9.8626	68	468.84	40.611	280	1930.5	113.13	780	...
2.7557	19	131.00	10.008	69	475.74	42.061	290	1999.5	114.58	790	...
2.9008	20	137.90	10.153	70	482.63	43.511	300	2068.4	116.03	800	...
3.0458	21	144.79	10.298	71	489.53	44.962	310	2137.4	117.48	810	...
3.1908	22	151.68	10.443	72	496.42	46.412	320	2206.3	118.93	820	...
3.3359	23	158.58	10.588	73	503.32	47.862	330	2275.3	120.38	830	...
3.4809	24	165.47	10.733	74	510.21	49.313	340	2344.2	121.83	840	...
3.6259	25	172.37	10.878	75	517.11	50.763	350	2413.2	123.28	850	...
3.7710	26	179.26	11.023	76	524.00	52.214	360	2482.1	124.73	860	...
3.9160	27	186.16	11.168	77	530.90	53.664	370	2551.1	126.18	870	...
4.0611	28	193.05	11.313	78	537.79	55.114	380	2620.0	127.63	880	...
4.2061	29	199.95	11.458	79	544.69	56.565	390	2689.0	129.08	890	...
4.3511	30	206.84	11.603	80	551.58	58.015	400	2757.9	130.53	900	...
4.4962	31	213.74	11.748	81	558.48	59.465	410	2826.9	131.98	910	...
4.6412	32	220.63	11.893	82	565.37	60.916	420	2895.8	133.43	920	...
4.7862	33	227.53	12.038	83	572.26	62.366	430	2964.7	134.89	930	...
4.9313	34	234.42	12.183	84	579.16	63.817	440	3033.7	136.34	940	...
5.0763	35	241.32	12.328	85	586.05	65.267	450	3102.6	137.79	950	...
5.2214	36	248.21	12.473	86	592.95	66.717	460	3171.6	139.24	960	...
5.3664	37	255.11	12.618	87	599.84	68.168	470	3240.5	140.69	970	...
5.5114	38	262.00	12.763	88	606.74	69.618	480	3309.5	142.14	980	...
5.6565	39	268.90	12.909	89	613.63	71.068	490	3378.4	143.59	990	...
5.8015	40	275.79	13.053	90	620.53	72.519	500	3447.4	145.04	1000	...
5.9465	41	282.69	13.198	91	627.42	73.969	510	...	147.94	1020	...
6.0916	42	289.58	13.343	92	634.32	75.420	520	...	150.84	1040	...
6.2366	43	296.47	13.489	93	641.21	76.870	530	...	153.74	1060	...
6.3817	44	303.37	13.634	94	648.11	78.320	540	...	156.64	1080	...
6.5267	45	310.26	13.779	95	655.00	79.771	550	...	159.54	1100	...
6.6717	46	317.16	13.924	96	661.90	81.221	560	...	162.44	1120	...
6.8168	47	324.05	14.069	97	668.79	82.672	570	...	165.34	1140	...
6.9618	48	330.95	14.214	98	675.69	84.122	580	...	168.24	1160	...
7.1068	49	337.84	14.359	99	682.58	85.572	590	...	171.14	1180	...
7.2519	50	344.74	14.504	100	689.48	87.023	600	...	174.05	1200	...

(*continued*)

Table 15   Metric stress or pressure conversions (*continued*)

The middle column of figures (in bold-faced type) contains the reading (in MPa or ksi) to be converted. If converting from ksi to MPa, read the MPa equivalent in the column headed "MPa." If converting from MPa to ksi, read the ksi equivalent in the column headed "ksi." 1 ksi = 6.894757 MPa. 1 psi = 6.894757 kPa

ksi		MPa	ksi		MPa	ksi		MPa	ksi		MPa
176.95	**1220**	...	226.26	**1560**	...	275.57	**1900**	...	324.88	**2240**	...
179.85	**1240**	...	229.16	**1580**	...	278.47	**1920**	...	327.79	**2260**	...
182.75	**1260**	...	232.06	**1600**	...	281.37	**1940**	...	330.69	**2280**	...
185.65	**1280**	...	234.96	**1620**	...	284.27	**1960**	...	333.59	**2300**	...
188.55	**1300**	...	237.86	**1640**	...	287.17	**1980**	...	336.49	**2320**	...
191.45	**1320**	...	240.76	**1660**	...	290.08	**2000**	...	339.39	**2340**	...
194.35	**1340**	...	243.66	**1680**	...	292.98	**2020**	...	342.29	**2360**	...
197.25	**1360**	...	246.56	**1700**	...	295.88	**2040**	...	345.19	**2380**	...
200.15	**1380**	...	249.46	**1720**	...	298.78	**2060**	...	348.09	**2400**	...
203.05	**1400**	...	252.37	**1740**	...	301.68	**2080**	...	350.99	**2420**	...
205.95	**1420**	...	255.27	**1760**	...	304.58	**2100**	...	353.89	**2440**	...
208.85	**1440**	...	258.17	**1780**	...	307.48	**2120**	...	356.79	**2460**	...
211.76	**1460**	...	261.07	**1800**	...	310.38	**2140**	...	359.69	**2480**	...
214.66	**1480**	...	263.97	**1820**	...	313.28	**2160**	...	362.59	**2500**	...
217.56	**1500**	...	266.87	**1840**	...	316.18	**2180**	...			
220.46	**1520**	...	269.77	**1860**	...	319.08	**2200**	...			
223.36	**1540**	...	272.67	**1880**	...	321.98	**2220**	...			

Table 16 Metric energy conversions

The middle column of figures (in bold-faced type) contains the reading (in J or ft · lb) to be converted. If converting from ft · lb to J, read the J equivalent in the column headed "J." If converting from J to ft · lb, read the equivalent in the column headed "ft · lb." 1 ft · lb = 1.355818 J

ft · lb		J	ft · lb		J	ft · lb		J	ft · lb		J
0.7376	**1**	1.3558	28.7649	**39**	52.8769	56.7923	**77**	104.3980	129.0734	**175**	237.2681
1.4751	**2**	2.7116	29.5025	**40**	54.2327	57.5298	**78**	105.7538	132.7612	**180**	244.0472
2.2127	**3**	4.0675	30.2400	**41**	55.5885	58.2674	**79**	107.1096	136.4490	**185**	250.8263
2.9502	**4**	5.4233	30.9776	**42**	56.9444	59.0050	**80**	108.4654	140.1368	**190**	257.6054
3.6878	**5**	6.7791	31.7152	**43**	58.3002	59.7425	**81**	109.8212	143.8246	**195**	264.3845
4.4254	**6**	8.1349	32.4527	**44**	59.6560	60.4801	**82**	111.1771	147.5124	**200**	271.1636
5.1629	**7**	9.4907	33.1903	**45**	61.0118	61.2177	**83**	112.5329	154.8880	**210**	284.7218
5.9005	**8**	10.8465	33.9279	**46**	62.3676	61.9552	**84**	113.8887	162.2637	**220**	298.2799
6.6381	**9**	12.2024	34.6654	**47**	63.7234	62.6928	**85**	115.2445	169.6393	**230**	311.8381
7.3756	**10**	13.5582	35.4030	**48**	65.0793	63.4303	**86**	116.6003	177.0149	**240**	325.3963
8.1132	**11**	14.9140	36.1405	**49**	66.4351	64.1679	**87**	117.9562	184.3905	**250**	338.9545
8.8507	**12**	16.2698	36.8781	**50**	67.7909	64.9055	**88**	119.3120	191.7661	**260**	352.5126
9.5883	**13**	17.6256	37.6157	**51**	69.1467	65.6430	**89**	120.6678	199.1418	**270**	366.0708
10.3259	**14**	18.9815	38.3532	**52**	70.5025	66.3806	**90**	122.0236	206.5174	**280**	379.6290
11.0634	**15**	20.3373	39.0908	**53**	71.8583	67.1182	**91**	123.3794	213.8930	**290**	393.1872
11.8010	**16**	21.6931	39.8284	**54**	73.2142	67.8557	**92**	124.7452	221.2686	**300**	406.7454
12.5386	**17**	23.0489	40.5659	**55**	74.5700	68.5933	**93**	126.0911	228.6442	**310**	420.3036
13.2761	**18**	24.4047	41.3035	**56**	75.9258	69.3308	**94**	127.4469	236.0199	**320**	433.8617
14.0137	**19**	25.7605	42.0410	**57**	77.2816	70.0684	**95**	128.8027	243.3955	**330**	447.4199
14.7512	**20**	27.1164	42.7786	**58**	78.6374	70.8060	**96**	130.1585	250.7711	**340**	460.9781
15.4888	**21**	28.4722	43.5162	**59**	79.9933	71.5435	**97**	131.5143	258.1467	**350**	474.5363
16.2264	**22**	29.8280	44.2537	**60**	81.3491	72.2811	**98**	132.8702	265.5224	**360**	488.0944
16.9639	**23**	31.1838	44.9913	**61**	82.7049	73.0186	**99**	134.2260	272.8980	**370**	501.6526
17.7015	**24**	32.5396	45.7288	**62**	84.0607	73.7562	**100**	135.5818	280.2736	**380**	515.2108
18.4390	**25**	33.8954	46.4664	**63**	85.4165	77.4440	**105**	142.3609	287.6492	**390**	528.7690
19.1766	**26**	35.2513	47.2040	**64**	86.7723	81.1318	**110**	149.1400	295.0248	**400**	542.3272
19.9142	**27**	36.6071	47.9415	**65**	88.1282	84.8196	**115**	155.9191	302.4005	**410**	555.8854
20.6517	**28**	37.9629	48.6791	**66**	89.4840	88.5075	**120**	162.6982	309.7761	**420**	569.4435
21.3893	**29**	39.3187	49.4167	**67**	90.8398	92.1953	**125**	169.4772	317.1517	**430**	583.0017
22.1269	**30**	40.6745	50.1542	**68**	92.1956	95.8831	**130**	176.2563	324.5273	**440**	596.5599
22.8644	**31**	42.0304	50.8918	**69**	93.5514	99.5709	**135**	183.0354	331.9029	**450**	610.1181
23.6020	**32**	43.3862	51.6293	**70**	94.9073	103.2587	**140**	189.8145	339.2786	**460**	623.6762
24.3395	**33**	44.7420	52.3669	**71**	96.2631	106.9465	**145**	196.5936	346.6542	**470**	637.2344
25.0771	**34**	46.0978	53.1045	**72**	97.6189	110.6343	**150**	203.3727	354.0298	**480**	650.7926
25.8147	**35**	47.4536	53.8420	**73**	98.9747	114.3221	**155**	210.1518	361.4054	**490**	664.3508
26.5522	**36**	48.8094	54.5796	**74**	100.3305	118.0099	**160**	216.9308	368.7811	**500**	677.9090
27.2898	**37**	50.1653	55.3172	**75**	101.6863	121.6977	**165**	223.7099			
28.0274	**38**	51.5211	56.0547	**76**	103.0422	125.3856	**170**	230.4890			

Table 17 Metric length and weight conversion factors

Unit	Inches to millimeters	Millimeters to inches	Pounds to kilograms	Kilograms to pounds
1	25.400 1	0.039 371	0.453 59	2.204 62
2	50.800 1	0.078 742	0.907 19	4.409 24
3	76.200 2	0.118 112	1.360 78	6.613 86
4	101.600 2	0.157 483	1.814 37	8.818 49
5	127.000 3	0.196 854	2.267 96	11.023 11
6	152.400 3	0.236 225	2.721 56	13.227 73
7	177.800 4	0.275 596	3.175 15	15.432 35
8	203.200 4	0.314 966	3.628 74	17.636 97
9	228.600 5	0.354 337	4.082 33	19.841 59
10	254.000 6	0.393 708	4.355 92	22.046 22

Table 18   Conversion of inches to millimeters

Inches	Millimeters	Inches	Millimeters	Inches	Millimeters	Inches	Millimeters
0.001	0.025	0.200	5.08	0.480	12.19	0.760	19.30
0.002	0.051	0.210	5.33	0.490	12.45	0.770	19.56
0.003	0.076	0.220	5.59	0.500	12.70	0.780	19.81
0.004	0.102	0.230	5.84	0.510	12.95	0.790	20.07
0.005	0.127	0.240	6.10	0.520	13.21	0.800	20.32
0.006	0.152	0.250	6.35	0.530	13.46	0.810	20.57
0.007	0.178	0.260	6.60	0.540	13.72	0.820	20.83
0.008	0.203	0.270	6.86	0.550	13.97	0.830	21.08
0.009	0.229	0.280	7.11	0.560	14.22	0.840	21.34
0.010	0.254	0.290	7.37	0.570	14.48	0.850	21.59
0.020	0.508	0.300	7.62	0.580	14.73	0.860	21.84
0.030	0.762	0.310	7.87	0.590	14.99	0.870	22.10
0.040	1.016	0.320	8.13	0.600	15.24	0.880	22.35
0.050	1.270	0.330	8.38	0.610	15.49	0.890	22.61
0.060	1.524	0.340	8.64	0.620	15.75	0.900	22.86
0.070	1.778	0.350	8.89	0.630	16.00	0.910	23.11
0.080	2.032	0.360	9.14	0.640	16.26	0.920	23.37
0.090	2.286	0.370	9.40	0.650	16.51	0.930	23.62
0.100	2.540	0.380	9.65	0.660	16.76	0.940	23.88
0.110	2.794	0.390	9.91	0.670	17.02	0.950	24.13
0.120	3.048	0.400	10.16	0.680	17.17	0.960	24.38
0.130	3.302	0.410	10.41	0.690	17.53	0.970	24.64
0.140	3.56	0.420	10.67	0.700	17.78	0.980	24.89
0.150	3.81	0.430	10.92	0.710	18.03	0.990	25.15
0.160	4.06	0.440	11.18	0.720	18.29	1.000	25.40
0.170	4.32	0.450	11.43	0.730	18.54		
0.180	4.57	0.460	11.68	0.740	18.80		
0.190	4.83	0.470	11.94	0.750	19.05		

Table 19   Conversion of millimeters to inches

Millimeters	Inches	Millimeters	Inches	Millimeters	Inches	Millimeters	Inches
0.01	0.0004	0.26	0.0102	0.51	0.0201	0.76	0.0299
0.02	0.0008	0.27	0.0106	0.52	0.0205	0.77	0.0303
0.03	0.0012	0.28	0.0110	0.53	0.0209	0.78	0.0307
0.04	0.0016	0.29	0.0114	0.54	0.0213	0.79	0.0311
0.05	0.0020	0.30	0.0118	0.55	0.0217	0.80	0.0315
0.06	0.0024	0.31	0.0122	0.56	0.0220	0.81	0.0319
0.07	0.0028	0.32	0.0126	0.57	0.0224	0.82	0.0323
0.08	0.0031	0.33	0.0130	0.58	0.0228	0.83	0.0327
0.09	0.0035	0.34	0.0134	0.59	0.0232	0.84	0.0331
0.10	0.0039	0.35	0.0138	0.60	0.0236	0.85	0.0335
0.11	0.0043	0.36	0.0142	0.61	0.0240	0.86	0.0339
0.12	0.0047	0.37	0.0146	0.62	0.0244	0.87	0.0343
0.13	0.0051	0.38	0.0150	0.63	0.0248	0.88	0.0346
0.14	0.0055	0.39	0.0154	0.64	0.0252	0.89	0.0350
0.15	0.0059	0.40	0.0157	0.65	0.0256	0.90	0.0354
0.16	0.0063	0.41	0.0161	0.66	0.0260	0.91	0.0358
0.17	0.0067	0.42	0.0165	0.67	0.0264	0.92	0.0362
0.18	0.0071	0.43	0.0169	0.68	0.0268	0.93	0.0366
0.19	0.0075	0.44	0.0173	0.69	0.0272	0.94	0.0370
0.20	0.0079	0.45	0.0177	0.70	0.0276	0.95	0.0374
0.21	0.0083	0.46	0.0181	0.71	0.0280	0.96	0.0378
0.22	0.0087	0.47	0.0185	0.72	0.0283	0.97	0.0382
0.23	0.0091	0.48	0.0189	0.73	0.0287	0.98	0.0386
0.24	0.0094	0.49	0.0193	0.74	0.0291	0.99	0.0390
0.25	0.0098	0.50	0.0197	0.75	0.0295	1.00	0.0394